Sample Preparation of Pharmaceutical Dosage Forms

Beverly Nickerson
Editor

Sample Preparation of Pharmaceutical Dosage Forms

Challenges and Strategies for Sample Preparation and Extraction

aapspress
Springer

Editor
Beverly Nickerson
Analytical Development
Pfizer Global Research and Development
Pfizer Inc., Groton, CT
USA
beverly.nickerson@pfizer.com

ISBN 978-1-4419-9630-5 e-ISBN 978-1-4419-9631-2
DOI 10.1007/978-1-4419-9631-2
Springer New York Dordrecht Heidelberg London

Library of Congress Control Number: 2011932984

Printed on acid-free paper

Springer is part of Springer Science+Business Media (www.springer.com)

To my wonderful husband Tom
and my terrific children Haley and Ashley
for their love, encouragement, and support

Preface

Analytical Chemistry is a corner stone of the drug development process. Analytical measurements and data underpin assessments and decisions that are made throughout the drug development process. Development and use of appropriate and robust analytical methods is critical to the ability to generate accurate and reliable analytical data. Sample preparation is an integral part of the analytical method and is often the most time-consuming portion of the method to perform. Developing appropriate and robust extraction and sample preparation methods can be challenging for pharmaceutical dosage forms due to the nature of the sample. Oftentimes method robustness and method transfer problems are the result of issues with the sample preparation portion of the method rather than the analysis portion of the method (e.g., HPLC chromatographic conditions).

This book is intended to serve as a resource for analysts in developing and troubleshooting sample preparation methods. These are critical activities in providing accurate and reliable data throughout the lifecycle of a drug product. This guide is divided into four sections. The first section, Chaps. 1 and 2, is an introductory section that discusses dosage form and diluent properties that impact sample preparation of pharmaceutical dosage forms and the importance of sampling considerations in generating data representative of the drug product batch. The second section of this book, Chaps. 3–5, discusses specific sample preparation techniques typically used with pharmaceutical dosage forms. The third section, Chaps. 6–9, discusses sample preparation method development for different types of dosage forms and includes information on addressing drug excipient interactions and post-extraction considerations (e.g., clarification, derivatization). It also includes discussions on method validation in Chap. 10, and applying Quality by Design (QbD) principles to sample preparation methods in Chap. 11. The last section, Chaps. 12–15, covers additional topics in sample preparation including automation, investigating aberrant potency results, and green chemistry considerations for sample preparation. The last chapter of this section discusses the ideal case where no sample preparation is required for sample analysis.

I would like to acknowledge my friends and colleagues in the pharmaceutical industry that I have worked with in supporting drug development candidates. Many of the issues we have dealt with involved various challenges with sample preparation and extraction of dosage forms. The prevalence of these issues, combined with the limited literature resources available on sample preparation of pharmaceutical dosage forms, prompted me to organize and co-teach a short course on "Sample Preparation/Extraction for Solid Oral Dosage Forms" at the 2006 American Association of Pharmaceutical Scientists (AAPS) Annual Meeting. The next step was the writing and editing of *Sample Preparation of Pharmaceutical Dosage Forms* to provide a comprehensive guide.

I sincerely acknowledge all the authors for their dedication, efforts, and valuable contributions to this work, which I trust readers will find to be a useful resource in developing and troubleshooting sample preparation methods for pharmaceutical dosage forms. I would also like to thank David De Antonis and Ling Zhang for their support and encouragement of my work on this volume and Thomas Bush for his review and proofreading of the manuscript. Last, but not least, I would like to thank my husband Tom and my children, Haley and Ashley, for their patience, understanding, and support during the time I have spent working on *Sample Preparation of Pharmaceutical Dosage Forms*.

Groton, CT Beverly Nickerson, Ph.D.

About the Editor

Dr. Beverly Nickerson received her Ph.D. in Analytical Chemistry at the University of North Carolina at Chapel Hill. After graduate school she worked in the Analytical Research and Development Department at Hoffmann-La Roche in Nutley, NJ. She later joined the Analytical Development Department at Pfizer in Groton, CT where she is currently an Associate Research Fellow. Her primary responsibilities include working as a member of cross-functional teams to develop drug candidates, supporting formulation development efforts, developing and validating methods, problem-solving and addressing analytical issues encountered during drug development, and writing reports and sections for regulatory submissions. Dr. Nickerson has worked on early stage and late stage development compounds as well as product enhancement projects. She served on the Executive Committee of the Analysis and Pharmaceutical Quality (APQ) Section of the American Association of Pharmaceutical Scientists (AAPS) during 2001 through 2005, including Chair of the APQ Section in 2004. Dr. Nickerson has published numerous articles in peer-reviewed journals, is author of several book chapters, and has presented at various scientific meetings.

Contents

Contributors

Edmund J. Bishop, Ph.D. Quality Laboratory Operations, Celgene Corporation, Summit, NJ, USA

Daniel Brannegan, M.S. Analytical Development, Pfizer Global Research and Development, Pfizer Inc., Groton, CT, USA

Xin Bu, Ph.D. Analytical Research and Development, Bristol-Myers Squibb Company, New Brunswick, NJ, USA

Sachin Chandran, Ph.D. Biopharmaceutics, Bristol-Myers Squibb Company, New Brunswick, NJ, USA

Hitesh P. Chokshi, Ph.D. Pharmaceutical and Analytical R&D, Hoffmann-La Roche Inc., Nutley, NJ, USA

Ivelisse Colón, Ph.D. Quality: CMC & QbD, Vertex Pharmaceuticals Inc., Cambridge, MA, USA

Eddie Ebrahimi, B.S. Analytical Development, Pfizer Global Research and Development, Pfizer Inc., Groton, CT, USA

Paul Ferguson, Ph.D. Research Analytics, Pfizer Global Research and Development, Pfizer Inc., Sandwich, UK

Mark Harding, Ph.D. Analytical Development, Pfizer Global Research and Development, Pfizer Inc., Sandwich, Kent, UK

Dawen Kou, Ph.D. Pharmaceutical and Analytical R&D, Hoffmann-La Roche Inc., Nutley, NJ, USA

Carlos Lee, Ph.D. Research Analytical, Pfizer Global Research and Development, Pfizer Inc., Groton, CT, USA

Yang (Angela) Liu, Ph.D. Analytical Development, Pfizer Global Research and Development, Pfizer Inc., Groton, CT, USA

K. Rick Lung, Ph.D. Global Technical Services, Operations, AstraZeneca Pharmaceuticals LP, Newark, DE, USA

Hua Ma, Ph.D. Pharmaceutical and Analytical R&D, Hoffmann-La Roche Inc., Nutley, NJ, USA

Gerald Manius, B.S. Pharmaceutical and Analytical R&D, Hoffmann-La Roche Inc., Nutley, NJ, USA

Beverly Nickerson, Ph.D. Analytical Development, Pfizer Global Research and Development, Pfizer Inc., Groton, CT, USA

Jackson D. Pellett, Ph.D. Small Molecule Pharmaceutical Sciences, Genentech, Inc., South San Francisco, CA, USA

George L. Reid, Ph.D. Analytical Development, Pfizer Global Research and Development, Pfizer Inc., Groton, CT, USA

Garry Scrivens, Ph.D. Analytical Development, Pfizer Global Research and Development, Pfizer Inc., Sandwich, UK

John Spirig, Ph.D. Analytical Research and Development, Bristol-Myers Squibb Company, New Brunswick, NJ, USA

Larry Taylor, Ph.D. Department of Chemistry, Virginia Polytechnic Institute and State University, Blacksburg, VA, USA

Jian Wang, Ph.D. Analytical Development, Pfizer Global Research and Development, Pfizer Inc., Groton, CT, USA

Qinggang Wang, Ph.D. Analytical Research and Development, Bristol-Myers Squibb Company, New Brunswick, NJ, USA

Gang Xue, Ph.D. Analytical Development, Pfizer Global Research and Development, Pfizer Inc., Groton, CT, USA

Shangdong Zhan, Ph.D. Pharmaceutical and Analytical R&D, Hoffmann-La Roche Inc., Nutley, NJ, USA

Lin Zhang, Ph.D. Department of Engineering, Thermo Scientific Portable Optical Analysis, Wilmington, MA, USA

Zhongli Zhang, Ph.D. Analytical Development, Pfizer Global Research and Development, Pfizer Inc., Groton, CT, USA

Section A
Introduction

Chapter 1
Properties That Impact Sample Preparation and Extraction of Pharmaceutical Dosage Forms

Beverly Nickerson

Abstract A significant portion of the time spent in testing and analyzing samples is spent on the sample preparation portion of the method. Developing appropriate extraction and sample preparation methods can be challenging for pharmaceutical dosage forms. An understanding of the steps involved in sample preparation and extraction as well as an understanding of the drug, dosage form, and diluent properties that impact sample preparation is critical in developing an adequate method. These steps and properties are discussed in detail.

1.1 Introduction

Accurate analytical data are critical in the pharmaceutical industry to ensure the quality and safety of the product. During drug development, this information is used to evaluate and select formulations for use in toxicology and clinical studies, to assess manufacturing processes and to assess the suitability and stability of clinical supplies. For marketed products, analytical data are used to evaluate the suitability and stability of the commercial product.

Development and use of robust analytical methods is critical in the ability to generate accurate analytical data. Sample preparation is an integral part of the analytical method. In a survey conducted by LC-GC (Majors 1991), responses indicated that approximately two-thirds of the time spent testing and analyzing samples was spent on the sample preparation portion of the method. In addition, issues related to sample preparation accounted for one-third of the errors generated while performing an analytical method.

B. Nickerson (✉)
Analytical Development, Pfizer Global Research and Development,
Pfizer Inc., 558 Eastern Point Road, Groton, CT 06340, USA
e-mail: beverly.nickerson@pfizer.com

B. Nickerson (ed.), *Sample Preparation of Pharmaceutical Dosage Forms: Challenges and Strategies for Sample Preparation and Extraction*, DOI 10.1007/978-1-4419-9631-2_1,

Some resources are available that discuss sample preparation and extraction by specific technique (Pawliszyn 1997; Thurman and Mills 1998) or for specific fields of application (Handley 1999; Mitra 2003). This work focuses on aspects of sample preparation for assay, content uniformity, and purity testing of pharmaceutical dosage forms. Sample preparation and extraction challenges and requirements for dosage forms include (1) achieving complete extraction of the drug and impurities without causing degradation; (2) using reasonable sample preparation methods and conditions (e.g., reasonable in terms of time, effort, and solvents); (3) final prepared samples must be compatible with the analysis method; (4) method must be rugged and robust enough to meet its intended purpose; and (5) meeting the time and resource constraints in developing the sample preparation method.

The key steps in the extraction and sample preparation of drug from the dosage form as well as the properties of the drug, dosage form, and solvent that affect extraction and sample preparation are discussed in this first chapter. Specific extraction techniques and sample preparation approaches used for various types of dosage forms are discussed in subsequent chapters of this book.

1.2 Sample Preparation of Pharmaceutical Dosage Forms

The general steps of sample analysis of a drug product are outlined in Fig. 1.1. The drug product batch may consist of hundreds to thousands or millions of individual dosage units. A representative sample of the batch must be taken for use in testing. Sampling and sampling considerations are discussed in detail in Chap. 2. Dosage units from the analytical sample are then selected and prepared for analysis as dosage forms typically cannot be introduced into the analysis equipment as is, although developments in the area of sample testing with no sample preparation are discussed in Chap. 15. Sample preparation can involve a number of steps including dispersion, particle size reduction (e.g., milling, grinding, homogenization), solubilization of the analytes of interest, derivatization, concentration, sample clean-up (e.g., removing interferences), and clarification (e.g., removing insoluble materials). The sample preparation steps required depend on the dosage form type and the end analysis technique. Once the sample preparation has been completed, the sample is then analyzed by the appropriate technique (e.g., chromatography, spectroscopy, titration) and data are available for analysis, interpretation, and decision making with respect to the drug product batch.

As illustrated in Fig. 1.2, the sample preparation steps required in a given method depend on the dosage form type being tested and the end analysis technique. For solution dosage forms (Fig. 1.2a), such as oral solutions and syrups, the drug is already dissolved in solution and uniformly distributed. In these cases, sample preparation is straight forward and typically requires only dilution of the formulation in a diluent (e.g., water or mobile phase) to make it compatible with the analysis method. In some cases, sample concentration, derivatization, or clean-up may be required. For dosage forms that are powders (e.g., powders for oral suspensions or

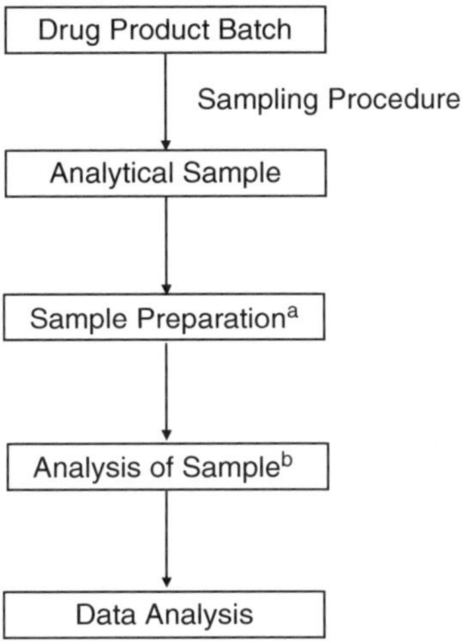

Fig. 1.1 General steps for sample preparation and analysis. (**a**) Sample preparation may include any of the following steps: disintegration/dispersion, particle size reduction (e.g., milling, grinding, homogenization), extraction and solubilization of the analytes of interest, derivatization, concentration, clean-up (e.g., remove interferences) and clarification (e.g., filtration to remove insoluble materials). (**b**) Analysis methods include chromatography, spectroscopy, titration, etc

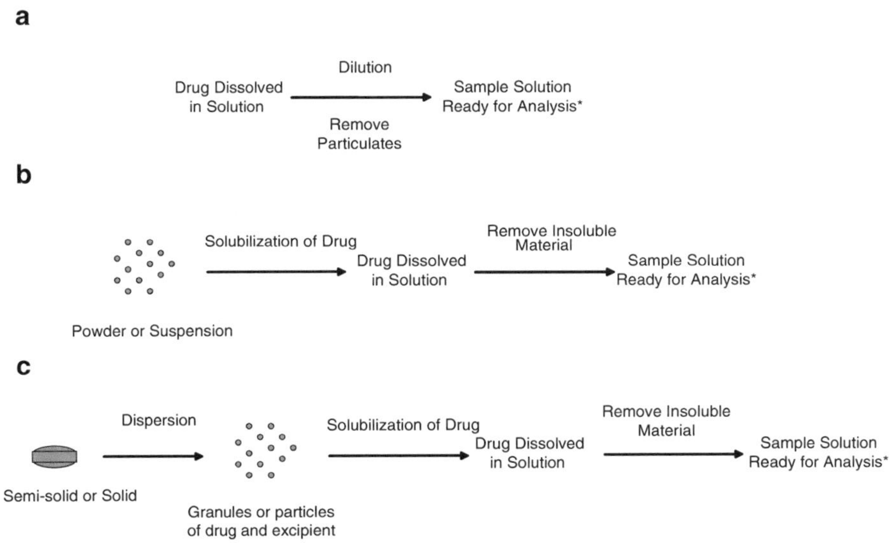

Fig. 1.2 General processes involved in sample preparation of dosage forms such as (**a**) solutions, (**b**) powders or suspensions, and (**c**) solid dosage forms. *Additional steps such as derivatization, sample concentration or sample clean-up may also be required prior to analysis

Table 1.1 Key API, diluent and dosage form properties that impact the (a) dosage form dispersion and (b) drug solubilization steps in sample preparation. The dispersion and solubilization steps are depicted schematically in Figure 1.2. Important components of the solubilization of drug step with respect to sample preparation include the extent of drug solubilization (e.g., total drug dissolved in solution) and the rate of drug solubilization

	(a) Parameters impacting dispersion of dosage forms	(b) Parameters impacting solubilization of drug	
		Extent of drug solubilization	Rate of drug solubilization
API Properties Impacting (a) and (b)		Solubility of API in diluent	Surface area/particle size Diffusion coefficient
Diluent Properties Impacting (a) and (b)	Ability of diluent to wet (solid-liquid contact angles, surface tension) and disperse dosage form Viscosity	Ability of diluent to solubilize API (solvent polarity) Ability of diluent to minimize drug-excipient interactions Volume of diluent	Volume of diluent Amount of API already dissolved
Dosage Form Properties Impacting (a) and (b)	Dosage form type (e.g., disintegrating or non-disintegrating) Excipients Manufacturing process Hardness/porosity	Drug-excipient interactions	Porosity
Other Factors Impacting (a) and (b)	Temperature (e.g., to liquefy semi-solid dosage forms) Particle size reduction techniques	Temperature	Agitation Temperature Time

lyophiles) or suspensions (Fig. 1.2b), the drug must be dissolved into solution and the final solution must be compatible with the end analysis technique. For semi-solid (e.g., creams, ointments), solid oral (e.g., tablets, capsules), and solid non-oral dosage forms (e.g., suppositories) (Fig. 1.2c), the dosage form must first be dispersed to allow efficient dissolution of the drug.

For all dosage form types except solutions, identification of an appropriate diluent is critical to ensuring dissolution and recovery of the drug from the dosage form. In addition, dispersion is important for all non-solution dosage forms. As shown in Table 1.1, the steps of solubilizing the drug and dispersing the dosage form depend on several properties of the API, dosage form, and diluent. Not all the parameters in Table 1.1 can be adjusted in sample preparation method development. For instance, dosage form type (e.g., non-disintegrating controlled release tablet) is selected based on the intended route of administration and dosing regime required to achieve efficacy. Excipients and manufacturing process are set in order to

Table 1.2 Key parameters in sample preparation method develop that impact (a) solubilization of drug from dosage forms (e.g., non-solution dosage forms) and (b) dispersion of dosage forms (e.g., solid oral dosage forms)

(a) Parameters impacting solubilization of drug		(b) Parameters impacting dispersion of dosage forms
Extent of drug solubilization	Rate of drug solubilization	
Diluent selection	Diluent selection	Diluent selection
Diluent volume	Particle size reduction techniques	Particle size reduction techniques
Time	Agitation	Agitation
	Temperature	Temperature
		Time

manufacture a stable and robust dosage form, not to make sample preparation easier. The analytical chemist is left with a subset of the parameters in Table 1.1 to use in method development and these are shown in Table 1.2. It is important, however, to understand how all the parameters in Table 1.1 affect sample preparation. If there is a change in the formulation or the manufacturing process, the impact on the sample preparation method will need to be evaluated and the method adjusted if necessary.

The key parameters to leverage in sample preparation method development are selection of the diluent, agitation conditions (e.g., shaking, sonication) including time, temperature and use of any mechanical particle size reduction techniques (e.g., grinding or homogenization). Selection of the diluent is critical to ensuring complete recovery of the drug. The solubility of the drug in the diluent must be high enough to ensure complete recovery. If not, no amount of agitation or particle size reduction can increase the recovery above this solubility limit. For non-solution dosage forms, not only is diluent selection critical but so is the means chosen to disperse the dosage form. If the dosage form remains intact, recovery of the drug may be slow or incomplete because the drug is not adequately exposed to the diluent. Dispersion of the dosage form may be performed using an appropriate diluent (e.g., water for immediate release tablets) or particle size reduction techniques (e.g., grinding). Agitation (e.g., shaking or sonication) is typically used to facilitate dispersion of the dosage form and to mix the sample solution to speed up the extraction process for all types of dosage forms. Heating may also be used to disperse semi-solid dosage forms (e.g., to melt the sample and form a solution).

The next sections of this chapter discuss details of the dissolution and dispersion steps and factors that influence these processes. Subsequent chapters of this book discuss specific extraction techniques and sample preparation approaches for specific types of dosage forms.

1.3 Properties That Impact Dispersion of Dosage Forms

As noted previously, extraction and sample preparation of drug from a semi-solid or solid dosage form typically involves two processes – dispersion of the dosage form and dissolution of the drug. Dispersion or disintegration can be defined as the breakup of the dosage form into smaller particles or granules when in contact with a liquid. If disintegration or dispersion does not occur, the drug will not be efficiently or completely extracted from the dosage form. The disintegration and dispersion process is influenced by properties of the dosage form and the extraction diluent. Disintegration mechanisms and the factors that influence dispersion are described below.

1.3.1 Disintegration Mechanisms

Dosage form factors that impact disintegration or dispersion include dosage form type, excipients used in the formulation, manufacturing process, and other factors (e.g., hardness/porosity for tablets). For immediate release and orally dispersive tablet formulations, disintegration occurs when the tablet is exposed to water due to the properties of the disintegrant in the formulation. Several different theories have been proposed to explain the mechanism of tablet disintegration and these have been summarized in a number of publications (Lowenthal 1972; Kanig and Rudnic 1984; Melia and Davis 1989; Guyot-Herman 1992). Most immediate release tablet formulations contain disintegrants, which play a critical role in the tablet disintegration process. Disintegrants appear to function by several different mechanisms, with each disintegrant type having a dominant mechanism or a combination of mechanisms. The two most commonly referenced mechanisms are wicking/capillary action and swelling. Wicking or capillary action is the ability of the disintegrant to draw water up into the porous network of the tablet. This leads to breakup of the intermolecular hydrogen bonding forces between the particles/granules in the formulation and results in tablet disintegration. The extent as well as the rate of wicking are important factors for disintegration. The swelling mechanism involves the swelling of the disintegrant after water uptake. This causes a build up in force and subsequent breakup of the dosage form. The extent and rate of swelling are important factors leading to disintegration.

In both the wicking/capillary action and swelling mechanisms of disintegration, water or solvent uptake is critical. Water or solvent uptake by a porous structure depends on the balance between several factors including capillary forces and viscous forces and is described by the Washburn equation in (1.1) (Washburn 1921):

$$l^2 = \left(\frac{\gamma}{\eta}\frac{\cos\theta}{2}\right)rt, \tag{1.1}$$

where

l = length of liquid penetration at time t,
γ = surface tension of the penetrating liquid,
η = viscosity of the penetrating liquid,
r = radius of capillary or pore size,
θ = solid–liquid contact angle, and
t = time.

1.3.2 *Factors That Impact Disintegration and Dispersion*

1.3.2.1 Solvent Properties

From (1.1), it is apparent that the water or solvent uptake is dependent on factors related to the dosage form (e.g., pore size) and factors related to the water or solvent as well (e.g., surface tension, viscosity, solid–liquid contact angle). Pore size is set by the formulation and manufacturing process. Therefore, during development of the extraction and sample preparation procedure, selection of the diluent is the key parameter to ensure wicking/solvent uptake since solvent selection impacts surface tension, liquid viscosity, and wettability of the solid by the liquid. Solvent selection is also critical to ensuring tablet disintegration after solvent update by disruption of forces holding the tablet together or by swelling of an excipient.

Before tablet disintegration can occur, the solvent must wet the surface of the dosage form. The degree of wetting is dependent on the contact angle, θ, the liquid makes with the solid surface. When θ is 0°, wetting is complete, while values of θ greater than or equal to 90° are indicative of poor wetting characteristics. A value of θ equal to 180° is indicative of non-wetting (the liquid is a spherical drop on the surface). In general, the lower the surface tension of a liquid, the smaller the contact angle on a given solid. In addition, the more polar the solid, the smaller the contact angle with the same solvent (Bummer 2000). The surface tension of a liquid can be reduced by adding a surfactant or wetting agent or by increasing temperature (Banakar 1992).

After the surface of the tablet is wetted, capillarity may occur in the tablet pores. Capillarity is the spontaneous movement of a liquid into a capillary or narrow tube due to surface forces. The greater the surface tension and the finer the capillary radius that exists, the higher the liquid will rise in the capillary. Capillarity will occur spontaneously in a cylindrical pore even if the contact angle is greater than 0°, but it will not occur at all if the contact angle becomes 90° or more (Bummer 2000).

For sample preparation and extraction considerations, unless a mechanical dispersion technique is used, a solvent that will wet the tablet surface, enter the pores of the tablet, and facilitate tablet dispersion is required.

1.3.2.2 Dosage Form Properties

Dosage form factors that impact tablet disintegration and dispersion include dosage form type, excipients used in the formulation, and the manufacturing process. Dosage form type obviously impacts dispersion as some types are disintegrating dosage forms (e.g., immediate release tablets, orally dispersive tablets) which are designed to disintegrate when exposed to water while others are non-disintegrating dosage forms (e.g., sustained release tablets).

Excipients are ingredients added to the API to enable manufacture of the dosage form. For immediate release tablets, disintegration occurs due to the properties of the disintegrant and therefore the disintegrant impacts tablet disintegration. Other types of excipients can also impact drug recovery. For example, excipients such as polymers that are used to optimize or modify drug release can impact drug extraction by making it difficult to disperse the dosage form or by trapping the drug. In addition, lubricants (may hinder tablet wetting), glidants (may hinder dissolution), diluents (may impact disintegration and dissolution), and binders (may have drug–excipient interactions) may also have an impact. Drug–excipient interactions are discussed in detail in Chap. 6.

During the manufacturing process, disintegrants may be added prior to granulation (intragranular – inside the granules) or during the lubrication step prior to compression (extragranular – outside the granules) or during both of these steps. It has been shown that extragranular formulations disintegrate more rapidly while intragranular formulations disintegrate into finer particles (Peck et al. 1990; Guyot-Herman 1992). The manufacturing process used for immediate release tablets will impact the disintegration process of the tablet and subsequent dissolution of the drug. Direct compression tablets will disintegrate into primary drug particles, while wet granulation tablets will disintegrate into granules consisting of drug and excipients (Carstensen 1977).

Additional dosage form properties may impact disintegration and dispersion. For example, tablet hardness is an important factor. As tablet hardness increases, the porosity or pore diameter throughout the tablet decreases. If the pore size is too small, a longer time will be required for water or solvent to penetrate the pores and disintegration times will therefore increase. On the other hand, if the pore size is too large and allows the tablet matrix to elastically yield as the disintegrant swells, there will be no generation of force to disintegrate the tablet (Guyot-Herman 1992). Thus, if there is a significant change in tablet hardness during the course of development, there could be an impact on the ability of the sample preparation method to adequately disperse and extract the active.

As discussed in detail in Chap. 7, solid oral dosage forms can be dispersed by finding a suitable diluent (e.g., water for immediate release tablets, other diluents for controlled release tablets). For capsule formulations, the capsule shell can be removed or conditions can be found to dissolve or rupture the capsule. For non-disintegrating solid oral dosage forms such as sustained or controlled release tablets, an appropriate solvent needs to be identified to disperse the tablet. Alternatively, mechanical means such as grinding or milling can be used to disperse tablet dosage forms. As discussed in Chap. 8, for solid, non-oral dosage forms (e.g., suppositories or patches), and semi-solid dosage forms (e.g., creams, ointments), appropriate diluents may be used to dissolve excipients and disperse the dosage form. Transdermal patches may also be cut into smaller pieces and heat can be used to liquefy suppositories. Agitation is typically used to facilitate dispersion and mixing of the sample solutions for all types of dosage forms.

1.4 Factors That Impact Dissolution and Solubilization of Drug in Dosage Forms

Dissolution or solubilization of API and components of interest is required during sample preparation of non-solution–type dosage forms. Dissolution models and the factors that influence dissolution are discussed below.

1.4.1 Dissolution Models

1.4.1.1 Pharmaceutical Solids

The diffusion layer theory is the best known model for transport-controlled dissolution (i.e., dissolution rate is controlled by the rate of diffusion of solute molecules across a diffusion layer). The diffusion layer theory accounts for the dissolution rates of most pharmaceutical solids and has been used to predict dissolution rates of drugs in powder form (Higuchi 1967; Stavchansky and McGinity 1990; Grant and Brittain 1995). In the diffusion layer model, which is graphically depicted in Fig. 1.3, interaction of the solvent with the surface of a drug particle produces an infinitesimally thin layer of saturated solution of drug (concentration $= C_S$) around the drug particle. At the solid–liquid interface, solid–solution equilibrium exits. With increasing distance, x, from the surface of the solid, the concentration of dissolved drug decreases from C_S (at $x=0$) to that in the bulk solution C (at $x=h$). The rate at which the drug diffuses across this layer, the diffusion layer, controls the dissolution rate. In addition, in a stirred solution, the flow velocity of the liquid dissolution medium increases from 0 at $x=0$ to the bulk value at $x=h$.

Dokoumetzidis and Macheras reviewed various equations that have been derived and proposed to describe dissolution based on the diffusion layer model (Dokoumetzidis and Macheras 2006). In 1897, Noyes and Whitney developed an equation to describe dissolution, or "The rate of solution of solid substances in their own solutions" (Noyes and Whitney 1897). The Noyes–Whitney equation is shown in (1.2). Bruner and von Tolloczko modified (1.2) to take into account the surface area of the substance and this equation is shown in (1.3) (Bruner and Tolloczko 1900; Dokoumetzidis and Macheras 2006). Nernst and Brunner later derived the Nernst–Brunner equation, (1.4) based on the diffusion layer model and Fick's second law (Nernst 1904; Brunner 1904; Dokoumetzidis and Macheras 2006):

$$\frac{\mathrm{d}C}{\mathrm{d}t} = k(C_S - C), \qquad (1.2)$$

$$\frac{\mathrm{d}C}{\mathrm{d}t} = k_1 S(C_S - C), \qquad (1.3)$$

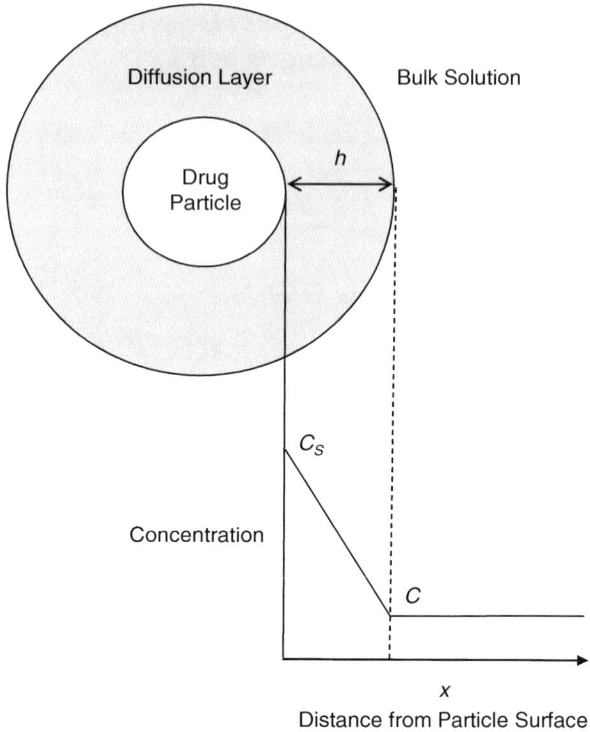

Fig. 1.3 Diffusion layer model describing the mechanism of dissolution of a solid into a solvent

$$\frac{\mathrm{d}C}{\mathrm{d}t} = \frac{DS}{Vh}(C_\mathrm{S} - C), \tag{1.4}$$

where

C_S = saturation concentration or saturation solubility,
C = concentration of drug in the bulk solution at time t,
k = a constant,
k_1 = a constant,
S = surface area,
D = diffusion coefficient,
V = volume of medium, and
h = thickness of the diffusion layer.

In 1931, Hixson and Crowell modified (1.3) to derive (1.5), the Hixson–Crowell cube root law, which relates time to the cube root of weight under sink conditions and accounts for the change in a particle's surface area during dissolution (Hixson and Crowell 1931; Dokoumetzidis and Macheras 2006):

$$w_0^{1/3} - w^{1/3} = k_2 t, \tag{1.5}$$

where

w_0 = initial weight of drug particle,
w = weight of the remaining undissolved drug particle at time t, and
k_2 = a constant.

1.4.1.2 Tablet Dosage Forms

Carstensen described dissolution of disintegrating direct compression tablets and wet-granulated tablets (Carstensen 1977). For disintegrating tablet dosage forms, disintegration is typically rapid and occurs first followed by drug dissolution. As the particles dissolve, the surface area of the drug decreases. The Hixson–Crowell cube root law as written in (1.6) describes the dissolution of primary drug particles after disintegration of direct compression tablets. This equation assumes that the drug is soluble in the dissolving solvent, that sink conditions exist, and that the solvent will cause disintegration of the tablet (Carstensen 1977):

$$m_0^{1/3} - m^{1/3} = K\left[t - t_1\right], \tag{1.6}$$

where

m_0 = original mass of drug in the tablet,
m = amount of drug not dissolved at time t,
$K = kSm_0^{1/3}/(\rho r_0)$,
t_1 = disintegration time,
k = intrinsic dissolution-rate constant,
S = drug solubility,
ρ = true density, and
r_0 = original radius of the particles.

For wet-granulated tablets exposed to a liquid, such as a dissolving solvent, tablets disintegrate into granules containing drug and excipient. These granules may be either porous or non-porous. For porous granules, drug diffusion into the bulk solution takes longer than penetration of the dissolving solvent into the granules. Tablet disintegration and drug diffusion into the bulk solution is therefore rate controlling. For wet-granulation tablets, the following equations by Carstensen describe the dissolution process for (a) granule→drug in solution (1.7) and (b) tablet→drug in solution (1.8) (Carstensen 1977):

$$\ln[(M/V) - C] = -k^*(t - t_{ii}) + \ln(M/V), \tag{1.7}$$

$$\ln[(M/V) - C] = -k''(t - t_i - t_{ii}) + \ln(M/V), \tag{1.8}$$

where

M = amount of drug in the tablet being dissolved,
V = volume of dissolving solvent,

C = concentration at time t,
t_i = disintegration time (tablet into granules),
t_{ii} = time required for solvent penetration into the granule,
k^* *and* k'' = apparent dissolution constants, which depend on the diffusion coefficient of the drug through the granule matrix and the radius of the granule (k^*/k'' is a function of surface area and porosity).

Carstensen notes that for poorly permeable granules, penetration of dissolving solvent into the granules is rate limiting and drug is dissolved from the granules according to the Higuchi square root law, which is shown below in (1.9) (Higuchi 1963; Carstensen 1977). In these cases, particle reduction techniques may speed up the extraction and sample preparation process:

$$Q = [KA\varepsilon t]^{1/2}, \tag{1.9}$$

where

Q = amount of drug dissolved per unit surface area (cm^2),
A = the fraction of drug in the tablet or granule,
ε = the porosity of the granules or dosage form mass,
t = time,
K = a proportionality constant and equals $2DS$, where D is the diffusion coefficient of the drug in the dissolving medium and S is the solubility of drug in the medium.

Carstensen notes that for an erosion tablet that does not disintegrate, and where the matrix erodes and releases drug, the erosion of the tablet is analogous to dissolution of a spherical particle. The disappearance rate of the tablet will follow the Hixson–Crowell cube root law, where m_0 is the amount of drug present in the dosage form at time 0 and m is the amount of drug still undissolved at time t (Hixson and Crowell 1931; Carstensen 1977). Some sustained release products are formulated by suspending drug in a film and grinding up the material, and in these cases, dissolution follows the Higuchi square root law (Carstensen 1977). For extraction and sample preparation of drug from erosion-based tablets and other types of sustained release formulations, mechanical means can be used to disperse the material and speed up drug recovery.

1.4.2 Leveraging Key Factors to Impact Dissolution During Sample Preparation

Mechanisms for dissolution of drug and drug particles are discussed above. Two aspects of dissolution are important for extraction and sample preparation – the extent and the rate of analyte dissolution. The extent of drug dissolution translates into drug recovery and is dependent on the properties of the API, dissolving or extraction solvent (e.g., diluent), and dosage form. Temperature and agitation also affect the rate of drug dissolution. All these factors are discussed below.

1.4.2.1 Extent of Dissolution

API Properties

The key limiting factor for drug dissolution from a dosage form is the solubility of the drug in the diluent. The equations in Sect. 1.4.1 show a dependence of dissolution on drug solubility in the solvent. Solubility is defined as the maximum amount of solute that can dissolve in a specific amount of solvent at a specific temperature. The solubility of a solid is dependent on the nature of both the solute (e.g., molecular size, functional groups/polarity, pK_a) and the selected dissolving solvent (e.g., polarity, pH, and buffer concentration) and the intermolecular interactions between the solute and the solvent.

Analyte functional groups and their interactions with a given solvent contribute to the overall solubility of the analyte and hence play a significant role in sample preparation/extraction. Functional groups can be classified as non-polar (hydrophobic), polar (hydrophilic), or ionic. In order for a solute to be solubilized by a solvent, the solvent must overcome the intermolecular interactions of the solute–solute molecules. In addition, the solvent molecules must be separated from each other by the solute molecules. This is likely to occur when the attractions between solute molecules and between solvent molecules are similar. If the attractions are different, then solute molecules will not separate from each other and the solvent molecules will not separate from each other and hence the solute will not dissolve (Burke 1984). In general, non-polar or hydrophobic dissolving/extraction solvents should be selected for non-polar/hydrophobic analytes and non-ionized analytes. Polar or hydrophilic dissolving/extraction solvents should be selected for polar/hydrophilic analytes and ionized analytes.

For drugs with ionizable functional groups, the pH of the solvent can be adjusted to effect ionization of the analyte (and hence polarity) and affect its solubility in the solvent as ionized groups are more soluble in aqueous and polar solvents, while non-ionized groups are soluble in non-polar solvents. Thus, when choosing a dissolving/extraction solvent for a compound with ionizable functional groups, the pK_a is important in that one can increase the solubility of the drug in polar dissolving/extraction solvents by having the pH of the dissolving solvent be at least two pH units above or below the pK_a on the side of the ionized form of the molecule, while solubility of the compound in non-polar solvents would be increased if the compound is maintained in a non-ionized form.

Solvent Properties

As discussed above, non-polar or hydrophobic solvents tend to dissolve non-polar/hydrophobic analytes and non-ionized analytes. Polar or hydrophilic solvents tend to dissolve polar/hydrophilic analytes and ionized analytes. A number of different solvent polarity classification schemes (e.g., Hildebrand Solubility Parameters, Hansen Solubility Parameters, Solvent-Selectivity Triangle) have been developed and have been discussed in various reviews (Snyder 1978; Burke 1984). In addition, there are programs (e.g., COSMOtherm, aspenONE) available that will give theoretically calculated estimates of solubility for analytes in different solvents (Klamt 1998). These classification schemes and programs provide a means to rank solvents with respect to their polarity and to identify solvents to maximize solubility for a given solute.

Dosage Form Properties

A significant dosage form factor that impacts the extent of drug dissolution is potential drug–excipient interactions. These interactions can affect the stability of the API and the performance of the formulation. In addition, drug–excipient interactions can affect the development of analytical methodology by impacting the conditions needed to achieve complete drug recovery in assay methods or by effecting dissolution tests. Physical interactions between a drug and an excipient include such interactions as adsorption and physical trapping or inclusion of drug by a non-soluble or gelling polymer excipient. These physical interactions can result in low recovery of the active during sample analysis and/or delayed drug release during dissolution testing. For sample preparation/extraction of drug from dosage forms with a potential for API to adsorb to excipients or become trapped by polymeric excipients, judicious selection of extraction solvent and sample preparation conditions is needed to minimize or eliminate these interactions. Otherwise, low drug recoveries may be obtained leading to inaccurate results. Drug–excipient interactions are discussed in detail in Chap. 6.

1.4.2.2 Rate of Dissolution

API Properties

As shown in the equations in Sect. 1.4.1, API-related factors that impact the rate of drug dissolution are API surface area (particle size) and diffusion coefficient. The dissolution rate will increase as the surface area of the solid increases. Therefore, solvation or dissolution rate can be increased by decreasing the particle size of the sample through crushing, grinding, milling, etc. to create increased surface area. In addition, smaller particles have a small diffusion boundary layer, resulting in faster transport of dissolved material from the particle surface (Randall 1995). Sample preparation strategies utilizing particle size reduction (e.g., grinding, ball mill) are discussed in Chap. 3. In some cases, however, particle size reduction may decrease (or fail to increase) the dissolution rate. This is caused by incomplete wetting of the solid as a result of increased adsorption of air to the particle surface and results in reduced effective surface area and decreased dissolution. The use of a surfactant in these cases may improve dissolution (Lantz 1990).

The dissolution rate will increase as the diffusion coefficient of the solute increases. The diffusion coefficient, D, is defined as shown in (1.10) (Hoener and Benet 1990). As shown in this equation, the diffusion coefficient is dependent on the solvent viscosity and temperature. Hence solvent selection and temperature will impact the rate of drug dissolution and hence extraction and sample preparation:

$$D = kT / (6\pi\eta r), \tag{1.10}$$

where

k = Boltzmann's constant,
T = absolute temperature,
η = viscosity of the solvent, and
r = radius of molecule in solution.

Solvent Properties

Solvent factors that impact the rate of drug dissolution/solubilization include solvent volume and amount of API already dissolved. Equations (1.2)–(1.4) show that dissolution rate is dependent on C, the concentration of drug in the bulk fluid. To remove this dependence for sample preparation considerations, a sufficient volume of solvent should be used to ensure sink conditions. In this case, C_s will be much greater than C. When C is less than 15% of C_s, sink conditions exist and C has a negligible effect on the dissolution rate of the solid (Stavchansky and McGinity 1990).

Dosage Form Factors

As shown in (1.7)–(1.9), porosity affects dissolution by affecting the rate at which the solvent penetrates the granules. Tablet porosity or hardness is determined during formulation development and is not a factor that is controlled during extraction and sample preparation method development. However, it is important to note that any changes in tablet porosity or hardness during the course of development may impact the sample preparation/extraction method and should therefore be assessed.

Miscellaneous Factors

Additional factors that impact the rate of drug dissolution and solubilization include temperature and agitation. The rate of dissolution generally increases as temperature increases because analyte solubility typically increases and analyte and solvent diffusion increases. This increase in diffusion speeds up the solvation process since the solute dissolved in the solvent will diffuse away from the undissolved sample particle at a faster rate, allowing more solute to dissolve. In addition, as temperature increases, solvent viscosity decreases (solvent can more readily penetrate pores of sample particles) and solvent surface tension decreases (solvent can more readily wet the sample) (Richter et al. 1996). There are, however, exceptions, as some polymers are more soluble at lower temperatures and use of higher temperatures may lead to drug–excipient interactions (e.g., entrapment of drug in the polymer matrix).

As shown in (1.4) the dissolution rate is dependent on the inverse of h, the thickness of the stationary layer of solvent around the drug particle. Thus, dissolution rate can be increased, by decreasing h. For extraction and sample preparation purposes, the value of h can be decreased by increasing the "stirring rate" or agitation of the solution (Hoener and Benet 1990). Agitation brings fresh solvent to the solute surface, so that more solute can dissolve and hence increase the rate of dissolution.

1.5 Summary

Properties of the API, dosage form, and diluent affect extraction and sample preparation. The API is chosen based on efficacy and toxicology considerations. Factors such as dosage form type, excipients, and manufacturing process are determined during formulation development and are typically not altered based on extraction and sample preparation considerations. It is important, however, to understand these factors and how they may impact dispersion of the dosage form during sample preparation. This is especially important if there is a change in any of these parameters during the course of development of the product and the impact of the change(s) on the extraction and sample preparation method should be assessed. Diluent selection, diluent volume, particle size reduction, agitation, temperature, and time are the key parameters to leverage during sample preparation. Sample preparation techniques as well as sample preparation strategies for different types of dosage forms, which leverage these variables, are discussed in subsequent chapters of this book.

References

Banakar UV (1992) Pharmaceutical dissolution testing. Marcel Dekker, New York

Bruner L, Tolloczko S (1900) Über die Auflösungsgeschwindigkeit Fester Körper. Z Phys Chem 35:283–290

Brunner E (1904) Reaktionsgeschwindigkeit in Heterogenen Systemen. Z Phys Chem 47(1):56–102

Bummer PM (2000) Chapter 20: Interfacial phenomena. In: Gennaro AR (ed) Remington: the science and practice of pharmacy, 20th edn. Lippincott Williams & Wilkins, Baltimore

Burke J (1984) Solubility parameters: theory and application. In: Jensen C (ed) The book and paper group annual, vol 3, part 2. The Hildebrand Solubility Parameter, The American Institute for Conservation, pp 13–58. http://cool.conservation-us.org/coolaic/sg/bpg/annual/v03/bp03-04.html. Accessed 9 Sept 2010

Carstensen JT (1977) Pharmaceutics of solids and solid dosage forms. Wiley, New York
Dokoumetzidis A, Macheras P (2006) A century of dissolution research: from Noyes and Whitney to the biopharmaceutics classification system. Int J Pharm 321:1–11
Grant DJW, Brittain HG (1995) Chapter 11: Solubility of pharmaceutical solids. In: Brittain HG (ed) Physical characterization of pharmaceutical solids. Marcel Dekker, New York
Guyot-Herman AM (1992) Tablet disintegration and disintegrating agents. STP Pharma Sci 2(6):445–462
Handley AJ (1999) Extraction methods in organic analysis. CRC Press, Boca Raton
Higuchi T (1963) Mechanism of sustained-action medication. Theoretical analysis of rate of release of solid drugs dispersed in solid matrices. J Pharm Sci 52:1145–1149
Higuchi WI (1967) Diffusional models useful in biopharmaceutics – drug release rate processes. J Pharm Sci 56:315–324
Hixson AW, Crowell JH (1931) Dependence of reaction velocity upon surface and agitation, I – theoretical considerations. Ind Eng Chem 23(8):923–931
Hoener B-A, Benet LZ (1990) Chapter 4: Factors influencing drug absorption and drug availability. In: Banker GS, Rhodes CT (eds) Modern pharmaceutics, 2nd edn. Marcel Dekker, New York
Kanig JL, Rudnic EM (1984) The mechanisms of disintegrant action. Pharm Technol 8:50–63
Klamt A (1998) COSMO and COSMO-RS. In: Schleyer PvR, Allinger L (eds) Encyclopedia of computational chemistry, vol 2. Wiley, New York
Lantz RA Jr (1990) Chapter 3: Size reduction. In: Lieberman HA, Lachman L, Schwartz JB (eds) Pharmaceutical dosage forms, vol 2, Tablets, 2nd edn. Marcel Dekker, New York
Lowenthal W (1972) Disintegration of tablets. J Pharm Sci 61(11):1695–1711
Majors RE (1991) An overview of sample preparation. LC GC 9(1):16–18, 20
Melia CD, Davis SS (1989) Review article: mechanisms of drug release from tablets and capsules. I: disintegration. Aliment Pharmacol Ther 3:223–232
Mitra S (2003) Sample preparation techniques in analytical chemistry. Wiley, Hoboken
Nernst W (1904) Theorie der Reaktionsgeschwindigkeit in Heterogenen Systemen. Z Phys Chem 47:52–55
Noyes AA, Whitney WR (1897) The rate of solution of solid substances in their own solution. J Am Chem Soc 19:930–934
Pawliszyn J (1997) Solid phase microextraction theory and practice. Wiley-VCH, New York
Peck GE, Bailey GJ, McCurdy VE, Banker GS (1990) Chapter 2: Tablet formulation and design. In: Liebeman HA, Lachman L, Schwartz JB (eds) Pharmaceutical dosage forms, vol 1 Tablets, 2nd edn. Marcel Dekker, New York
Randall CS (1995) Chapter 6: Particle size distribution. In: Brittain HG (ed) Physical characterization of pharmaceutical solids. Marcel Dekker, New York
Richter BE, Jones BA, Ezzell JL, Porter NL, Avdalovic N, Pohl C (1996) Accelerated solvent extraction: a technique for sample preparation. Anal Chem 68(6):1033–1039
Snyder LR (1978) Chapter II: Solvent selection for separation processes. In: Perry ES, Weissberger A (eds) Techniques of chemistry, vol XII, Separation and purification, 3rd edn. Wiley, New York
Stavchansky SA, McGinity JW (1990) Chapter 6: Bioavailability in tablet technology. In: Lieberman HA, Lachman L, Schwartz JB (eds) Pharmaceutical dosage forms, vol 2 Tablets, 2nd edn. Marcel Dekker, New York
Thurman EM, Mills MS (1998) Solid-phase extraction principles and practice. Wiley, New York
Washburn EW (1921) The dynamics of capillary flow. Phys Rev 17(3):273–283

Chapter 2
Sampling Considerations

Dawen Kou, Hua Ma, Edmund J. Bishop, Shangdong Zhan, and Hitesh P. Chokshi

Abstract Accurate analytical data for pharmaceutical dosage forms are dependent on adequate sample preparation and appropriate sample analysis methods. In addition, in order for the results to truly reflect the quality of the overall batch, the sample tested must be representative of the batch. Sampling is the first step and a critical aspect of the overall analysis process. Many analysts, however, are not as familiar with sampling techniques and principles as they are with analytical techniques or even sample preparation techniques. This chapter covers some fundamental sampling considerations and strategies relevant to pharmaceutical dosage forms. Both the theoretical aspects and applications in the development, manufacture, and quality control of pharmaceutical products are discussed. Examples of issues caused by sampling bias/errors are also given.

2.1 General Considerations

Sampling as a concept and practice is certainly not limited to chemical, physical, or biological analysis. Any work that deals with measurement of part of a population has to deal with sampling. Clinical trials, for example, rely on results from a limited number of patients (samples) to demonstrate statistical significance for the target population. The general theory of sampling and statistics is well discussed by Cochran in the book "Sampling Techniques" (Cochran 1977).

Major contributions to the development of the sampling theory in chemical testing were made by Gy, Ingamells, Visman, and Benedetti-Pichler, among others.

D. Kou (✉)
Pharmaceutical and Analytical R&D, Hoffmann La-Roche Inc.,
340 Kingsland Street, Nutley, NJ 07110, USA
e-mail: dawen.kou@roche.com

B. Nickerson (ed.), *Sample Preparation of Pharmaceutical Dosage Forms: Challenges and Strategies for Sample Preparation and Extraction*, DOI 10.1007/978-1-4419-9631-2_2,

A discussion on their individual work can be found in a review article (Kratochvil et al. 1984). Many types of chemical tests are destructive testing, which means that the samples are destroyed during testing. It is impractical to test 100% of the products, otherwise there will be no products left. Even for non-destructive tests, 100% testing, especially for a large lot, can be time and cost prohibitive. Therefore, sampling is inevitable before testing can begin.

2.1.1 Gy's Theory of Sampling

Widely recognized as a leading expert in the field of sampling, Pierre Gy published his first paper on sampling in 1950 (Gy 1950; Gy 2004e). His work has spanned the last 50 plus years and evolved into a comprehensive and relatively complete theory, with the publication of three books and more than 200 papers by Gy, and more by others (Gy 1979, 1998; Minkkinen 2004; Petersen et al. 2005). The latest version of the theory was published in 2004 in five installments (Gy 2004a, b, c, d, e). Gy's theory was originally developed for mining and geochemical applications, but many concepts and principles established in the theory are universal and applicable to other practical fields and scientific disciplines. Some of the concepts and principles are briefly discussed in this section.

Sampling can be understood as a process of mass reduction to obtain a representative portion of the whole. A sampling plan must address both the qualitative and quantitative aspects of sampling. The qualitative question is how: e.g., what sampling tools and techniques are used, how they are used, and from what locations. The quantitative question is how much: e.g., what is the sample size (number of units) and what should be the unit sample weight.

Sampling, sample preparation, and sample analysis errors all contribute to the total error in analytical results, as shown in (2.1), where σ^2 is the variance:

$$\sigma^2_{\text{total}} = \sigma^2_{\text{sampling}} + \sigma^2_{\text{sample preparation}} + \sigma^2_{\text{sample analysis}} \tag{2.1}$$

It was reported that errors caused by sampling bias/errors can be 100 times that of sample analysis related bias/errors. (Gy 1998). It is obvious that minimizing the errors caused by sampling is critical for accurate test results representative of the batch.

Gy's theory provides different mathematical treatments for two models of sampling: sampling from zero-dimensional objects and sampling from one-dimensional objects. The dimension in this case does not refer to physical dimensions but order, particularly the order in time. There is the absence of order in time in zero-dimensional objects such as a stationary batch of materials, while time is the order or dimension in one-dimensional objects such as materials in a flow stream or on a manufacturing conveyor belt.

The different types of errors present in sampling from zero-dimensional objects are summarized in Fig. 2.1 (Gy 2004a). Total sampling error (TSE) includes two

FIRST CASE: SAMPLING OF ZERO-DIMENSIONAL OBJECTS

AT EACH SAMPLING SUB-STAGE

□□□□□

Selection PROBABILITY can be either UNIFORM or NOT.
The total sampling error *TSE* is the sum of two, and only two, components:

□□□□□

SELECTION WITH UNIFORM PROBABILITY.
ERRORS RESULT FROM THE TWO FORMS OF HETEROGENEITY OF THE MATERIAL:

"CORRECT SAMPLING ERRORS" *CSE*

□□□□□

SELECTION WITH NON-UNIFORM PROBABILITY. ADDITIONAL ERRORS ARE GENERATED BY WRONG DESIGN OR USE OF INCORRECT SAMPLING EQUIPMENT OR METHODS:

"INCORRECT SAMPLING ERRORS" *ISE*

$$TSE = CSE + ISE$$

CORRECT SAMPLING ERROR *CSE*
The Correct Sampling Error *CSE* is the sum of two, and only two, components:

□□□□□

COMPONENT RESULTING FROM THE CONSTITUTIONAL HETEROGENEITY OF THE MATERIAL
"FUNDAMENTAL SAMPLING ERROR" *FSE*

□□□□□

COMPONENT RESULTING FROM THE DISTRIBUTIONAL HETEROGENEITY OF THE MATERIAL
"GROUPING AND SEGREGATION ERROR" *GSE*

$$CSE = FSE + GSE$$

INCORRECT SAMPLING ERROR *ISE*
The Incorrect Sampling Error *ISE* is the sum of three, and only three, components:

□□□□□

GEOMETRICAL DELIMITATION OF THREE-DIMENSIONAL INCREMENTS
"INCORRECT DELIMITATION ERROR" *IDE*

□□□□□

EXTRACTION OF MATERIAL-INCREMENTS from THREE-DIMENSIONAL INCREMENTS
"INCORRECT EXTRACTION ERROR" *IEE*

□□□□□

PREPARATION AND HANDLING OF MATERIAL-INCREMENTS AND SAMPLE
"INCORRECT PREPARATION ERROR" *IPE*

$$ISE = IDE + IEE + IPE$$

RECAPITULATION : Zero-Dimensional Objects
TOTAL SAMPLING ERROR, *TSE*

$$TSE = CSE + ISE = (FSE + GSE) + (IDE + IEE + IPE)$$

Fig. 2.1 Errors present during sampling from zero-dimensional objects. (Reproduced from Gy (2004a), with permission from Elsevier)

SECOND CASE: SAMPLING OF ONE-DIMENSIONAL OBJECTS

A flowing lot L is a one-dimensional object. To all intents and purposes, however, it can be regarded as a sequence of Zero-Dimensional Objects.
These are the adjacent Potential-Increments centred on the Point-Increments.
Sampling of L is then a two-step process:

1. Selection, on the time axis, of immaterial, extension-less, Point-Increments ,
2. Extraction of the Material-Increments about the Point-Increments or *Materialization*

SELECTION OF EXTENSION-LESS POINT-INCREMENTS.
This selection generates the:
"POINT-SELECTION ERROR" ***PSE***

MATERIALIZATION OF THE POINT-INCREMENTS

THIS IS A ZERO-DIMENSIONAL PROBLEM THAT GENERATES
"THE MATERIALIZATION SAMPLING ERROR" ***MSE***

TSE = PSE + MSE

The materialization can be either CORRECT or INCORRECT. IN THE GENERAL CASE, MSE is therefore the SUM OF TWO, AND ONLY TWO, MAIN COMPONENTS (SEE ABOVE):

MSE = CSE + ISE

AT EACH SAMPLING SUB-STAGE

THE TOTAL SAMPLING ERROR TSE IS THEREFORE THE SUM OF SIX COMPONENTS BELONGING TO THREE CLASSES

TSE = PSE + CSE + ISE = PSE + (FSE + GSE) + (IDE + IEE + IPE)

Fig. 2.2 Errors present during sampling from one-dimensional objects. (Reproduced from Gy (2004a), with permission from Elsevier)

types of errors: Correct sampling error (CSE) and incorrect sampling error (ISE). CSE can be further divided into fundamental sampling error (FSE) and grouping and segregation error (GSE). ISE can be further divided into incorrect delimitation error (IDE), incorrect extraction error (IEE), and incorrect preparation/processing error (IPE) due to contamination, loss of material, alteration in composition, and involuntary and deliberate faults. CSEs are not true errors but the inherent variance or heterogeneity of the material. CSE is zero if the lot is absolutely homogeneous, which is rarely the case in real-world situations. CSE cannot be further reduced while ISE can be minimized with a proper sampling plan.

Figure 2.2 (Gy 2004a) is a summary of the errors in sampling from one-dimensional objects. In addition to the errors in zero-dimensional objects, the TSE when sampling

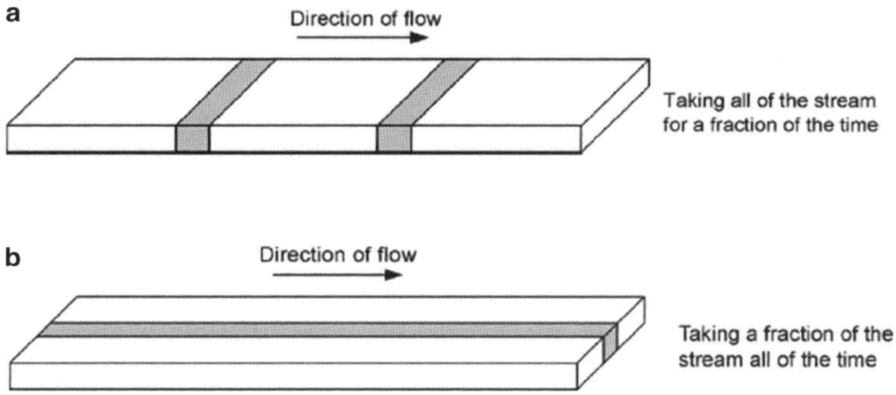

Fig. 2.3 (**a**) The correct way and (**b**) incorrect way of sampling from a flow stream. (Reproduced from Gy (2004a), with permission from Elsevier)

one-dimensional objects also includes point selection error (PSE). As illustrated in Fig. 2.3 (Gy 2004a), in sampling from a one-dimensional flow stream, the right way is to take the entire stream for a fraction of the time. The wrong way is to sample a fraction of the stream all of the time. The correct approach when applied to sampling from powder blends is also known as the "golden rule of sampling": (1) samples should be collected in motion and (2) samples should be collected from the powder bed at small time intervals throughout the entire powder stream rather than at a pre-selected site at all times (Allen 1997).

2.1.2 Primary Sampling and Secondary Sampling

Generally speaking, sampling can also be classified as primary sampling and secondary sampling, which if not distinguished can lead to confusion when the term sampling is used in various situations (Gy 2004a). Primary sampling (or field sampling) can be defined as sampling from one or more lots or batches. Secondary sampling (or lab sampling) refers to sampling from the samples received in the lab from primary sampling. Secondary sampling can also be considered as the first step of sample preparation.

2.2 Strategies for Primary Sampling

The appropriate strategies for primary sampling are dependent on the objective of the application and the nature of the samples. For example, sampling in mining is to get a representative assay of an often extremely large and heterogeneous mass (e.g., tens of thousands of tons of ores). Mixing, particle size reduction (via comminuting), and composite sampling are often necessary to provide a suitable sample. This strategy is applicable to sampling for materials of greatly heterogeneous composition, size, properties, and distribution. Pharmaceutical products are in units that appear identical to each other, be it tablets, capsules, solution in vials, or some other forms. Therefore, primary sampling for pharmaceutical dosage forms resembles sampling for other types of uniform products more than sampling for materials of great heterogeneity.

2.2.1 Acceptance Sampling

Acceptance sampling is based on statistics. The US military was the first major organization to apply acceptance sampling plans for procurement, starting from the period of World War II. The sampling plans were further developed by academia, and subsequently adopted widely outside of the military for sampling, testing, and making dispositions (acceptance or rejection) of a lot or batch of products.

If a test leads to a binary result (pass or fail), then it is sampling by attributes. If a test leads to a continuous measurement, then it is sampling by variables. The military standard MIL-STD-105 (the latest version 105E in 1989) was developed for sampling by attributes (MIL-STD-105E 1989). Its civilian equivalents are ANSI/ASQ Z1.4 and ISO Standard 2859, adopted by the American National Standards Institute (ANSI) in 1971 and by the International Organization for Standardization (ISO) in 1974, respectively (ANSI/ASQ Z1.4 2008). The MIL-STD-144 was developed for sampling by variables, with its civilian counterparts being ANSI/ASQ Z1.9 and ISO 3951 (ANSI/ASQ Z1.9 2008). The military standards 105 and 144 were withdrawn in 1995 and 1999 for economical reasons, since the civilian standards serve the same purposes.

Acceptable quality level (*AQL*) is defined in MIL-STD-105E as "The maximum percent defective (or the maximum number of defects per hundred units) that, for purposes of sampling inspection, can be considered satisfactory as a process average." In other words, a lot with a defective rate below *AQL* is to be accepted. Rejectable quality level (*RQL*) or lot tolerance percentage defective (*LTPD*) is the defect level above which a lot is to be rejected. For example, a sampling plan with an *AQL* of 1% and an *RQL* of 5% means that a lot should be accepted if its defective rate is below 1%, and should be rejected if the defective rate is above 5%. A sampling plan provides the means to achieve the desired *AQL* and *RQL*, by specifying the appropriate sample number (n) and acceptance number (a). A lot with defects above a among the n samples will be rejected. However, there are risks or uncertainties in achieving the target *AQL* and *RQL* by using a specific sampling plan. The type one risk (a.k.a. producer's risk, or R_p) is the risk of rejecting a good lot with a defective rate equal to *AQL*. The type two risk is the risk of accepting a bad lot (consumer's risk or R_c) at *RQL/LTPD*.

An acceptance sampling plan is best understood through an operating curve (*OC*). Figure 2.4 (Breunig and King 1962) is an example of an *OC* curve of an acceptance sampling-by-variables plan, where the four key parameters are specified as follows: the $AQL = 10\%$, the $RQL = 20\%$, the risk level of rejecting a good lot, $R_p = 5\%$, and the

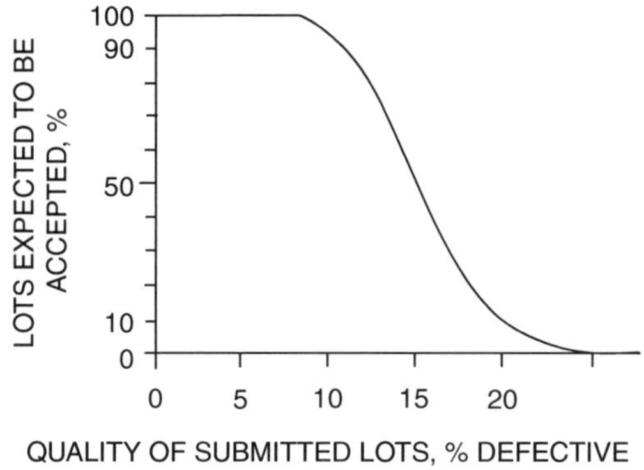

Fig. 2.4 Typical *OC* curve from MIL-STD-414, illustrating $AQL = 10\%$; $RQL = 20\%$; $R_p = 5\%$; $R_c = 10\%$; $n = 85$. (Reproduced from Breunig and King (1962), with permission from Wiley InterScience)

risk level of accepting a bad lot, $R_c = 10\%$. Then the corresponding number of samples (n) needed is determined to be 85, according to the established statistical tables in MIL-STD-144 (1957). As shown by the *OC* curve, using this sampling plan, there is a 95% of certainty that a lot with 10% of *AQL* (a good lot) will be accepted, or 5% of risk that such a lot will be rejected. The same plan also provides a 90% of certainty that a lot with 20% of *RQL* (a bad lot) will be rejected, or 10% of risk that the lot will be accepted. The statistical chance of accepting a lot with a quality level between *AQL* and *RQL* can also be assessed by the corresponding point on the *OC* curve.

2.2.2 The Square Root of N Plus One Rule

In the square root of *N* plus one rule, *N* is the number of units in the lot. The number of the samples to be taken from the lot is simply calculated as $\sqrt{N} + 1$. For example, a lot of products are stored in 1,000 drums. The number of drums to sample from is calculated to be $\sqrt{1{,}000} + 1 = 33$.

The square root of *N* plus one rule has been applied to several situations. One is to use the rule to calculate the number of containers to sample from, while the actual sample size (number of samples) is determined by a statistical method such as the ANSI/ASQ Z1.4. Another is to use the rule to calculate the number of containers to sample from, and then to create a composite sample from the selected containers. A third scenario is to use the rule as a sampling plan, and to accept a lot with zero defects and reject with one or more defects.

Although without official reference, the origin of the rule has been traced to sampling from agricultural products in the 1920's (Izenman 2001). Unlike the acceptance sampling plans discussed above, the square root of *N* plus one rule is not based on statistics. Because of this, its validity has been a matter of debate, questioned by some but defended by others (Quackenbush and Rund 1967; Saranadasa 2003; Torbeck 2009). Saranadasa compared the rule with a statistical method based on the Edgeworth approximation (Saranadasa 2003). It was determined that for a lot size larger than 30 units and in normal distribution, the rule provides "that at least 90% of the time, the 95% confidence interval would cover the population mean."

Despite the different views, and largely thanks to its simplicity, the square root of *N* plus one sampling rule is widely used in various industries, including pharmaceuticals. In the FDA's Investigations Operations Manual (IOM), it states in Chap. 4 Sampling, Sect. 4.7.3.2 Random Sampling for Adulteration Violations: "a general rule is to collect samples from the square root of the number of cases or shipping containers but not less than 12 or more than 36 subs in duplicate. If there are less than 12 containers, all should be sampled" (US FDA 2010).

2.3 Strategies for Secondary Sampling

What constitutes an appropriate secondary sampling plan is largely dependent on the objective of the analysis. For example, if the analysis is to determine the homogeneity of a suspension, the sampling plan must cover its spatial distribution. If the analysis is to determine the uniformity of dosage units (UDU), then a sufficient number of samples should be taken to achieve statistical significance. For dissolution testing, the samples should be taken across the entire time span of dissolution process. A proper sampling strategy is also dependent on the matrix of the drug product. In general, solutions are more homogeneous than solids and suspensions. Sampling and testing for homogeneity, which is necessary for suspensions, for example, is not needed for solutions. This section is not intended to provide a complete and comprehensive review of the secondary sampling requirements for all types of pharmaceutical products, but instead to offer an analysis of the commonality and differences among various tests.

2.3.1 Sampling and Compendial Methods

Compendial methods, e.g., USP methods, specify the number of units to be used for a specific test. This type of sampling is secondary sampling. To analysts working in the lab, these units are often called samples. However, USP considers all the units in a single test a single sample and the test a singlet determination. The sampling and acceptance criteria in USP are meant to be valid for the sample, not the lot. It is clearly stated in the USP: "At times, compendial standards take on the character of statistical procedures, with multiple units involved and perhaps a sequential procedural design to allow the user to determine that the tested article meets or does not meet the standard. The similarity to statistical procedures may seem to suggest an intent to make inference to some larger group of units, but in all cases, statements about whether the compendial standard is met apply only to the units tested. Repeats, replicates, statistical rejection of outliers, or extrapolations of results to larger populations are neither specified nor proscribed by the compendia; such decisions are dependent on the objectives of the testing (USP 2009)."

USP is meant to be prescribed standards and are not a set of acceptance specifications. Torbeck wrote an excellent article explaining the purpose and importance of USP standards being absolute and not negotiable (Torbeck 2005). Using USP tests alone as release tests for a large-size batch has been recognized as inadequate (Breunig and King 1962; Murray et al. 1995; Tsong et al. 1995). A batch passing USP testing once does not mean that it can pass USP testing at any time. USP testing is not intended to replace primary sampling and additional testing procedures and manufacturing specifications. Instead, the latter should be used to meet the former. As stated in the USP: "The manufacturer's release specifications, and compliance with good manufacturing practices generally, are developed and followed to assure that the article will indeed comply with compendial standards until its expiration date when stored as directed (USP 2009)."

The pharmaceutical industry has largely adopted a zero acceptance policy regarding USP test failures. If any sample from a lot at any time within expiry date fails to meet USP standards, the lot will be considered defective. Therefore, it is advisable for manufacturers to develop statistical sampling plans and release specifications that are more stringent than USP standards. Bergum has proposed a statistical procedure to construct acceptance limits for multiple stage tests that will provide an acceptable probability (e.g., 95%) for a given sample to meet USP requirements with a chosen level of confidence (e.g., 90%) (Bergum 1990). Following this seminal paper and subsequent work (Bergum and Utter 2000, 2003), Bergum and Li (2007) also proposed acceptance limits for the new ICH USP 29 content uniformity test (USP <905> Uniformity of dosage units 2009), which became official on 1 January 2007. The validity and value of this statistics-based approach is being increasingly recognized. The American Society for Testing and Materials (ASTM) recently adopted this approach to develop ASTM E2709-09 "Standard practice for demonstrating capability to comply with a lot acceptance procedure" (ASTM 2009).

2.3.2 *Average and Variability*

Replicate units are used to test for the average and variability of dosage units. In general, for a homogenous sample, using averages can provide a more accurate result. For assays and impurities, replicate units are prepared and analyzed, and a single final result is reported instead of the individual results of the replicates.

In tests for variability (e.g., UDU), multi-stage sampling, testing, and acceptance criteria are adopted. As defined in the latest version of USP <905>, for UDU testing, at first 10 units are selected and tested, and evaluated against the acceptance criteria. If the criteria are not met, then an additional 20 units are sampled, tested, and evaluated against the acceptance criteria for 30 units. The acceptance value (AV) is calculated using (2.2) (USP <905> Uniformity of dosage units 2009):

$$\mathrm{AV} = \left|M - X\right| + k_s, \tag{2.2}$$

where X is the mean of individual sample values, M is the reference value ($M = X$ if $98.5 \le X \le 101.5$; $M = 98.5$ if $X < 98.5$; $M = 101.5$ if $X > 101.5$), k is the

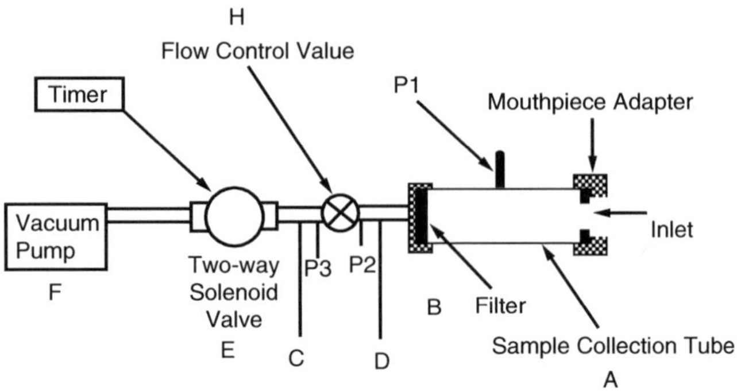

Fig. 2.5 Schematic diagram of a dosage unit sampling apparatus (DUSA) used for dry powder inhalers. (Reproduced from USP <601> (2009), with permission from the United States Pharmacopeia)

acceptability constant ($k=2.4$ if $n=10$, and $k=2.0$ if $n=30$), and s is the sample standard deviation. The calculated acceptance value is then compared to the maximum allowed acceptance value ($L_1=15.0$). The maximum allowed range for deviation of each dosage unit tested from the calculated value of M is (1 ± 0.01) (L_2) M, where $L_2=25.0$. All values in (2.2) and the acceptance criteria are in percentage of label claim. The acceptance criteria are based on statistical tolerance levels, taking into account both the average and variability of the individual results.

A similar approach is applied to inhalers and pre-metered dosage units labeled with a named inhalation device. In this case, specialized sampling apparatus are used at specific operating conditions. As an example, a schematic diagram of sampling apparatus for drug powder inhalers is shown in Fig. 2.5. The dosage unit sampling apparatus (DUSA) should operate at a pressure drop of 4 kPa for the duration of time to allow 4 L of air, as specified in USP <601>, to be drawn through the mouthpiece of the inhaler (USP <601> Aerosols, nasal sprays, metered-dose inhalers, and dry powder inhalers 2009).

Delivered-dose uniformity requires the test of ten inhalers, with one dose from each inhaler, for stage 1 test. The stage 1 criteria are that no less than nine of the ten doses are in the range of 75–125% of the specified target-delivered dose and none is outside of 65–135%. If the results fail to meet the stage 1 criteria, then 20 additional inhalers are tested for stage 2. The stage 2 criteria are that no more than three out of the 30 values are outside 75–125% and none is outside of 65–135%. Inhalers containing multiple doses need to be tested for delivered-dose uniformity over the entire contents. First, ten doses from one inhaler are tested as stage 1 and evaluated. For a dry powder inhaler, the ten doses are three at the beginning, four in the middle [$(n/2)-1$ to $(n/2)+2$, where n is the number of doses on the label], and three at the end. If results from the ten doses do not meet the acceptance criteria, two additional inhalers are selected, and ten doses from each are tested as stage 2. The stage 1 and 2 criteria are the same as described in delivered-dose uniformity test.

Dissolution testing is an example of testing both mean and variability. Dissolution also involves what can be called tertiary sampling, as samples are drawn from the dissolution medium in which a dosage unit is dissolving at pre-determined time intervals to generate a dissolution profile. USP <711> adopts a three-stage procedure for dissolution testing, with 6, 6, and 12 units for each stage, respectively (USP <711> Dissolution 2009). Tsong et al. argued that the USP dissolution acceptance criteria, when for product release, are dependent on the average value of all units and do not reject a lot or batch that has a large percentage of tablets that dissolve with less than the label specification Q. They proposed an acceptance rule based on a sampling-by-variables approach, which is shown to provide tighter control on the percentage of tablets that dissolve with less than Q (Tsong et al. 1995).

2.3.3 OOS and Re-sampling

If the analytical result of a sample is out of specifications (OOS), this will trigger an analytical investigation. If the investigation finds no error in the original analysis, the next step will be a retest of the original sample and/or re-sampling and analysis of new sample preparations.

Retesting uses the original homogenous sample (e.g., a solution formulation) or sample preparation (e.g., a composite sample for assay) and serves to verify or invalidate the original analysis. The maximum number of repeated testing allowed should be based on statistics and be pre-defined in a Standard Operating Procedure (SOP). It should not be adjusted depending on the repeat testing results obtained, because this would lead to the unacceptable practice of "testing into compliance."

Re-sampling and new sample preparations serve to verify or invalidate the original sample preparation. Re-sampling preferably should be from the same (primary) sample. Only when there are no sufficient sample units available, should a new primary sample be taken from the lot. Detailed procedures regarding the investigation of OOS, retesting, re-sampling, and reporting (as well as applicability of statistics on outlier values) can be found in a recently published FDA Guidance for Industry (US FDA 2006).

2.4 Sampling in Various Stages of Development and Manufacturing

Sampling and testing finished products can only verify the quality of the products, not create or improve it. To achieve the desired quality in final products, the individual steps of the development and manufacturing process must be well designed, properly carried out, and have achieved the pre-defined acceptance criteria. To that end, appropriate sampling and testing plans and acceptance criteria must be in place for every stage of manufacturing, from powder blends, to in-process dosage units, to finished units, and from the development batch, to the validation batch, to routine manufacturing batches.

2.4.1 Blend Uniformity Analysis

Sampling from blends for blend uniformity analysis (BUA) can be more challenging than sampling for finished products. The starting materials (drug substance and excipients) are inherently heterogeneous and require extensive processing to achieve acceptable homogeneity. Even after blend uniformity is demonstrated during mixing, the components are subject to segregation upon subsequent storage and handling, and may become less homogenous (Muzzio et al. 2003). For this reason, the FDA has taken the position that the blend uniformity acceptance criteria need to be more stringent than the content UDU criteria in order to allow room for segregation in powder blends and still meet the product specifications (US FDA 1994, 1996, 1999).

The FDA has published a series of documents related to BUA sampling and testing criteria (US FDA 1994, 1996, 1999, 2003), and the latest was a draft "Guidance for powder blends and finished dosage units – stratified in-process dosage unit sampling and assessment" (US FDA 2003). After being finalized, it will represent the agency's current thinking on stratified sampling strategies and acceptance criteria appropriate during various stages of development and manufacturing. Here stratified sampling means that the samples are taken from pre-determined time intervals and pre-selected locations. In-process samples are defined as dosage units before coating and packaging.

The guideline addresses three important questions regarding sampling: sampling location, sample size (number), and unit sample weight. It recommends the unit sample weight from blends to be 1–3× of the weight of the final dosage form, but also allows up to 10× the weight with adequate scientific justification. This is a noteworthy change from the strict 1–3× rule set forth by Judge Alfred M. Wolin in the court decision (The United States of America vs. Barr Laboratories et al. 1993). That decision was intended to prevent excessive large sample sizes that may mask localized heterogeneity. However, it has been reported that sampling bias inherent in certain sampling techniques and the limits of small sample weight led to powder samples not representative of the bulk blend (Berman and Planchard 1995; Garcia et al. 1995; Berman et al. 1996). Such evidences appeared to have been taken into consideration by the draft Guidance. Regarding sampling locations, samples should be taken from areas where poor mixing can occur leading to extreme high and low results. The sample sizes appropriate for blends and dosage units are described in Figs. 2.6 and 2.7 (US FDA 2003).

To verify that powder blends have achieved adequate mix, sampling and testing at every stage of development is needed. First, for the development batch,

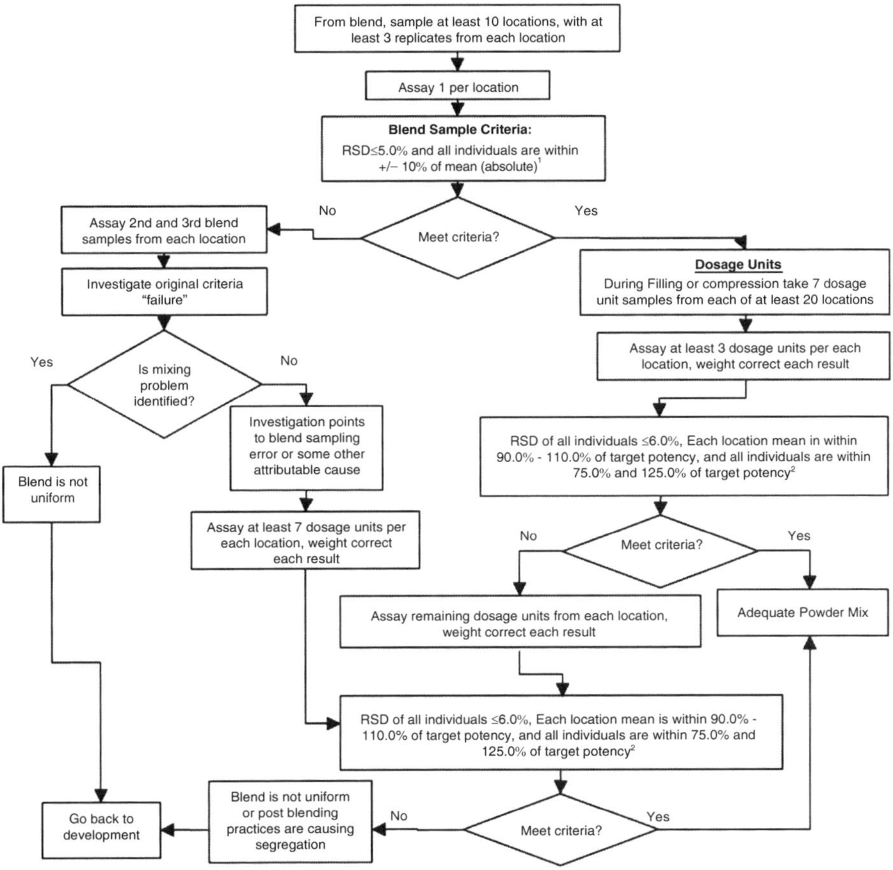

[1] Examples of "mean +/– 10% (absolute)" are: If the mean strength = 95%, then the interval is 95% +/– 10%; thus, all individuals must fall within 85.0% to 105.%. If the mean strenth = 103.0%, then the interval is 103.0% +/– 10.0%; thus all individuals must fall within 93.0% to 113.0%.

[2] When comparing individual dosage units to 75.0% - 125.0% of target strength, use the *as is* results (not corrected for weight).

Fig. 2.6 Sampling, testing, and manufacturing verification criteria for blend and dosage unit uniformity

correlation between powder mix uniformity and in-process dosage unit uniformity and finished product uniformity must be established. Twenty samples per location from at least ten locations are recommended for dosage units sampling. After the correlation is demonstrated, the next step is verification of manufacturing criteria using the exhibit and process validation batch. The detailed steps for sampling, testing, and acceptance criteria are illustrated in Fig. 2.6. Three samples per location from at least ten locations are recommended for powder blends, and seven samples per location from at least 20 locations are recommended for dosage units. Only when both the blend samples and in-process samples meet their criteria, is the powder mix considered adequate for the exhibit and validation batch. It is noted that

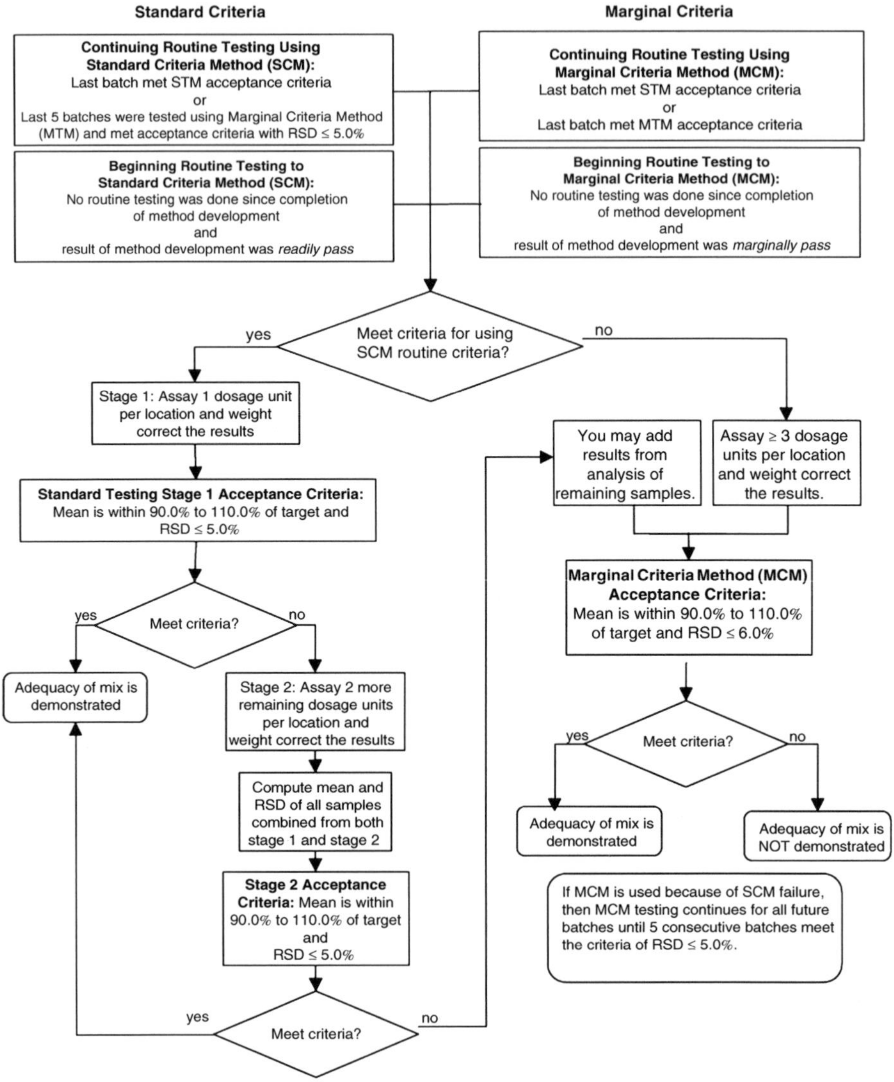

Fig. 2.7 Sampling and acceptance criteria for routine manufacturing batch

the recommended acceptance criteria for powder blends are more stringent than the acceptance criteria for dosage units.

The sampling strategies for routine manufacturing batch testing and corresponding acceptance criteria are shown in Fig. 2.7. Ten sampling locations during capsule filling or tablet compression should be identified to represent the routine manufacturing batch. Three samples are collected from each sampling location. A two-stage test and two types of acceptance criteria, standard criteria method (SCM) and marginal criteria method (MCM), are used to assess the results.

2.4.2 *Sampling Errors/Bias*

Errors or bias in blend sampling can be influenced by multiple factors, such as the type of sampling device, the sampling technique, the properties of the powder blend, the sampling location, and the sample weight. Berman wrote an excellent review article in which the causes of sampling errors were examined in detail (Berman 2001).

The conventional tool of sampling from powder blends is a sampling thief probe that is inserted into a powder bed. It is well documented that thief probes tend to disturb the powder bed during insertion by dragging the particles in the upper layer deeply into the powder bed, causing potential local segregation (Harwood and Ripley 1977; Chang et al. 1996; Garcia et al. 1998; Muzzio et al. 1999). As a result, the collected powder samples may not reflect the true quality of the mixture at a given location. Coarse particles can be preferentially sampled from the top of the blender when a thief probe is maintained in a vertical position rather than at an acute angle. Moreover, there are many types of sampling thieves. Berman reported that sampling from the same blend using two different sampling thieves led to different results (Berman et al. 1996).

Particle flowability also plays a role in powder sampling via thief probes. If certain components in a formulation are more free flowing than others, they can be collected more selectively in a sampling thief. If the excipients are more free flowing than the active pharmaceutical ingredient, this can lead to an assay lower than the true value. A higher assay can be obtained vice versa. Static charges on bulk blends can also lead to sampling errors. Blends sometimes have to be grounded for days for the static to completely discharge. Samples drawn from the bottom of the container may be more compressed than the top portions. In general, the lower the drug concentration in the blend, the smaller the sample weight, the more likely the analysis is subject to sampling errors.

Sampling bias was demonstrated in the following real example (Berman and Planchard 1995). In an attempt to validate the manufacturing process of two new but lower strengths (at X mg and 2X mg) of an existing tablet product (at 4X mg), multiple samples were taken from the blends in the V-blender and the hoppers in X mg and 2X mg sample weight. As shown in Fig. 2.8, despite outstanding content uniformity of the tablets at all three strengths, the blend uniformity failed to meet the acceptance criteria. A thorough investigation found no analytical error, and the low assays in the blend samples were most likely caused by sampling bias as discussed in the preceding paragraphs. Nonetheless, the manufacturing process could not be validated due to the limits of sampling techniques and procedures allowed in that circumstance.

2.5 Sampling and Process Analytical Technology

Process analytical technology (PAT) has been gaining a lot of momentum in recent years in the pharmaceutical industry. It is aimed to provide a mechanism to design, analyze, and control the operation processes through real-time measurement of

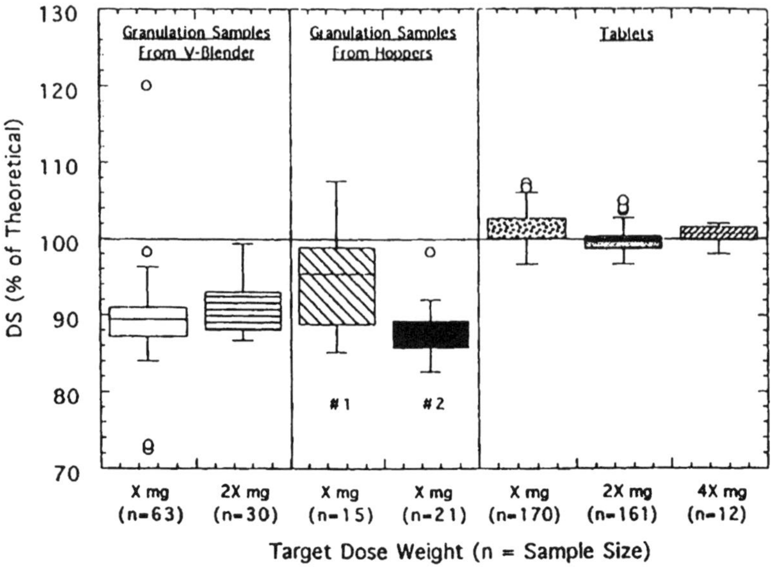

Fig. 2.8 Uniformity of a blend and uniformity of dosage units made from the blend. (Reprinted from Berman and Planchard (1995), with permission of the publisher (Taylor & Francis Group, http://www.informaworld.com))

critical process parameters (CPP), which affect critical quality attributes (CQA). PAT is closely related to quality by design (QbD), another major initiative in recent years by the FDA. PAT relies heavily on spectroscopy-based technologies, such as near infrared (NIR), Raman, and light-induced fluorescence (LIF). These techniques are non-destructive and avoid some of the problems associated with the use of sampling thieves mentioned above.

Online process monitoring of pharmaceutical unit operations can be achieved through collecting and analyzing real-time data (spectra) via an NIR sensor attached to the unit blender, fluidized bed, granulator, etc. In this approach, measurement is carried out over time through one or more sampling points where the sensor(s) are attached. Thus, sampling still plays an important role in PAT with respect to obtaining accurate determinations on critical material attributes and/or process end-points. Determination of the optimal sampling location and beam size for an NIR sensor is essential in order to achieve proper and robust process control. For pharmaceutical powder blending unit operations, several studies have discussed the potential impact of sampling location on the blending variability determination. Portillo et al. concluded from a blending simulation study that the sampling locations can dramatically offset blending variance distributions (Portillo et al. 2006). Ma and Anderson reported after analyzing NIR chemical images collected from a small-scale mixing process of a model pharmaceutical powder system that the areas near the blender edges demonstrated higher blending composition variation than the center (Ma and Anderson 2008). The results indicate that erroneous blending uniformity determination could be reached if the NIR sensor is not placed at the optimal sampling locations. Li et al. studied the effect of beam size on real-time determination of powder blend homogeneity by an online NIR sensor, and suggested that excessively large beam size could lead to underestimation of blend heterogeneity (Li et al. 2007). Green et al. investigated the sampling effects on method accuracy in the monitoring of moisture content in fluid bed dryers (Green et al. 2005).

2.6 Conclusions

This chapter has discussed some theoretical aspects of sampling and how they are applied in the analysis of pharmaceutical dosage forms. Sampling is an important aspect of cGMP procedures and quality control. Every stage of development and manufacturing relies on appropriate sampling plans to obtain representative and sufficient samples in order to make correct decisions. Sampling errors can lead to failure to validate a good batch. Sampling is an integral part of compendial methods, where entire analytical procedures and acceptance criteria are closely related to the stages of sampling and the number of samples for each stage. Sampling also plays an important role in the application of new technologies such as PAT. Anyone wishing to be well rounded in the overall process of pharmaceutical development and manufacturing will benefit from these fundamental understandings about sampling and may wish to learn more about this important subject.

References

21CFR Parts 210 and 211 (1996) Current good manufacturing practice: amendment of certain requirements for finished pharmaceuticals: proposed rule. Federal Register 61(87):20104–20105

Allen T (1997) Particle size measurement powder sampling and particle size measurement, 5th edn. Chapman & Hall, London, pp 1–4

American Society for Quality (2008a) ANSI/ASQ Z1.4 Sampling procedures and tables for inspection by attributes. ASQ, Milwaukee, WI

American Society for Quality (2008b) ANSI/ASQ Z1.9 sampling procedures and tables for inspection by variables for percent nonconforming. ASQ, Milwaukee, WI

American Society for Testing and Materials (2009) ASTM E2709-09 standard practice for demonstrating capability to comply with a lot acceptance procedure. ASTM, West Conshohocken, PA

Bergum JS (1990) Constructing acceptance limits for multiple stage tests. Drug Dev Pharm 16(14):2153–2166

Bergum JS, Li H (2007) Acceptance limits for the new ICH USP 29 content-uniformity test. Pharm Tech, October 2, 2007, www.pharmtech.com

Bergum JS, Utter ML (2000) Process validation. In: Chow SC (ed) Encyclopedia of biopharmaceutical statistics. Marcel Dekker, New York, pp 422–439

Bergum JS, Utter ML (2003) Statistical methods for uniformity and dissolution testing. In: Nash RA, Watchter AH (eds) Pharmaceutical process validation. Marcel Dekker, New York, pp 667–697. doi:667

Berman J (2001) The compliance and science of blend uniformity analysis. PDA J Pharm Sci Tech 55(4):209–222

Berman J, Planchard JA (1995) Blend uniformity and unit dose sampling. Drug Dev Ind Pharm 21(11):1257–1283

Berman J, Schoeneman A, Shelton JT (1996) Unit dose sampling – a tale of two thieves. Drug Dev Ind Pharm 22(11):1121–1132

Breunig HL, King EP (1962) Acceptance sampling of finished pharmaceutical products. J Pharm Sci 51(12):1187–1194

Chang RK, Shukla J, Buehler J (1996) An evaluation of a unit-dose compacting sample thief and a discussion of content uniformity testing and blending validation issues. Drug Dev Ind Pharm 22(9&10):1031–1035

Cochran WG (1977) Sampling techniques, 3rd edn. Wiley, New York, NY

Garcia T, Elsheimer B, Tarczynshi F (1995) Examination of components of variance for a production scale, low dose powder blend and resulting tablets. Drug Dev Ind Pharm 21(18): 2035–2045

Garcia TP, Taylor MK, Pande GS (1998) Comparison of the performance of two sample thieves for the determination of the content uniformity of a powder blend. Pharm Dev Tech 3(1):7–12

Green RL, Thurau G, Pixley NC, Mateos A, Reed RA, Higgins JP (2005) In-line monitoring of moisture content in fluid bed dryers using near-IR spectroscopy with consideration of sampling effects on method accuracy. Anal Chem 77:4515–4522

Gy PM (1950) Formule donnant la masse minimale d'echantillon (A formula for the minimum sample mass), unpublished paper

Gy PM (1979) Sampling of particulate materials – theory and practice. Elsevier, Amsterdam

Gy PM (1998) Sampling for analytical purposes. Wiley, New York, NY

Gy PM (2004a) Sampling of discrete materials – a new introduction to the theory of sampling I: quantitative approach. Chemo and Intel Lab Sys 74:7–24

Gy PM (2004b) Sampling of discrete materials II: quantitative approach – sampling of zero-dimensional objects. Chemo and Intel Lab Sys 74:25–38

Gy PM (2004c) Sampling of discrete materials III: quantitative approach – sampling of one-dimensional objects. Chemo and Intel Lab Sys 74:39–47

Gy PM (2004d) Part IV: 50 years of sampling theory – a personal history. Chemo and Intel Lab Sys 74:49–60

Gy PM (2004e) Part V: annotated literature compilation of Pierre Gy. Chemo and Intel Lab Sys 74:61–70

Harwood CF, Ripley T (1977) Errors associated with the thief probe for bulk powder sampling. J Powder Bulk Solids Tech 11:20–29

Izenman FC (2001) Statistical and legal aspects of the forensic study of illicit drugs. Stat Sci 16(1):35–57

Kratochvil B, Wallace D, Taylor JK (1984) Sampling for chemical analysis. Anal Chem 56:113R–129R

Li W, Johnson AC, Bruce R, Rasmussen H, Worosila GD (2007) The effect of beam size on real-time determination of powder blend homogeneity by an on-line near infrared sensor. J Pharm Biomed Anal 43:711–717

Ma H, Anderson CA (2008) Characterization of powder blends by NIR chemical imaging. J Pharm Sci 97(8):3305–3320

MIL-STD-105E (1989) Military standard sampling procedures and tables for inspection by attributes. Supt. of Documents, US Government Printing Office, Washington, DC

MIL-STD-144 (1957) Military standard sampling procedures and tables for inspection by variables for percent defective. Supt. of Documents, US Government Printing Office, Washington, DC

Minkkinen P (2004) Practical applications of sampling theory. Chemo and Intel Lab Sys 74:85–94

Murray M, Uraizee S, Sakr A (1995) Preliminary investigation of the suitability of the USP uniformity of dosage units tests for evaluating the uniformity of powder blends and their corresponding tablets. Pharm Ind 57:256–262

Muzzio FJ, Roddy M, Brone D, Alexander AW, Sudah O (1999) An improved powder-sampling tool. Pharm Tech 23(4):92–110

Muzzio FJ, Goodridge CL, Alexander A, Arratia P, Yang H, Sudah O, Mergen G (2003) Sampling and characterization of pharmaceutical powders and granular blends. Int J Pharm 250:51–64

Petersen L, Minkkinen P, Esbensen KH (2005) Representative sampling for reliable data analysis: theory of sampling. Chemo and Intel Lab Sys 77:261–277

Portillo P, Muzzio FJ, Ierapetritou MG (2006) Characterizing powder mixing processes utilizing compartment models. Int J Pharm 320:14–22

Quackenbush FW, Rund RC (1967) The continuing problem of sampling. J Assoc Off Anal Chem 50:997–1006

Saranadasa H (2003) The square root of N plus one sampling rule – how much confidence do we have? Pharm Tech 27(5):50–62

The United States of America vs. Barr Laboratories, et al. (1993) 812, F. Supp. 458, District of New Jersey

Torbeck LD (2005) In defense of USP singlet testing. Pharm Tech 29(2):105–106

Torbeck LD (2009) Statistical solutions: square root of (N) + 1 sampling plan. Pharm Tech 33(10):108

Tsong Y, Hammerstrom T, Lin K, Ong TE (1995) Dissolution test acceptance sampling plans. J Biopharm Stat 5(2):171–183

United States FDA (2010) Investigations operations manual, collection technique, Sect. 4.3.7.2 Random sampling http://www.fda.gov/ICECI/Inspections/IOM/ucm122525.htm#4.3.7.2. Accessed 8 Aug 2010

United States Pharmacopeia 32 – National Formulary 27 (2009) US Pharmacopeial Convention, Rockville, MD, http://www.uspnf.com/uspnf/pub/index?usp=32&nf=27&s=2&officialOn=December%201,%202009. Accessed 8 Aug 2010

US FDA (1994) Division of field investigations, guide to inspections of oral solid dosage forms pre/post approval issues for development and validation. United States FDA, Silver Spring, MD

US FDA (1999) CEDR, Office for generic drugs, draft guidance for industry ANDAs: blend uniformity analysis (withdrawn in 2002). United States FDA, Silver Spring, MD

US FDA (2003) CEDR, Draft guidance for industry: powder blends and finished dosage units - stratified in-process dosage unit sampling and assessment. United States FDA, Silver Spring, MD

US FDA (2006) CEDR, Guidance for industry: investigating out-of-specification (OOS) test results for pharmaceutical production. United States FDA, Silver Spring, MD

USP <601> Aerosols, nasal sprays, metered-dose inhalers, and dry powder inhalers (2009) United States Pharmacopeia 32 – National Formulary 27, http://www.uspnf.com/uspnf/pub/index?usp=32&nf=27&s=2&officialOn=December%201,%202009. Accessed 8 Aug 2010

USP <711> Dissolution (2009) United States Pharmacopeia 32 – National Formulary 27, http://www.uspnf.com/uspnf/pub/index?usp=32&nf=27&s=2&officialOn=December%201,%202009. Accessed 8 Aug 2010

USP <905> Uniformity of dosage units (2009) United States Pharmacopeia 32 – National Formulary 27, http://www.uspnf.com/uspnf/pub/index?usp=32&nf=27&s=2&officialOn=December%201,%202009. Accessed 8 Aug 2010

Section B
Extraction and Sample Preparation Techniques

Chapter 3
Agitation and Particle Size Reduction Techniques

Beverly Nickerson and K. Rick Lung

Abstract This chapter discusses techniques used to facilitate disintegration or dispersion of dosage forms. Shaking, stirring, vortexing, and sonication are common agitation techniques used to facilitate dispersion, mixing, and extraction of drug from various types of dosage forms. In many cases, particle size reduction techniques are used to increase the speed and efficiency of dosage form disintegration or dispersion. These techniques include grinding, milling or blending, homogenization, and sonication. Each of these techniques will be discussed.

3.1 Introduction

Chapter 1 discussed parameters that can be leveraged in sample preparation method development. Two of these parameters are agitation conditions and use of particle size reduction techniques. Agitation and particle size reduction can be critical for nonsolution dosage forms, to facilitate dispersion of the sample in order to enhance extraction of the drug. If the dosage form remains intact, recovery of the drug may be slow or incomplete because the drug is not adequately exposed to the diluent.

This chapter discusses techniques used to facilitate disintegration or dispersion of dosage forms. Shaking, stirring, vortexing, and sonication are common agitation techniques used to facilitate dispersion, mixing, and extraction of drug from various types of dosage forms. Particle size reduction techniques that are used to increase the speed and efficiency of dosage form disintegration or dispersion include grinding, milling or blending, homogenization, and sonication. Each of these techniques is

B. Nickerson (✉)
Analytical Development, Pfizer Global Research and Development,
Pfizer Inc., 558 Eastern Point Road, Groton, CT 06340, USA
e-mail: beverly.nickerson@pfizer.com

B. Nickerson (ed.), *Sample Preparation of Pharmaceutical Dosage Forms: Challenges and Strategies for Sample Preparation and Extraction*, DOI 10.1007/978-1-4419-9631-2_3,

discussed below. Additional discussion and examples of the use of these techniques in the preparation of solid oral dosage forms is covered in Chap. 7 and in Chap. 8 for select nonsolid dosage forms.

3.2 Agitation Techniques

Agitation is used to facilitate dispersion, mixing, and extraction of drug from various types of dosage forms. For dosage forms such as solutions and suspensions, agitation may be used to mix the dosage form with diluent to extract the drug or make the sample solution compatible with the analysis technique. For semisolid and solid dosage forms, agitation is used to disperse or disintegrate the dosage and extract the drug, although in some cases particle size reduction techniques (discussed in Sect. 3.3) may also be required. Agitation is also important to ensure a homogenous sample solution prior to analysis.

Agitation of the sample solution can be performed in a number of ways and the more common techniques include shaking, stirring, vortexing, and sonication. In a survey published by Majors in 2002, a little over 30% of respondents reported working on the analysis of pharmaceutical compounds. In this same survey, ~30% of respondents reported using mixing, ~35% reported using vortexing, and ~45% reported using sonication in sample preparation methods. More than 50% of respondents reported using multiple sample preparation techniques in a single method (Majors 2002). Agitation techniques are relatively cost-effective and simple to use. Each of these techniques is discussed below.

3.2.1 Shaking

Shaking sample solutions is a common practice in preparing samples of pharmaceutical dosage forms. Shaking may be performed manually or by using a mechanical shaker. Manual shaking is simple to perform, but can be labor-intensive for high sample numbers and in some cases can lead to analyst-to-analyst variability and impact the robustness of the method. As an example, Kirschbaum reported analyst-to-analyst variability in shaking sample flasks that resulted in up to a 3% difference in assay results (Kirschbaum 1989). Additional examples of issues with manual shaking are discussed in Case Studies 1 and 3 in Chap. 13.

Mechanical shakers have the advantage over manual shaking in the ability to shake multiple samples at a consistent speed. Sample preparation methods using mechanical shaking typically specify a speed (rpm: revolutions per minute; opm: oscillations per minute or cycles per minute) and time. Different types of shakers are commercially available (e.g., from Thermo Fisher Scientific, Waltham, MA; VWR International, West Chester, PA; IKA Works, Inc., Wilmington, NC) that can be used with different types of sample containers (e.g., volumetric or Erlenmeyer flasks, beakers, bottles, test tubes, microwell plates, and separatory flasks) and some

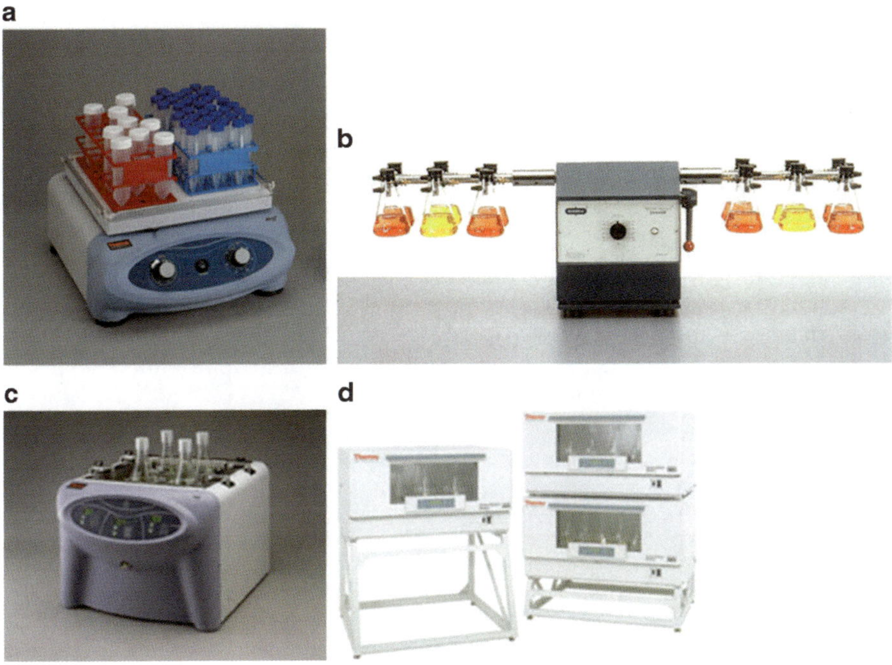

Fig. 3.1 Examples of (**a**) reciprocating shaker, (**b**) wrist-action® shaker, (**c**) water bath orbital shaker, (**d**) incubated stackable orbital shakers. (Figure 3.1 (**a, c, d**) reproduced with permission from Thermo Fisher Scientific. Figure 3.1 (**b**) reproduced with permission from Burrell Scientific)

models include incubation, refrigeration, or water bath options. Multiple samples can be shaken at one time to facilitate sample throughput and some models are stackable to save on laboratory space. Different types of shakers commonly used with sample preparation of pharmaceutical dosage forms include the types described below (Cole-Parmer Technical Library (a)) and examples of some of these are shown in Fig. 3.1.

Reciprocating shaker: Samples are placed on a platform that is moved in a back-and-forth motion to mix the sample solutions.
Orbital shaker: Samples are placed on a platform that is moved in a circular orbit to mix the sample solutions.
Reciprocating orbital shaker or *dual-action shaker*: Samples are placed on a platform that is moved in both an orbital and reciprocating motion.
Wrist-action® shaker: Samples are held in place on a horizontal rod and the flasks are moved in a motion that mimics the side-to-side action of hand mixing. The shaking amplitude and speed can be adjusted.
Rotating shaker: Samples are spun at adjustable angles.
Rocking shaker: Samples are placed on a platform that rocks in a "see-saw" motion.
Water bath shakers: Samples are placed on a platform inside a water bath for agitation at a controlled temperature.

As described above, shakers have different modes of action and all modes may not give equivalent results for a given drug product and sample preparation method. It is important to evaluate specific shaker types for a given application, particularly for agitation critical (e.g., viscous solutions, difficult to disperse dosage forms) sample preparation methods. If warranted, the shaker type should be specified in the method. As an example, Cain reported a comparison of orbital shaking versus reciprocating shaking for a composite sample preparation of 20 tablets for an IR dosage form. Based on a Design of Experiment study, it was shown that shaking speed (150–250 cycles per minute) and bottle shape (round or square base) were not significant factors. Extraction was constant at full recovery (approximately 96%) using a reciprocating shaker when shaking from 10 to 20 min. Extraction, however, increased from approximately 85% in 10 min to approximately 95% in 20 min when using the orbital shaker (Cain 2007).

3.2.2 Stirring

A magnetic stirrer contains an electric motor that spins a magnet that is under a chemical resistant flat surface. Alternatively, a set of stationary electromagnets may be used to create a rotating magnetic field. A flask containing the sample, diluent, and a magnetic stir bar is placed on the surface of the stirrer. When the magnetic stirrer is turned on, the rotating magnetic field in the unit causes the stir bar to spin and mix the solution. A stir bar retriever can be used to remove the stir bar from the solution after agitation is complete.

Magnetic stirrers and stirring hot plates are available from many vendors (e.g., IKA Works, Inc., Wilmington, NC; Thermo Fisher Scientific, Waltham, MA; VWR International, West Chester, PA) in a variety of sizes and configurations and some examples are shown in Fig. 3.2. Stir plates may be either analog or digital. Analog magnetic stirrers are economical, but are not designed to provide exact control of speed. In cases where stir speed control (and temperature control for stirring hotplates) is critical, a digital unit with electronic feedback controls is used. Digital stir plates and digital stirring hotplates allow programming to control speed, time, and temperature. Multiposition stir plates are commercially available and simultaneously stir a number of flasks either at the same speed or different speeds, depending upon the particular unit. Stirring hot plates and multiposition stirring hotplates allow stirring and heating at the same time. For viscous solutions, a stirrer with a large drive magnet, heavy duty motor, and capacity to accommodate long stir bars is typically required (Cole-Parmer Technical Library (b)). Magnetic stir bars are available in a variety of shapes (e.g., round, octagonal, starburst, double) and sizes (e.g., micro to large) and come with a PTFE (polytetrafluoroethylene) coating for chemical resistance.

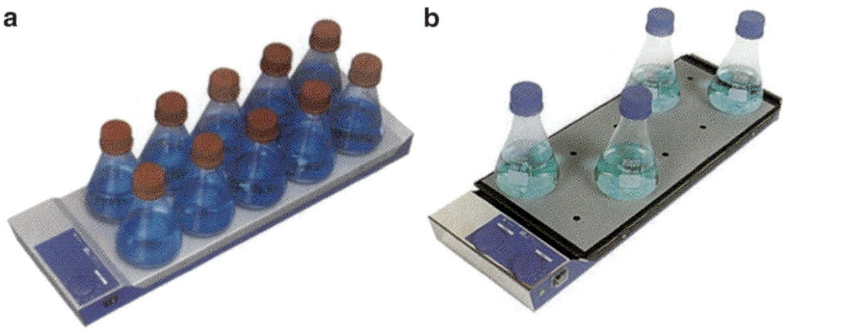

Fig. 3.2 Example of (**a**) multiposition magnetic stirrer and (**b**) multiposition stirring hot plate. (Reproduced with permission from IKA Works, Inc.)

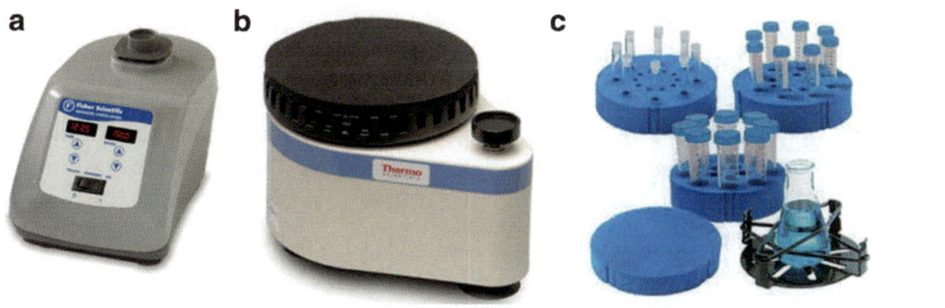

Fig. 3.3 Examples of (**a**, **b**) vortex mixers and (**c**) accessories. (Reproduced with permission from Thermo Fisher Scientific)

3.2.3 Vortexing

A vortex mixer or vortexer is another instrument commonly used to mix sample solutions in various types of containers (e.g., volumetric or Erlenmeyer flasks, beakers, test tubes, eppendorf tubes, microwell plates) and is available from a number of different vendors (e.g., IKA Works, Inc., Wilmington, NC; VWR International, West Chester, PA; Thermo Fisher Scientific, Waltham, MA). Some examples of vortex mixers are shown in Fig. 3.3. Mixing is achieved through the use of a rapid circular or orbital motion. The unit consists of an electric motor, drive shaft, and rubber cup that is attached slightly off-center to the shaft. The motor and drive shaft cause the rubber cup to move rapidly in a circular or orbital motion that when in contact with a sample container causes the solution to form a vortex. The vortexer has a heavy base and rubber feet to provide stability during use. There are fixed speed vortexers and analog vortexers that allow variable speeds (100–3,200 rpm) for gentle to vigorous mixing of sample solutions. Vortexers can be set to run continuously or only when a sample container is pressed down on the rubber cup (touch mode).

3.2.4 Sonication

Sonication can be considered both an agitation technique and a particle size reduction technique. Details of sonication are discussed below in Sect. 3.3.

3.3 Particle Size Reduction Techniques

Many of the traditional "shake flask" extraction techniques work well for dosage forms such as immediate release tablets, suspensions, powders for oral suspension, and lyophiles. In these cases, diluent can be added to the sample and the sample solution can be agitated (e.g., shaking, stirring) to extract and dissolve the drug substance. Agitation alone is typically not sufficient to disperse nondisintegrating dosage forms, such as extended release tablets. In these cases, several techniques are available to mechanically disperse solid samples such as nondisintegrating tablet formulations or tablet formulations that do not disintegrate readily. For example, the use of hydroxypropyl methylcellulose in sustained release formulation has been reviewed in the literature (Hogan 1989). HPMC formulated extended release tablets are designed to swell into a gel matrix and retain the active ingredient. As the HPMC gel matrix tablet passes through the GI tract, the HPMC matrix is slowly eroded into colloidal particles while the active ingredient is slowly released in the upper, mid-, and lower GI track. An extended release tablet formulated with HPMC will likely swell up quickly when it comes into contact with a typical extraction solvent that contains water. When such types of tablets are extracted in the laboratory using conventional agitation or sonication technology, it may take as much as 20 h to completely dissolve the gel matrix. Similarly, some enteric coatings that are designed to act as a protective barrier in the acidic environment commonly found in the stomach may cause extraction problems when traditional shake-flask methods of extractions are used. Sustained release formulations that are based on a polymeric film coating that controls the rate of diffusion through the coating can also be more difficult to extract than conventional immediate release tablets (Heller 1980; Tarvainen et al. 2004).

In order to speed up the sample extraction process, particle size reduction techniques are often necessary to effectively disperse a dosage form and speed up extraction and dissolution of the drug. In a survey published by Majors in 2002, respondents reported using various particle size reduction techniques in sample preparation: ~15% of respondents reported using blending, ~20% reported using homogenization, and ~45% reported using sonication (Majors 2002). These particle size reduction techniques are discussed in detail below. Specific applications of these techniques to solid oral dosage forms are discussed in Chap. 7.

3.3.1 Grinding

Grinding, which relies on pressure and friction, is a commonly used technique to disperse and reduce the particle size of solid dosage forms down to a fineness of approximately 10-μm diameter. Manual grinding is commonly performed using a mortar and pestle and this is one of the oldest ways of preparing samples for a composite assay of pharmaceutical tablets. Mortars and pestles made from different materials (e.g., porcelain, agate, glass) are available and grinding is typically performed dry (without water or solvent) but can also be performed wet (with water or solvent). Individual dosage units or a composite of multiple dosage units can be ground to a coarse or fine powder and either the entire amount or a portion of the powder is then used in subsequent steps of sample preparation.

Malleable samples can be frozen in liquid nitrogen to make them brittle prior to grinding. As an example, Zuo and colleagues froze nicotine chewing gum in liquid nitrogen, then cut the gum into small pieces and ground them in a mortar prior to extract the drug in diluent using sonication (Zuo et al. 2004). In some cases, crushing samples instead of grinding them to a fine powder is sufficient to improve recovery of the drug (Choi and Dong 2005; Oliveira et al. 2009). For example, Choi and Dong describe the evaluation of different agitation techniques, shaking, and sonication, with intact and crushed tablets for the sample preparation of a controlled-release tablet formulation. Since the API was soluble in aqueous and organic diluents and a polymer excipient present in the formulation was soluble in methanol, methanol was used as the sample diluent. Recovery of ≥99% of the API was achieved with intact tablets when using sonication for 30 min, while similar recoveries were achieved with crushed tablets with either 5 min of sonication or with 10 min of shaking (Choi and Dong 2005).

Manual grinding is simple, but can be labor-intensive for high numbers of samples and lead to analyst-to-analyst variability in the grinding technique. Mechanical mortar grinders (from e.g., Fritsch GmbH, Germany; Retsch Inc., Newton, PA) can be used to automate and standardize the grinding process and are cited in a number of papers (Nelson et al. 2006; Chen et al. 2009, 2010). An example of a mechanical mortar grinder is shown in Fig. 3.4. Mortars and pestles of different materials (e.g., chrome or stainless steel, porcelain, agate) are available for these instruments. Mechanical mortar grinders can be used dry or with solvent. In addition, liquid nitrogen can also be used to cool the mortar and pestle for thermally labile samples. While mechanical mortar grinders reduce manual labor, they may not speed up overall sample preparation since the mechanical unit still needs to be cleaned manually after each sample. Additional limitations of grinding samples include the potential loss of sample during transfer if the entire powder is required for analysis, potential segregation of particles leading to inaccurate results (see end of Sect. 3.3.2), and sample handling issues for high potency or hazardous compounds.

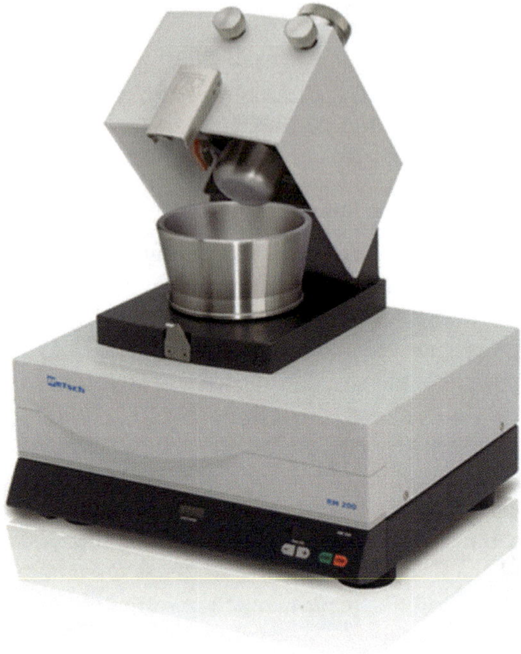

Fig. 3.4 Example of a mechanical mortar grinder. (Reproduced with permission from Retsch Inc.)

3.3.2 Milling or Blending

Samples may be milled or dispersed using various types of mills, such as ball mills, freezer mills, and knife mills or blenders, to obtain submicron particle sizes. Examples of different mills are shown in Fig. 3.5 and are discussed below.

3.3.2.1 Ball Mills

Ball mills (from e.g., Fritsch GmbH, Germany; Retsch Inc., Newton, PA) use impact and friction to pulverize hard, medium-hard, and brittle samples to approximately 10-μm-sized particles. The sample and single or multiple balls made from stainless steel or other materials (e.g., Teflon) are placed in the milling chamber. The milling chamber is then mechanically shaken, rotated, or vibrated. The ball(s) impact and pulverize the sample against the side of the milling chamber. Milling can be done dry or with solvent in the milling chamber. Heat may be generated during milling and is a potential concern for thermally labile compounds. The milling chamber may be placed in liquid nitrogen for several minutes prior to milling or dry ice can be added to the chamber to keep thermally labile compounds cool during milling or to make malleable or elastic samples brittle prior to milling. A cryogenic ball mill is also commercially available (Retsch Inc., Newton, PA; see section below on Freezer Mills). Other limitations of ball milling are the limited amount of solvent

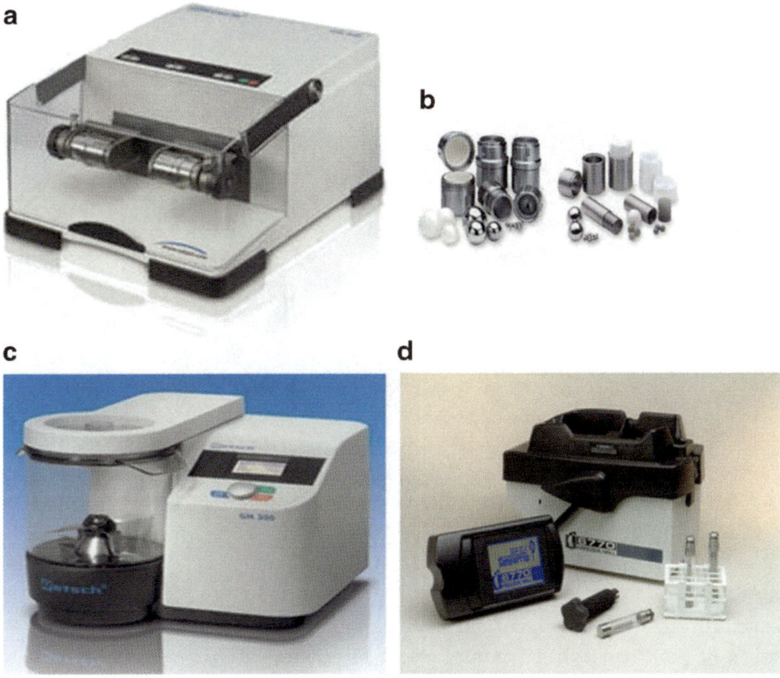

Fig. 3.5 Example of (**a**) a ball mill with (**b**) milling chambers and balls, (**c**) a knife mill, and (**d**) a freezer mill. (Figure 3.5(**a–c**) reproduced with permission from Retsch Inc. Figure 3.5(**d**) reproduced with permission from SPEX SamplePrep, LLC)

that can be used in the milling chamber (<25 mL) and the need for manual manipulations and potential loss of sample in transferring the milled sample from the chamber to a flask for subsequent sample preparation steps. The technique can therefore be labor-intensive for preparation of large numbers of samples as it is not easily amenable to automation.

As an example, Kok and Debets reported the use of a ball mill extraction method for several tablet dosage forms. In their work, they used milling vials that were constructed in house to allow use of up to 35 mL of diluent and enabled centrifugation of the sample solution in the milling vial itself. Different numbers and sizes of balls made of two different materials were evaluated for their effectiveness in milling intact tablets. The use of high frequency milling resulted in gray-colored pulverized samples due to stainless steel particles from the balls/vials and this discoloration was not observed at lower milling frequencies. A milling rate of 15 Hz for 2 min was chosen with the use of one stainless steel ball (12 mm for 18 and 35 mL vials; 9 mm for 6 mL vial) to achieve adequate wet milling and extraction of intact tablet samples. The milling and extraction portion of the method was completed in 2 min followed by 5 min of centrifugation in the milling vial. This was considerably shorter than the 15–30 min required per sample for manual grinding and weighing of the ground sample. Recoveries of drug for low- and high-dose products were

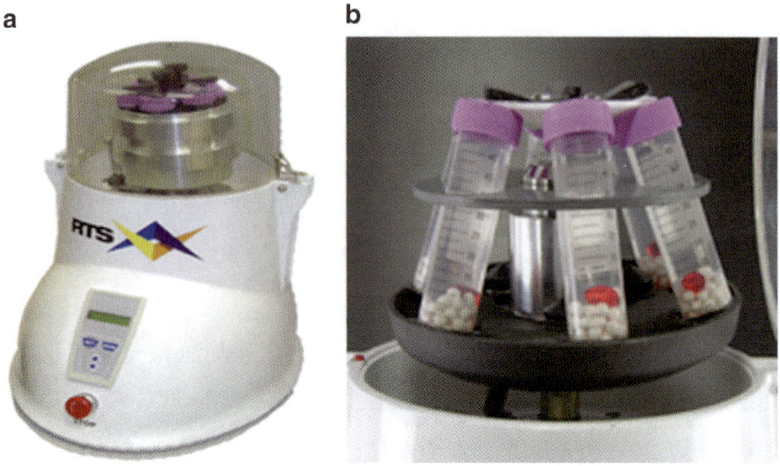

Fig. 3.6 RTS SolidPrep System. (Reproduced with permission from RTS Life Science)

comparable to the conventional extraction method. In the case of a tablet dosage form containing high levels of HPMC, the ball mill was found to provide good extraction of the drug, while manual grinding led to low recoveries due to smearing of the HPMC during grinding (Kok and Debets 2001).

The RTS SolidPrep system (RTS Life Science, Fall River, MA), shown in Fig. 3.6, uses beads instead of balls to mill samples. The instrument uses wet milling by rapidly shaking samples in disposable tubes filled with an extraction solvent and a measured amount of inert ceramic beads. The rapid reciprocating agitation results in both a milling and vortexing action that causes rapid disintegration and homogenization of the sample. Up to five samples can be processed at one time in volumes ranging from 50 to 125 mL. Fish and Pollard reported results in using this instrument on several solid oral dosage forms: a film-coated tablet formulation that took 30 min to prepare using magnetic stirring took only 45 s using the RTS SolidPrep system; a compressed tablet formulation that required 20 min to prepare using high shear homogenization, needed only 2 min using the RTS SolidPrep System; and a modified release tablet formulation that took 40 min to prepare by manual grinding and shaking took only 40 s with the RTS SolidPrep system (Fish and Pollard 2009).

3.3.2.2 Knife Mills and Blenders

Knife mills (from e.g., Retsch Inc., Newton, PA) use blades and a cutting mechanism to mill samples that are not amenable to grinding or ball milling, such as sugar or gelatin-coated tablets (ThomasNet News 2004a, b) or samples with high fat, oil, or moisture content that tend to clump during grinding or ball milling. Knife mills reduce sample particle size down to a fineness of approximately 300 μm or less and homogenize samples in a few seconds to a few minutes. Blenders (from e.g., Waring Laboratory, Torrington, CT; Kinematica AG, Lucerne, Switzerland) also use blades and a cutting mechanism to reduce samples to smaller particle size. A number of USP monographs report the use of blenders to prepare samples (USP Monograph for Amoxicillin Intramammary Infusion 2010; USP Monograph for Amphotericin B Cream 2010; USP Monograph for Erthromycin Delayed-Release Capsules 2010; USP Monograph for Erthromycin Delayed-Release Tablets 2010; USP Monograph for Erthromycin Topical Gel 2010; USP Monograph for Hydroxyzine Hydrochloride Tablets 2010; USP Monograph for Methoxsalen Capsules 2010; USP Monograph for Nifedipine Extended-Release Tablets 2010; USP Monograph for Nystatin Cream 2010; USP Monograph for Nystatin Lozenges 2010; USP Monograph for Nystatin Oral Suspension 2010; USP Monograph for Troleandomycin Capsules 2010).

3.3.2.3 Freezer Mills

Cryogenic or freezer mills (from e.g., Fritsch GmbH, Germany; Retsch Inc., Newton, PA; SPEX SamplePrep, LLC, Metuchen, NJ) use impact and friction to mill samples that are difficult to pulverize at room temperature by milling them in the presence of liquid nitrogen. Freezing samples in liquid nitrogen will make malleable or elastic samples brittle and amenable to pulverization. Freezing thermally labile samples will prevent or minimize degradation of thermally labile samples during milling and will also preserve volatile components in the sample. Samples are placed in the milling chamber along with a metal ball or rod (i.e., impactor) and the milling chamber is placed in liquid nitrogen within the freezer mill. After the sample is chilled (10–15 min), milling begins. The chamber is moved rapidly back and forth causing the impactor to pulverize the sample against the metal end of the milling chamber. Alternatively, the milling chamber is placed within a magnetic coil holder that is then submerged in liquid nitrogen within an insulated tub in the freezer mill. The magnetic coil causes the impactor to move back and forth pulverizing the sample against the metal end of the milling chamber. Typically, the freezer mill is programmed to perform several milling cycles with pauses in between the milling cycles to minimize heating of the sample. References are available that cite the use of a freezer mill in preparing pharmaceutical dosage forms for testing (e.g., USP Monograph for Beta Carotene Capsules 2010).

3.3.2.4 Additional Considerations

A potential disadvantage of using grinding or milling is segregation of particles or materials in the sample that can result in inaccurate results if only a portion of the ground or milled material is used in the subsequent steps of the procedure (Greco 1982, 1983, 1985). As an example Doucette described assay results obtained for composite samples of hydralazine hydrochloride tablets, which were mechanically milled and manually ground to fine powders. Samples were prepared with a mill and a portion of the material was dissolved and analyzed. These samples had assay values approximately 4% lower than samples that were manually ground or samples that were prepared with the mill where the entire sample was quantitatively transferred using water. The author demonstrated that the milled sample material that adhered to the walls of the bowl, blades, and cover after milling was significantly higher in drug content than the material in the powder bed of the mill (Doucette 1987). In another example, Kirschbaum reported 5–10% lower assay results for amitriptyline hydrochloride tablets due to drug adsorption (electrostatic attraction) on the surfaces of the mortar and pestle used to grind the samples (Kirschbaum 1989).

3.3.3 Sonication

Sonication, also referred to as ultrasonic extraction, can be considered both a particle size reduction technique and an agitation technique. This technique is used for a variety of purposes including cleaning, mixing, dispersing, degassing, cell disruption, and DNA shearing. Because of its ability to disperse, mix, and dissolve samples, it is widely used in the sample preparation of pharmaceutical dosage forms. Sonication can be performed by using a sonication bath or a sonication probe or horn. Instruments are available from multiple vendors (e.g., Branson Ultrasonics Corporation, Danbury, CT; VWR International, West Chester, PA; Qsonica, LLC, Newton, CT) and some sonication baths have heaters for temperature control. Examples are shown in Fig. 3.7.

Sonication is used with intact tablets as well as dispersed tablets (e.g., crushed) and other coarse, granular material. The sample along with diluent in a flask is placed in an ultrasonic bath and subjected to ultrasonic radiation. Alternatively, an

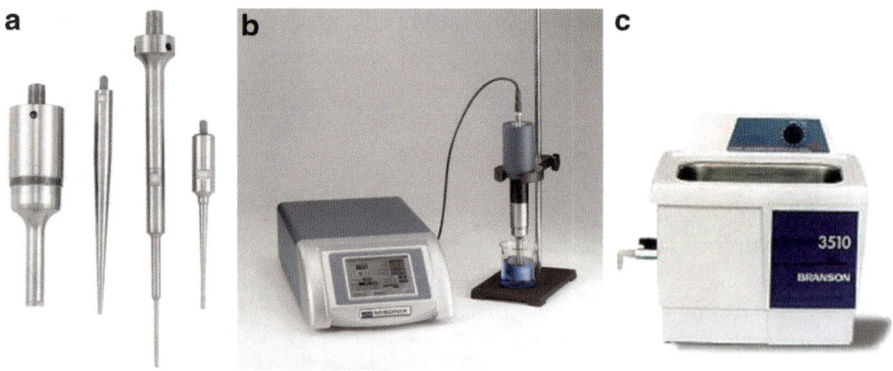

Fig. 3.7 Example of (**a**) ultrasonic probes, (**b**) ultrasonic horn and processor and (**c**) ultrasonic bath. (Figure 3.7(**a, b**) reproduced with permission from Qsonica, LLC. Figure 3.7(**c**) reproduced with permission from Branson Ultrasonics Corporation)

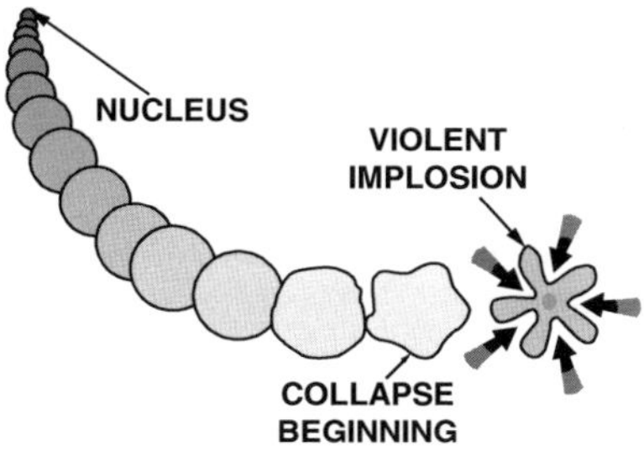

Fig. 3.8 Growth and collapse of a cavitation bubble. (Reproduced with permission form Branson Ultrasonics Corporation)

ultrasonic probe may be placed into the sample and diluent mixture. Ultrasonic energy is generated by a piezoelectric transducer at 20–40 kHz. This energy causes high and low pressure waves within the solvent. "The liquid is compressed during the high-pressure phase of the wave cycle, then pulled apart during the low pressure phase. As the pressure in the liquid is reduced during the low pressure phase, cavities grow from microscopic nuclei to a maximum critical diameter … During the subsequent high-pressure phase, these cavities are compressed and implode. The released energy is powerful, but … localized on a microscopic scale. This process is called "cavitation" " (Branson Ultrasonics Corporation 2000). Cavitation is depicted in Fig. 3.8.

As noted by the survey conducted by Majors, sonication is a popular sample preparation technique – used by 45% of the respondents in the survey (Majors 2002). There are limitations in using a sonicator for sample preparation and extraction methods and these include improper use of sonicators and variability in ultrasonic energy within a given bath and between baths that may lead to method transfer and robustness issues. For example, when a large number of volumetric flasks are placed into an ultrasonic bath, the energy applied to some of the samples may become too diffused to effectively fragment the dosage form and dissolve the drug substance. Therefore, a sample preparation method that works well in an R&D laboratory involving the processing of a small number of samples may fail in a QC laboratory when a much larger number of samples are put in the same type of ultrasonic bath. This may be more of an issue with controlled-release or nondisintegrating dosage forms where agitation may be critical for dispersion of the dosage form and extraction of the active ingredient. Some authors recommend avoiding the use of sonication for the reasons described above (Lee 2004; Chap. 7, Sect. 7.5.1 of this book).

An example of bath-to-bath variability is shown in Table 3.1. During method development for a high- dose immediate release tablet formulation, tablets were prepared in two different diluents using two different sonication baths. When using a diluent of 25% water/75% methanol, complete recovery was achieved for duplicate preparations with 15 min of sonication in Bath 2, but took 60 min in Bath 1. Using 100% methanol as the diluent reduced, but did not eliminate the differences between the baths as 15 min was required with Bath 2 and 30 min with Bath 1.

A number of factors affect sonication: surface tension of the solvent (addition of a small amount of surfactant increases cavitation), temperature of the bath, ultrasonic

Table 3.1 Recovery results for a tablet formulation using two different diluents and two different sonication baths

	Drug recovery (%) in 25% water/75% methanol		Drug recovery (%) in 100% Methanol	
Time (min)	Sonication Bath 1	Sonication Bath 2	Sonication Bath 1	Sonication Bath 2
15	95.4	100.2	99.6	100.2
	89.5	100.0	100.1	100.0
30	98.0		100.0	
	98.5		100.2	
45	86.3			
	100.4			
60	100.3			
	100.2			

frequency (20 vs. 40 kHz), power input, container position in the bath, level of solution in the bath, load in the bath (number of containers, size/shape of containers), and particulates in the solution in the bath (Branson Ultrasonics Corporation 2007). Many laboratories fill a sonication bath with only 1/2–1 in. of water and place the sample container on the bottom of the bath to increase sonication intensity (Choi and Dong 2005). These practices can adversely affect the bath performance (e.g., affect the system frequency and effectiveness) and components (e.g., cause overheating and decrease the life-time of heaters and transducers) (Branson Ultrasonics Corporation 2007) and may lead to method robustness issues and variability between baths. As mentioned previously, some authors recommend avoiding the use of sonication for these reasons (Lee 2004; Chap. 7, Sect. 7.5.1 of this book).

Recommendations by Branson (Branson Ultrasonics Corporation 2000) for effectively using a sonication bath include the following points:

- Operate only when liquid (water, not solvents) is in the bath. Fill the bath to the recommended level with sample containers and trays already in the bath. If the bath is not filled to the appropriate level, this will affect the system frequency leading to decreased effectiveness and potential damage to the unit (e.g., protects heaters and transducers from overheating).
- Frequently change the solution in the bath. Solutions that become contaminated with particles that settle on the bottom of the bath will decrease ultrasonic activity.
- Do not place sample containers directly on the bottom of the bath. Containers on the bottom of the bath will decrease cavitation and will damage the transducer because they reflect the ultrasonic energy back to the transducer. Containers should be placed in an open mesh basket or in an insert tray or be suspended in the solution to position the container in the optimal zone of the bath and to allow the ultrasonic waves to penetrate around the containers.
- Degas the solution in the bath by sonicating for 5–10 min to enhance cavitation prior to sonicating samples.

There are several ways to test whether a sonication bath is functioning (i.e., cavitating) properly. Two of these are the frosted glass slide test and the aluminum foil test that are described below. In addition, one can use an ultrasonic energy or cavitation meter (from e.g., Megasonics, Lake Oswego, OR) to map the energy intensity and frequency of various positions in the bath.

Frosted glass slide test: Take a frosted glass slide, wet it, and draw an "X" on the wet slide with a number 2 pencil. Insert the slide with the "X" into the solution in the bath and turn the bath on. The "X" should begin to disappear right away and all the pencil lead should be removed within 10 s (Branson Ultrasonics Corporation 2007).

Aluminum foil test: Cut a piece of aluminum foil to a size that will fit within the bath. Suspend the foil within the solution in the bath and turn on the bath for approximately 10 min. The foil should be uniformly wrinkled and perforated (Branson Ultrasonics Corporation 2007).

Sonication often results in heating of sample solutions. If using volumetric flasks, sonicated sample solutions should be allowed to cool to room temperature before they are diluted to volume. Since sonication induces localized areas of high temperature, this may adversely affect thermally labile compounds. As an example, Doyle reported a study evaluating potential degradation of APIs in two formulations when using a high power sonication probe (130 W, 40 kHz) to prepare the sample. In the case of Formulation 1, a degradation product was observed that ranged from 0.12 to 0.65% depending on the sonication power setting and the sonication time. An increase in temperature and loss of solvent due to evaporation was also observed at some conditions. For Formulation 2, two degradation products were observed and varied depending upon the sonication conditions (Doyle 2004). Choi and Dong reported an example where sonication of crushed tablets for 30 min yielded good recoveries of two APIs in the formulation and their respective impurities but created a degradation product (dihydroxy derivative) of the one of the APIs. Using vortexing and shaking created no artifact peak and gave >97% recovery of the two APIs (Choi and Dong 2005).

3.3.4 Homogenization

Homogenization is a technique that breaks down a sample into smaller parts and blends them to make them more uniform in texture and consistency. High sheer homogenizers, such as the example shown in Fig. 3.9(a), use a set of rotating blades (rotor and stator) with speeds up to 40,000 rpm, combined with wet grinding, shredding, and shearing to break up the sample in the presence of a diluent. The mechanism of dispersing a sample is depicted in Fig. 3.9(b). Homogenizers also provide vigorous mixing that enhances sample contact with the solvent, thereby facilitating sample extraction. Different models of handheld and small benchtop units (from e.g., Kinematica, Inc., Bohemia, NY; IKA Works, Inc., Wilmington, NC; Fischer Scientific, Pittsburgh, PA) perform homogenization in solutions as low as 0.1 mL and up 2,500 mL. Homogenizer probes can vary in size and blade shape as shown in Fig. 3.9(c).

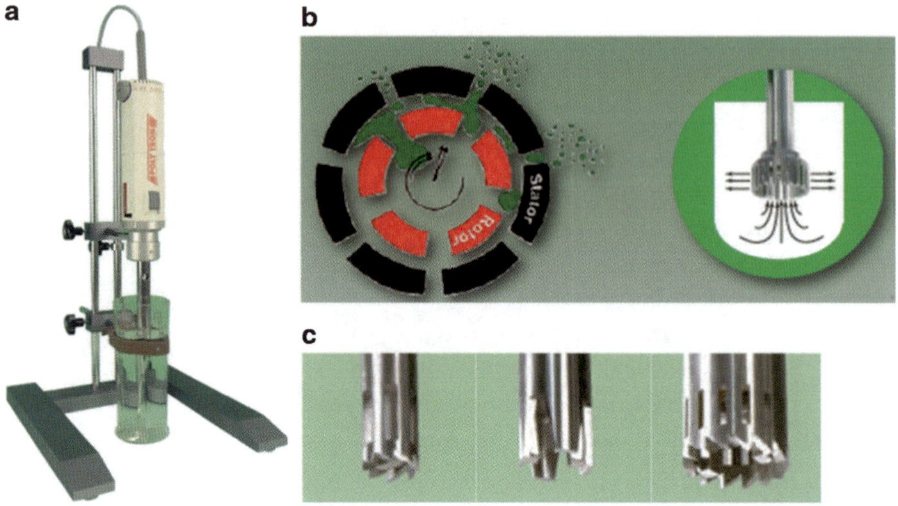

Fig. 3.9 (**a**) Example benchtop homogenizer. (**b**) Schematic representation of the homogenization process. (**c**) Examples of different size and shape homogenizer probes. (Reproduced with permission from Kinematica, Inc.)

Several USP monographs (USP Monograph for Bupropion Hydrochloride Extended-Release Tablets 2010; USP Monograph for Metformin Hydrochloride Extended-Release Tablets 2010; USP Monograph for Pseudoephredine Hydrochloride Extended-Release Tablets 2010) and papers in the literature (Shamrock et al. 2000; Höller et al. 2003; Toro et al. 2004; Lee et al. 2007; Nickerson et al. 2008) report the use of homogenization in sample preparation of pharmaceutical dosage forms. As an example, Lee and colleagues developed a sample preparation method for a challenging immediate release spray dried dispersion tablet formulation that took 30 min to prepare using a homogenizer instead of 5.5 h using shaking and sonication. This tablet formulation contained polymers that gelled during sample preparation causing challenges with drug recovery. A homogenization method was developed using the TPWII (Tablet Processing Workstation II, Sotax Corporation, Hopkinton, MA formerly Caliper LifeSciences) that involved homogenizing a tablet in 100 mL of diluent (80% acetonitrile/20% water) using eight 15 s pulses at 10 krpm. Additional diluent was added to bring the volume to 150 mL to achieve the desired final sample concentration and then 5–10 s pulses at 10 krpm were used to mix the sample solution. The use of a homogenizer significantly reduced the time to prepare the tablet samples (Lee et al. 2007).

Disadvantages of using homogenization include generation of heat and the introduction of oxygen, which can be problematic for heat-labile or oxygen-sensitive compounds. In addition, the technique can be labor-intensive as one sample is prepared at a time and requires cleaning of the homogenizer in between samples. Automated sample preparation systems utilizing homogenization (e.g., Tablet

Processing Workstation and Content Uniformity Testing System, Sotax Corporation, Hopkinton, MA) are available and reduce analyst hands-on labor. These systems are discussed in detail in Chap. 12.

3.4 Additional Considerations

Many sample preparation procedures for complex dosage forms involve the use of more than one type of agitation and/or particle size reduction technique. As an example, Dong and Pace describe the development of a sample preparation procedure for the assay of water-soluble vitamins in a multivitamin tablet formulation that is challenging due to the different solubilities of the many analytes in the formulation. The sample preparation method that was developed uses grinding, sonication, and stirring. First, one tablet is ground in a mortar and pestle and the sample is transferred to a flask and 10 mL of diluent is added. The sample solution is then sonicated for 2 min and 90 mL of a second diluent is added. The sample solution is stirred for 1 min and then sonicated in a 40°C bath for 5 min. The sample solution is then filtered and analyzed by HPLC (Dong and Pace 1996).

3.5 Summary

Agitation and particle size reduction techniques are commonly used to prepare samples of pharmaceutical dosage forms for analysis. These techniques are relatively cost-effective and can be simple to use, but there are limitations to some of these techniques as discussed above. It is important for analysts to understand these limitations and ensure that they do not adversely impact the performance of their sample preparation methods.

Acknowledgments The authors would like to thank Jack Howie for discussions and information related to sonication and John Warzeka for the data in Table 3.1.

References

Branson Ultrasonics Corporation (2000) B-117 Optimizing your ultrasonic cleaner. http://www.bransonic.com/literature.asp. Accessed 8 October 2010

Branson Ultrasonics Corporation (2007) Bransonics tabletop ultrasonic cleaners frequently asked questions. http://www.bransonic.com/faq.asp. Accessed 8 October 2010

Cole-Parmer Technical Library (a). Shaker motion descriptions. http://www.coleparmer.com/techinfo/techinfo.asp?htmlfile=ShakerMoTerms.htm&id=708. Accessed 7 October 2010

Cole-Parmer Technical Library (b). Stirrers and hot plates selection guide/selecting stirrers and hot plates: five important factors. http://www.coleparmer.com/techinfo/techinfo.asp?htmlfile=SelectingStirrersHP.htm&ID=696. Accessed 7 October 2010

Cain L (2007) Case study: analytical aspects of QbD. Presented at the 47th Annual Land O-Lakes Pharmaceutical Analysis Conference on Quality by Design: A New Frontier for the Analytical Chemist. July 30–August 1

Chen P, Atkinson R, Wolf WR (2009) Single-laboratory validation of a high-performance liquid chromatographic-diode array detector-fluorescence detector/mass spectrometric method for simultaneous determination of water-soluble vitamins in multivitamin dietary tablets. J AOAC Int 92(2):680–687

Chen P, Wolf WR, Castanheira I, Sanches-Silva A (2010) A LC/UV/Vis method for determination of cyanocobalamin (VB12) in multivitamin dietary supplements with on-line sample clean-up. Anal Methods 2:1171–1175

Choi CK, Dong MW (2005) Sample preparation for HPLC analysis of drug products, Chap. 5. In: Ahuja S, Dong MW (eds) Handbook of pharmaceutical analysis by HPLC, vol 6. Academic, London

Dong MW, Pace JL (1996) A rapid HPLC method for multivitamin analysis. LC-GC 14:794–803

Doucette Y (1987) Variability of assay results for hydralazine hydrochloride tablets due to the use of electric mill commotion during sample preparation. Drug Dev Ind Pharm 13(2):289–302

Doyle PJ (2004) Automated sonication and sample degradation; a study of two pharmaceutical formulations. Caliper LifeSciences Application Note. http://www.jasco.hu/konyvtar/Prelude-App-Note-Degradation-AP13.pdf. Accessed 7 October 2010

Fish M, Pollard M (2009) Preparing solid samples reliably and rapidly. Pharm Technol Eur 21(12):35–38

Greco GT (1982) I. Segregation of active constituents from tablet formulations during grinding: theoretical considerations. Drug Dev Ind Pharm 8(4):565–578

Greco GT (1983) II. Segregation of active constituents from tablet formulations during grinding: significance to pharmaceutical analysis. Drug Dev Ind Pharm 9(6):971–988

Greco GT (1985) III. Segregation of active constituents from tablet formulations during grinding: effects on coated tablet formulations. Drug Dev Ind Pharm 11(9&10):1889–1899

Heller J (1980) Controlled release of biologically active compounds from bioerodible polymers. Biomaterials 1(1):51–58

Hogan J (1989) Hydroxypropyl methylcellulose sustained release technology. Drug Dev Ind Pharm 15:975–999

Höller U, Brodhad C, Hofman P, Spitzer V (2003) Automated determination of selected water-soluble vitamins in tablets using a bench-top robotic system coupled to reversed phase (RP-18) HPLC with UV detection. J Pharm Biomed Anal 31(1):151–158

Kirschbaum JJ (1989) Inter-laboratory transfer of HPLC methods: problems and solutions. J Pharm Biomed Anal 7(7):813–833

Kok SJ, Debets AJJ (2001) Fast sample preparation for analysis of tablets and capsules: the ball-mill extraction method. J Pharm Biomed Anal 26:599–604

Lee C, Gallo J, Arikpo W, Bobin V (2007) Comparison of extraction techniques for spray dried dispersion tablet formulations. J Pharm Biomed Anal 45(4):565–571

Lee YC (2004) Method validation for HPLC analysis of related substances in pharmaceutical drug products, Chap. 3. In: Chan CC, Lam H, Lee YC, Zhang X-M (eds) Analytical method validation and instrument performance verification. Wiley, Hoboken, NJ

Majors RE (2002) Trends in sample preparation. LC-GC 20(12):1098–1113

Nelson BC, Sharpless KE, Sander LC (2006) Improved liquid chromatography methods for the separation and quantification of biotin in NIST standard reference material 3280: multivitamin/multielement tablets. J Agric Food Chem 54:8710–8716

Nickerson B, Arikpo WB, Berry MR, Bobin VJ, Houck TL, Mansour HL, Warzeka J (2008) Leveraging elevated temperature and particle size reduction to extract API from various tablet formulations. J Pharm Biomed Anal 47(2):268–278

Oliveira PR, Bernardi LS, Mendes C, Cardoso SG, Sangoi MS, Silva MAS (2009) Liquid chromatographic determination of norfloxacin in extended-release tablets. J Chromatogr Sci 47:739–744

Shamrock WF, Reilly K, Lloyd DK (2000) Automated sample preparation of Roxifiban tablets: transfer of a manual method to an automated workstation. J Pharm Biomed Anal 21(6):1225–1232

Tarvainen M, Peltonen S, Mikkonen H, Elovaara M, Tuunaineñ M, Paronen P, Keolainen J, Sutinen R (2004) Aqueous starch acetate dispersion as a novel coating material for controlled release products. J Control Release 96:179–191

ThomasNet News (2004a) Knife Mill Grinds Coated Tablets for Analysis. http://news.thomasnet.com/fullstory/Knife-Mill-grinds-coated-tablets-for-analysis-31439. Accessed 19 October 2010

ThomasNet News (2004b) Optimal Grinding of Coated tablets with the Grindomix GM200 from Retsch. http://news.thomasnet.com/fullstory/Knife-Mill-grinds-coated-tablets-for-analysis-31439. Accessed 19 October 2010

Toro I, Dulsat JF, Fabregas JL, Claramunt J (2004) Development and validation of a fully automated method for the chromatographic determination of content uniformity of drug tablets. J Pharm Biomed Anal 36(1):57–63

USP Monograph for Amoxicillin Intramammary Infusion (2010) In: United States Pharmacopeia, vol. 33 Reissue. USP-NF Online. http://www.uspnf.com

USP Monograph for Amphotericin B Cream (2010) In: United States Pharmacopeia, vol. 33 Reissue. USP-NF Online. http://www.uspnf.com

USP Monograph for Beta Carotene Capsules (2010) In: United States Pharmacopeia, vol. 33 Reissue. USP-NF Online. http://www.uspnf.com

USP Monograph for Bupropion Hydrochloride Extended-Release Tablets (2010) In: United States Pharmacopeia, vol. 33 Reissue. USP-NF Online. http://www.uspnf.com

USP Monograph for Erthromycin Delayed-Release Capsules (2010) In: United States Pharmacopeia, vol. 33 Reissue. USP-NF Online. http://www.uspnf.com

USP Monograph for Erthromycin Delayed-Release Tablets (2010) In: United States Pharmacopeia, vol. 33 Reissue. USP-NF Online. http://www.uspnf.com

USP Monograph for Erthromycin Topical Gel (2010) In: United States Pharmacopeia 2010, vol. 33 Reissue. USP-NF Online. http://www.uspnf.com

USP Monograph for Hydroxyzine Hydrochloride Tablets (2010) In: United States Pharmacopeia, vol. 33 Reissue. USP-NF Online. http://www.uspnf.com

USP Monograph for Metformin Hydrochloride Extended-Release Tablets (2010) In: United States Pharmacopeia, vol. 33 Reissue. USP-NF Online. http://www.uspnf.com

USP Monograph for Methoxsalen Capsules (2010) In: United States Pharmacopeia, vol. 33 Reissue. USP-NF Online. http://www.uspnf.com

USP Monograph for Nifedipine Extended-Release Tablets (2010) In: United States Pharmacopeia, vol. 33 Reissue. USP-NF Online. http://www.uspnf.com

USP Monograph for Nystatin Cream (2010) In: United States Pharmacopeia, vol. 33 Reissue. USP-NF Online. http://www.uspnf.com

USP Monograph for Nystatin Lozenges (2010) In: United States Pharmacopeia, vol. 33 Reissue. USP-NF Online. http://www.uspnf.com

USP Monograph for Nystatin Oral Suspension (2010) In: United States Pharmacopeia, vol. 33 Reissue. USP-NF Online. http://www.uspnf.com

USP Monograph for Pseudoephredine Hydrochloride Extended-Release Tablets (2010) In: United States Pharmacopeia, vol. 33 Reissue. USP-NF Online. http://www.uspnf.com

USP Monograph for Troleandomycin Capsules (2010) In: United States Pharmacopeia, vol. 33 Reissue. USP-NF Online. http://www.uspnf.com

Zuo Y, Zhang L, Wu J, Fritz JW, Medeiros S, Rego C (2004) Ultrasonic extraction and capillary gas chromatography determination of nicotine in pharmaceutical formulations. Anal Chim Acta 526:35–39

Chapter 4
Liquid–Liquid and Solid-Phase Extraction Techniques

Beverly Nickerson and Ivelisse Colón

Abstract Liquid-liquid extraction and solid-phase extraction are classical sample preparation techniques that have been used with various types of samples. The fundamentals of these two techniques, as well as several microextraction techniques based on the same principles, are described in this chapter. Application of these techniques to the sample preparation of pharmaceutical dosage forms for analysis is also discussed.

4.1 Introduction

Solvent extraction in one form or another is typically performed to render a pharmaceutical dosage form amenable to analysis. In its simplest form solvent extraction is performed by adding a single diluent (a single solvent or a solvent mixture) to the sample to extract the drug and analytes of interest. In other cases, two phases are required to partition the analyte and matrix components to achieve a sample solution compatible with the analysis technique. In this case, the two phases can both be liquid, as in liquid–liquid extraction (LLE), one phase can be solid and the other liquid, as in solid-phase extraction (SPE), or one phase can be a gas and the other a liquid, as in gas chromatography (GC) headspace analysis.

GC headspace analysis is performed to determine residual solvents in active pharmaceutical ingredients (API) as well as in dosage forms that use organic solvents in the manufacturing process (e.g., non-aqueous film coating). GC headspace analysis has been discussed in the literature (Snow and Slack 2002; Slack et al. 2003; Kolb and Ettre 2006) and will not be discussed in this chapter. LLE and

B. Nickerson (✉)
Analytical Development, Pfizer Global Research and Development,
Pfizer Inc., 558 Eastern Point Road, Groton, CT 06340, USA
e-mail: beverly.nickerson@pfizer.com

B. Nickerson (ed.), *Sample Preparation of Pharmaceutical Dosage Forms: Challenges and Strategies for Sample Preparation and Extraction*, DOI 10.1007/978-1-4419-9631-2_4,

SPE are techniques that are used to extract API and impurities from dosage forms and to remove interfering matrix components from sample solutions prior to analysis. An overview of LLE and SPE as well as liquid-phase microextraction (LPME) and solid-phase microextraction (SPME) are provided below along with applications of these techniques to pharmaceutical products.

4.2 Liquid–Liquid Extraction

4.2.1 Fundamentals

LLE is a classical sample preparation technique used to extract components of interest, to clean up or remove interfering matrix components, to preconcentrate analytes or to perform a solvent exchange to make the sample solution compatible with the analysis technique. The details of LLE have been described in various references (Holden 1999; Cantwell and Losier 2002; Wells 2003), so only a brief overview is provided here.

LLE uses two immiscible liquid phases to perform an extraction and separation. The two phases are typically an aqueous phase and an organic phase. The analyte of interest must have greater solubility in one of these phases than in the other phase as the analyte will distribute itself between the two phases based on its relative solubility in each solvent. Solvent miscibility charts are available in the literature (Sadek 2002) and water-immiscible solvents commonly used in LLE of pharmaceutical dosage forms include chloroform, ether, *n*-heptane, hexanes, isooctane, and methylene chloride.

In LLE, the sample is placed in a separator (e.g., separatory funnel, test tube, vial). If the sample is already in solution, then an immiscible solvent is added. If the sample is not dissolved, then the aqueous and immiscible organic diluents are added. The solution is thoroughly mixed (e.g., by vortexing or shaking) to provide maximum surface contact between the two phases to allow components to partition between the phases. Overly vigorous mixing, however, can lead to the formation of an emulsion. If not broken up, the emulsion can prevent separation and adequate removal of the phase of interest. After mixing, the two phases are allowed to separate. The phase with the higher density will form the bottom layer. This process is illustrated in Fig. 4.1. The phase that contains the drug or components of interest is removed. Typically, a second or multiple additional extractions are made with fresh volumes of extracting solvent in order to ensure adequate recovery of the analyte(s) or adequate removal of interfering components. The phases containing the analytes are combined. If the analyte of interest is in the organic phase, the solution can be filtered through a drying agent (e.g., anhydrous sodium sulfate) to remove residual water in the organic phase. The sample solution can then be analyzed or be diluted to an appropriate concentration and analyzed. In some cases, the organic sample solution is evaporated to concentrate the sample or is evaporated

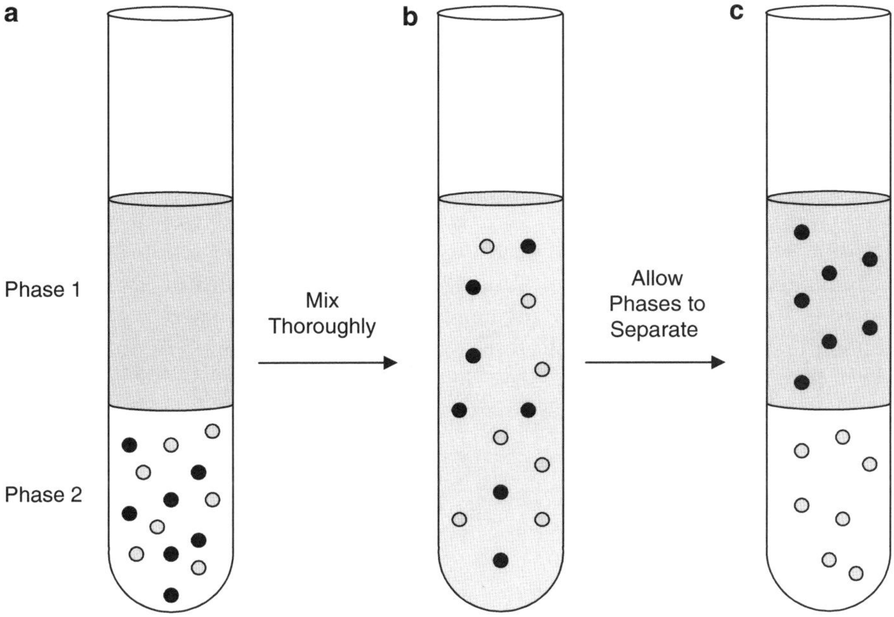

Fig. 4.1 Schematic demonstrating the principles of liquid–liquid extraction

to dryness and the residue is reconstituted with an appropriate diluent compatible with the analysis method (e.g., mobile phase for chromatographic analysis).

A *back extraction* may also be performed as part of the LLE procedure. In this case, after the two phases are mixed and separated, the analyte of interest has partitioned from the aqueous phase into the organic phase, leaving polar interferences in the aqueous phase. A portion of fresh aqueous phase is modified (e.g., pH of aqueous phase is adjusted) to increase the solubility of the analyte of interest in the aqueous phase. The aqueous phase is then added to the organic phase to partition the analyte of interest back into the aqueous phase to separate the analyte from components that were co-extracted (e.g., non-polar interferences) into the organic phase.

During LLE, an analyte will distribute itself between the two phases based on its relative solubility in each solvent. This distribution between the phases at equilibrium is described by the Nernst distribution or partition law, which is shown in (4.1) (Holden 1999). Equation (4.1) applies to two immiscible phases, E and O, which can either be a solid and a liquid, two liquids, or a gas and a liquid. In the case of LLE, the two phases are liquids. K_D is dependent on the particular analyte, solvent system, and temperature. The higher the solubility of the analyte in the extraction solvent, phase E, the higher will be the value of K_D:

$$K_{\mathrm{D}} = \frac{[X]_E}{[X]_O}, \tag{4.1}$$

where

K_D = distribution or partition coefficient,
$[X]_E$ = concentration of analyte X in phase E, the solvent used to extract the analyte from the original matrix, and
$[X]_O$ = concentration of analyte X in phase O, the phase that originally contained the analyte.

Another term used in LLE is the distribution ratio, D, which is used for compounds that dissociate in solution. The distribution ratio of a given compound is the ratio of the sum of the concentrations of all the species of the compound in phase E to the sum of all the species of the compound in phase O. For example, compound HS might dissociate to H+ and S− in phase O but exist as HS in phase E. The distribution ratio for HS is shown in (4.2) (Holden 1999):

$$D = \frac{[\mathrm{HS}]_E}{[\mathrm{HS}]_O + [\mathrm{S}^-]_O}. \tag{4.2}$$

Selection of the solvent system (e.g., pH of aqueous phases for phase O, type of immiscible organic solvent for phase E) is critical to achieving an acceptable LLE procedure. Solvent system considerations include a number of factors, such as solubility of components, solubility of the two phases in each other, distribution coefficients, recovery, density, interfacial tension, chemical reactivity, viscosity, safety, and cost (Holden 1999) and each of these are discussed below.

Solubility of Analyte(s): The analyte(s) of interest must have greater solubility in one of the phases than in the other phase as the analyte(s) will distribute itself between the two phases based on its relative solubility in each solvent. For acidic and basic analytes, pH of the aqueous layer is an important consideration. The pH of an aqueous sample solution may be adjusted to deionize the analyte and increase solubility and extractability of the analyte into the organic phase. And then for a back extraction, the pH of the aqueous phase is adjusted to increase analyte solubility in the aqueous phase in order to extract the analyte out of the organic phase and into the aqueous phase. Another consideration is the "salting out" effect. High salt concentrations in the aqueous phase can be used to increase partitioning of a water-soluble analyte into the organic phase (Majors and Slack 1997).

Solvent Immiscibility and *Solvent Solubility*: The solvents selected for an LLE procedure need to be immiscible in order to form two phases. Although the two phases are immiscible, they will have some solubility in each other and become saturated with each other when mixed (e.g., chloroform is 0.815% soluble in water; water is 0.056% soluble in chloroform) and it is desirable to minimize this solubility (Wells 2003).

Distribution Coefficient: As shown in (4.1), a high distribution coefficient, K_D, is desirable to extract an analyte into phase E.

Recovery: Since LLE involves an equilibrium partitioning of components between two phases, multiple extractions may be required to achieve desired recoveries of the component(s) of interest as discussed below.

Density: The solvent with the higher density will form the bottom layer, while the solvent with the lower density will form the top layer after phase separation.

Interfacial Tension and *Viscosity*: Low interfacial tension and low viscosity can lead to the formation of an emulsion when the phases are mixed and require a long time for the emulsion to disperse and achieve phase separation.

Chemical Reactivity: The solvents selected for an LLE procedure should not react or cause degradation of the components of interest in the sample.

Safety and *Cost*: Large quantities of solvent(s) are typically used in LLE. Analyst safety due to exposure and handling of these solvents and the cost of purchasing and disposing of the solvents should be considered.

Since LLE involves an equilibrium partitioning of components between two phases, multiple extractions may be required to achieve adequate recoveries of the component(s) of interest. The recovery of an analyte in one extraction step can be calculated using (4.3) (Wells 2003) while the total recovery of an analyte using multiple extractions steps can be calculated using (4.4) (Wells 2003). For compounds that dissociate, D may be substituted for K_D in (4.3) and (4.4) (Wells 2003):

$$\%R_X = \frac{100K_D}{K_D + (V_O / V_E)}, \tag{4.3}$$

$$\%R_X = \left\{1 - \left[\frac{1}{1 + K_D(V_E / V_O)}\right]^n\right\} \times 100, \tag{4.4}$$

where

$\%R_X$ = percent recovery of analyte X,
K_D = distribution or partition coefficient,
V_O = volume of the original sample in phase O,
V_E = volume of the extraction solvent, and
n = number of extractions.

Based on (4.4), the total amount of analyte extracted in multiple extractions depends upon the distribution coefficient, K_D, and the ratio of the volumes of the two phases used, V_E / V_O. As shown in Table 4.1 for a given volume of V_O and V_E, the % recovery increases as K_D increases. For a given value of K_D, % recovery increases as the ratio of V_E / V_O increases. In addition, more analyte will be recovered by using multiple aliquots of solvent (e.g., 3×25 mL) compared to one aliquot of the total solvent volume (e.g., 1×75 mL) with the difference being more significant at lower K_D values. However, it is not desirable to do more than four to five extractions as the amount recovered with each successive extraction approaches zero asymptotically (Holden 1999; Rossi and Miller 2003).

Table 4.1 Comparison of % recovery of component X ($\%R_X$) as a factor of K_D, V_O, V_E and the number of extractions performed

V_O	75	25	25	25	25
V_E	1×75 mL	1×75 mL	1×25 mL	2×25 mL	3×25 mL
V_E/V_O	1	3	1	1	1
K_D	$\%R_X$	$\%R_X$	$\%R_X$	$\%R_X$	$\%R_X$
0.1	9.1	23.1	9.1	17.0	24.9
0.5	33.3	60.0	33.3	55.6	70.4
1	50.0	75.0	50.0	75.0	87.5
5	83.3	93.8	83.3	97.2	99.5
10	90.0	96.7	90.0	99.2	99.9
50	98.0	99.3	98.0	100.0	100.0
100	99.0	99.7	99.0	100.0	100.0
500	99.8	99.9	99.8	100.0	100.0

Use of a separatory funnel (or test tube, vial) for LLE is suitable when the distribution ratio is favorable ($D > 5$) for one component in the solution and unfavorable for the other components ($D < 0.0001$). Other types of apparatus or systems should be used when the analyte distribution ratio is unfavorable (e.g., Soxhlet extraction, continuous extraction, countercurrent extraction) or when the analyte is likely to be in the vapor phase (e.g., bubbler extraction system) or to be in a solid form (e.g., impinger extraction system) (Majors and Slack 1997; Holden 1999).

A potential limitation of LLE is the formation of an emulsion that is a suspension of tiny droplets of one phase mixed in the other phase. An emulsion may form when the two phases are vigorously mixed and it may require a long time for the emulsion to disperse and achieve phase separation. The addition of salt, a salt solution or a small amount of a different organic solvent, or the use of centrifugation may break up an emulsion. Alternatively, coalescence may be achieved by creating turbulence on the droplet surfaces by passing the solution through a bed of glass wool or by stirring with a glass rod (Majors and Slack 1997).

Another potential limitation of LLE is that non-polar, water-immiscible organic solvents do not extract very polar or highly charged analytes well. In order to extract these types of components from the aqueous phase, a variation of LLE, salting out liquid–liquid extraction (SALLE), can be used. In this technique, a high concentration of inorganic salt is added to an aqueous and water-miscible organic solvent (e.g., acetone, methanol, acetonitrile) to cause the formation of two phases. In this way, LLE can be performed to extract a polar component from the aqueous phase into the organic phase (Majors 2009).

In addition, using LLE includes the use of large quantities of solvent that results in exposure of the analyst to these solvents and the cost to purchase and dispose of these solvents. Factors to consider in making a particular LLE procedure "more green" and environmentally friendly are discussed in Chap. 14. LLE can also be labor intensive as it typically involves many steps and manipulations. Since mixing of the two phases in a separatory funnel is performed manually, there can be method robustness or ruggedness issues and method transfer issues due to analyst-to-analyst variability in the thoroughness of mixing the phases, which may result in low or variable recoveries. Because of these many limitations, LLE is being replaced by other techniques such as SPE (Sect. 4.4) and SPME (Sect. 4.5). In addition, a number of miniaturized modes of LLE have been developed to facilitate automation, speed up extractions, and reduce solvent consumption and these are briefly described in Sect. 4.3.

There are several advantages to using LLE. It is a versatile technique with large linear sample capacity that can be used to extract an analyte of interest or to remove interfering components in the sample matrix (Cantwell and Losier 2002). If the analyte of interest is extracted into the organic phase, the solution can be evaporated to increase the concentration of the analyte or it can be evaporated to dryness and the residue dissolved in a diluent compatible with the analysis method. In addition, extraction efficiency is not dependent on the original analyte concentration, so it is amenable for trace-level analytes (Holden 1999).

4.2.2 *Pharmaceutical Applications*

Almost 40% of respondents in a survey reported that they use LLE for sample preparation (Majors 2002). LLE has been used to extract and/or remove interferences from sample matrices for various types of pharmaceutical formulations. Sample cleanup may be critical if spectroscopic methods are used to quantitate the sample without a chromatographic separation to separate excipient interferences from the active component or degradants. Some applications cited in USP monographs are presented in Table 4.2. The analytes of interest may be extracted into either the aqueous or the organic layer depending upon the relative solubilities of the analytes and interfering components. In some cases, the analyte of interest is extracted first into one phase and then back extracted into the other phase.

As a simple example, consider the USP monograph assay method for Thioridazine Oral Suspension. Thioridazine is practically insoluble in water and very soluble in chloroform (USP Reference Tables: Description and Solubility – T 2010). In this method, a volume of Thioridazine Oral Suspension is placed in a separator with water. The mixture is made alkaline with ammonium hydroxide and mixed. This pH adjustment is performed to keep thioridazine neutral and maximize the difference in solubility of thioridazine in chloroform vs. the aqueous phase. The sample is then extracted 6 times with chloroform. The extracts are dried by filtering through anhydrous sodium sulfate, combined and diluted to obtain a solution of the desired concentration. The final sample solution is analyzed by UV spectroscopy (USP Monograph for Thioridazine Oral Suspension 2010).

In another example, USP monographs describe an LLE method for assay of flurandrenolide ointment and flurandrenolide cream. Flurandrenolide is practically

Table 4.2 USP Monographs employing LLE sample preparation procedures

Pharmaceutical Dosage Form	USP 2010 References
Oral solutions	USP Monograph for Haloperidol Oral Solution
	USP Monograph for Levocarnitine Oral Solution
	USP Monograph for Doxylamine Succinate Oral Solution
	USP Monograph for Mesoridazine Besylate Oral Solution
	USP Monograph for Mibolerone Oral Solution
	USP Monograph for Thioridazine Hydrochloride Oral Solution
	USP Monograph for Valproic Acid Oral Solution
Syrups	USP Monograph for Chlorpromazine Hydrochloride Syrup
	USP Monograph for Promazine Hydrochloride Syrup
	USP Monograph for Docusate Sodium Syrup
Oral suspensions	USP Monograph for Erythromycin Estolate and Sulfisoxazole Acetyl Oral Suspension
	USP Monograph for Simethicone Oral Suspension
	USP Monograph for Thioridazine Oral Suspension
	USP Monograph for Chlorothiazide Oral Suspension
	USP Monograph for Mebendazole Oral Suspension
Tablets	USP Monograph for Apomorphine Hydrochloride Tablets
	USP Monograph for Hyoscyamine Tablets
	USP Monograph for Norethindrone Acetate Tablets
	USP Monograph for Norethindrone Acetate and Ethinyl Estradiol Tablets
	USP Monograph for Phenmetrazine Hydrochloride Tablets
	USP Monograph for Carbinoxamine Maleate Tablets
	USP Monograph for Codeine Phosphate Tablets
	USP Monograph for Codeine Sulfate Tablets
	USP Monograph for Guanethidine Monosulfate Tablets
	USP Monograph for Hydrocodone Bitartrate Tablets
	USP Monograph for Dexchlorpheniramine Maleate Tablets
Capsules	USP Monograph for Amantadine Hydrochloride Capsules
	USP Monograph for Docusate Calcium Capsules
	USP Monograph for Chlordizepoxide Hydrochloride and Clidinium Bromide Capsules
Suppositories	USP Monograph for Morphine Sulfate Suppositories
	USP Monograph for Acetaminophen Suppositories
Ointments	USP Monograph for Alclometasone Dipropionate Ointment
	USP Monograph for Flurandrenolide Ointment
	USP Monograph for Neomycin Sulfate Ointment
	USP Monograph for Undecylenic Acid Ointment
Creams	USP Monograph for Dexamethasone Phosphate Cream
	USP Monograph for Flurandrenolide Cream
	USP Monograph for Mafenide Acetate Cream
	USP Monograph for Tolnaftate Cream
Gels	USP Monograph for Tolnaftate Gel

insoluble in water, soluble in methanol, and freely soluble in chloroform (USP Reference Tables: Description and Solubility – F 2010). In the USP monograph method, methanolic sodium chloride and hexane are used to extract hydrophobic excipients and additives in the formulation into the organic phase. Chloroform is then added to the aqueous phase containing flurandrenolide to extract the drug for analysis. Sodium chloride is used to increase the extraction of flurandrenolide from the aqueous phase to the organic phase (e.g., salting out effect). The procedure effectively extracts flurandrenolide from the drug product but does involve many steps and manipulations as shown in the method description below.

> A quantity of flurandrenolide cream or ointment is placed in a separator with hexanes and methanolic sodium chloride. After mixing, the phases are allowed to separate and the aqueous phase is removed. The aqueous layer is extracted again with hexanes. Then each hexanes phase is washed twice with methanolic sodium chloride. All the aqueous phases are combined and extracted 4 times with chloroform. Each chloroform extract is dried by filtering through anhydrous sodium sulfate and combined. An internal standard is added and the sample solution is evaporated on a steam bath, then with nitrogen to dryness. The residue is reconstituted in mobile phase, filtered and analyzed by liquid chromatography (USP Monograph for Flurandrenolide Cream 2010; USP Monograph for Flurandrenolide Ointment 2010).

In this next example, LLE is used to remove an interfering matrix component in a liquid formulation to enable determination of a stabilizing agent in the formulation by cation exchange HPLC. A proprietary compound, compound A, is formulated as a liquid topical formulation. One of the excipients in the formulation is cetearyl octanoate, a fatty ester (an oil), which is used as a vehicle. This excipient needs to be removed by LLE prior to analysis of the sample because it is incompatible with the mostly aqueous HPLC mobile phase used for analysis. In the sample preparation procedure, an aliquot of the liquid sample is transferred to a centrifuge tube. Mobile phase (aqueous buffer with EDTA, pH 2.3/methanol, 95:5, v/v) is added and the solution is mixed by vortexing and then centrifuged to facilitate separation of the two phases. The drug also precipitates under these conditions. The top phase, the oil layer (cetearyl octanoate), and the precipitated drug are removed by vacuum aspiration. A portion of the aqueous layer is filtered and injected onto the HPLC system for analysis and determination of the stabilizing agent.

In another example, LLE is used to remove an interfering matrix component in a formulation to enable determination of a low-level degradation product. Compound B, a proprietary API in early stage development, was formulated as a sprayed dried dispersion (SDD) to enhance the solubility of the API. The API in the form of the SDD was then formulated into tablets. One of the degradation products of the API is a potential genotoxic impurity (PGI), which needs to be controlled at the parts per million level. An LC/MS/MS quantitation method was developed in order to determine this PGI at low parts per million levels in the SDD and tablets and an LLE method was developed to remove the SDD polymer from the sample matrix in order to make the sample solution amenable to mass spectroscopic analysis. A sample of SDD or ground tablet is dissolved in dichloromethane/ethanol (50/50) and then the aqueous phase (0.5% formic acid in water/ethanol, 90/10) is added. The solution is mixed by vortexing and shaking and the phases are separated by centrifugation. The aqueous layer is sampled and analyzed by LC/MS/MS. The method is linear from 10 to 120 ppm with a lower limit of quantitation of 14.5 ppm.

4.3 Liquid-Phase Microextraction Techniques

Lord and Pawliszyn wrote that "Microextraction is defined as an extraction technique where the volume of the extracting phase is very small in relation to the volume of the sample, and extraction of analytes is not exhaustive (Lord and Pawliszyn 2000)." LPME techniques have advantages over traditional LLE that include use of less solvent, potential for automation, online coupling with the analysis method, and potential for high throughput. LPME has been used extensively for drug analysis in biological and environmental samples. LPME techniques have been reviewed in the literature (Pedersen-Bjergaard and Rasmussen 2005, 2008; Wille and Lambert 2007; Xu et al. 2007; Lee et al. 2008; Nerin et al. 2009; Kataoka 2010). As noted by Kataoka (2010), LPME techniques can be classified into two groups: single-drop microextraction (SDME) and membrane-assisted LPME. A brief summary of these techniques is provided below.

4.3.1 Single-Drop Microextraction

SDME with two phases entails suspending a drop of organic solvent, the acceptor phase, from the tip of a microsyringe needle while in an aqueous solution containing the analyte of interest, the donor phase. The analyte is extracted into the organic droplet by diffusion. After extraction, the droplet is pulled back into the microsyringe and can then be directly injected into the analysis system (e.g., GC, HPLC). In a three-phase system, the analyte is extracted from the aqueous phase (the donor phase) into an organic phase and then is back extracted into an aqueous phase (the acceptor phase). SDME can also be performed by suspending an organic droplet in the headspace above an aqueous sample solution for analysis of volatile and semivolatile components. Advantages of SDME compared to LLE include significantly reduced solvent usage and the ability to analyze the droplet directly (e.g., no need for solvent evaporation). SDME has been used with a variety of analysis techniques including molecular spectroscopy, electrochemical, chromatography, electrophoresis, atomic spectroscopy, and mass spectroscopy. The limitations of SDME are potential dislodgment and instability of the droplet. Careful manual manipulation during the sample preparation is therefore required and solvent selection is critical (Xu et al. 2007; Kataoka 2010).

As an example, Daneshfar et al. used a three-phase SDME system (referred to as single-drop liquid–liquid–liquid microextraction (LLLME) by the authors) with

Fig. 4.2 Schematic illustration of LLLME device, (1) clamp, (2) microsyringe, (3) acceptor phase, (4) organic phase, (5) donor phase (sample), (6) stirring bar, (7) magnetic stirrer (reproduced from Daneshfar et al. (2009) with permission from Wiley)

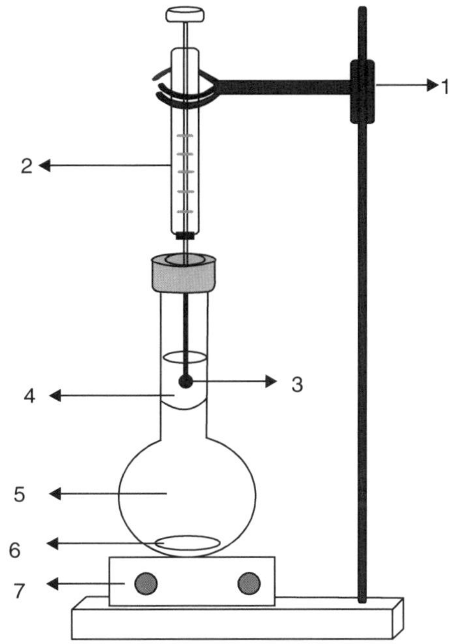

HPLC-UV analysis to determine an anti-malaria drug, chloroquine, in tablets and human urine samples. The drug was extracted from a basic aqueous phase (donor phase) into an organic phase and then back extracted into an acidic aqueous phase (acceptor phase) as shown in Fig. 4.2. Several method parameters were evaluated and optimized including type and volume of organic solvent, volume of aqueous acceptor phase, composition of aqueous donor and acceptor phases, stir rate, and extraction time. The mean extraction recovery for chloroquine tablet samples was 98.0–101.0% and 95.0–96.6% for chloroquine spike urine samples (Daneshfar et al. 2009). The method is described below.

> A composite sample of tablets was pulverized to a fine powder and a portion was dissolved in water with shaking and heating. A 5-mL aliquot of the aqueous tablet sample solution or urine sample (diluted 2:3 with 0.5 M sodium hydroxide) and an internal standard was placed in a volumetric flask (solution pH 12). A 250-μL aliquot of the organic phase, cyclohexane:2-ethyl-1-hexanol (1:1, v/v), was added to the volumetric flask on top of the aqueous phase. A microsyringe was used to suspend a 7-μL microdrop of 0.02 M phosphoric acid (pH 2) in the organic phase. The pH of the microdrop was selected to fully protonate the analyte in order to prevent it from re-extracting into the organic phase. The solution in the volumetric flask was stirred for 35 min to allow extraction of the drug. After extraction the microdrop was pulled back into the microsyringe and injected onto the HPLC-UV system for analysis (Daneshfar et al. 2009).

This example demonstrates some of the advantages of SDME: low solvent usage and the ability to directly analyze the droplet without drying or preconcentration.

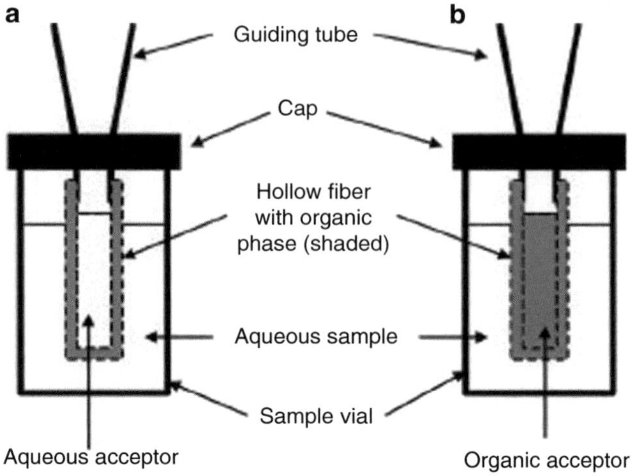

Fig 4.3 Principle of (**a**) three- and (**b**) two-phase LPME (reprinted from Lee et al. (2008) with permission from Elsevier)

4.3.2 Membrane-Assisted Microextraction

One of the major limitations of SDME is dislodgement and instability of the organic droplet. In membrane-assisted microextraction, a membrane (e.g., hollow fiber, membrane bags, flat-sheet membranes) is used as a support to immobilize the organic phase in the pores of the membrane. In hollow fiber LPME (HF-LPME), a polymeric membrane (e.g., polypropylene or other porous hydrophobic polymer) in the form of a hollow fiber is used as the support and can be operated in a two-phase or three-phase mode. The hollow fiber is soaked in the organic solvent (few seconds), which fills the pores of the fiber (with 15–20 μL) via capillary action. One end of the fiber is sealed and the lumen of the fiber is filled with an aqueous acceptor phase (2–30 μL) for a three-phase system. The fiber is then placed in the aqueous sample solution (50 μL–1 L). The analyte in the sample solution passes into the organic phase in the fiber via diffusion, then is back extracted into the aqueous acceptor phase in the lumen of the fiber. After extraction, the acceptor phase is sampled and analyzed (e.g., HPLC, GC, CE, MS). In the case of a two-phase system, the lumen is filled with the same organic solvent as contained in the fiber. A schematic of two-phase and three-phase HF-LPME is shown in Fig. 4.3. The high donor-to-acceptor phase volume ratio allows for high analyte enrichments without sample preconcentration or evaporation. Additional advantages of HF-LPME include the use of very small quantities of organic solvent, the ability to use larger volumes of extracting solvent compared to SDME, no carry over between samples as the hollow fibers are used only once, prevention of large molecules and particles from getting into the extraction solvent and the elimination of emulsion formation (Pedersen-Bjergaard and Rasmussen 2005, 2008; Kataoka 2010).

As an example, Yamini et al. used a three-phase HF-LPME system to extract and preconcentrate salbutamol and terbutaline from pharmaceutical tablet samples, urine samples, and environmental water samples. The following method parameters were evaluated and optimized: composition and volume of donor phase and acceptor phases, type and concentration of carrier in donor phase to increase enrichment of salbutamol and terbutaline in the organic phase, organic solvent type, stir rate, and extraction time (Yamini et al. 2006). A summary of the HF-LPME method is described below:

> Hollow fibers were cut into segments and one end was heat-sealed. Twenty-four microliter of acceptor phase was injected into the hollow fiber using a syringe and the syringe was left in the fiber. The fiber was inserted in organic solution (20% Aliquat 336 in diethyl ether) for 10 s to impregnate the pores and excess organic solvent was rinsed off with water for 10 s. The fiber was bent to create a "U" shape and was placed in 11-mL of sample solution (donor phase consisting of aqueous solution of tablet extract, environmental water sample or urine) and the solution was stirred for 60 min. After extraction the fiber was removed, the sealed end was cut and the syringe was used to remove the acceptor phase and inject it onto the HPLC system for the tablet and water samples or was transferred to a vial for HPLC-MS analysis for the urine samples (Yamini et al. 2006).

In this example, the addition of a carrier in the organic phase is required to extract salbutamol and terbutaline, hydrophilic drugs with high aqueous solubility, from the aqueous donor phase into the organic phase. The drug molecule is negatively charged in the high pH aqueous donor phase. At the interface with the organic phase, it forms a neutral ion pair with the carrier Aliquat 336 (tri-octyl methyl ammonium chloride; water insoluble), releases Cl^- and diffuses across the membrane into the organic phase. At the interface with the acceptor phase, a high concentration of anions (Br^-) in the acceptor phase will cause displacement of the drug from the ion pair and the negatively charged drug will diffuse into the aqueous acceptor phase while the carrier picks up Br^-. Excess Br^- in the acceptor phase drives the mass transfer. Recovery of salbutamol and terbutaline in the tablet samples was 96.8% (0.53% RSD) and 93.9% (3.01% RSD), respectively. The limit of detection of salbutamol and terbutaline in the tablet samples was 2.5 and 0.5 ng/mL, respectively (Yamini et al. 2006).

4.4 Solid-Phase Extraction

The use of SPE is now more common in analytical laboratories to overcome some of the drawbacks of LLE. In SPE, partitioning of the analytes occurs between a solid and a liquid phase (sometimes referred to as liquid–solid extraction) and therefore possesses mechanisms of retention analogous to chromatography. Samples are exposed to an adsorbent phase for either retention of the analytes of interest to increase their concentration or to remove matrix interferences. Following the chromatography analogy, SPE can be conducted in reversed-phase mode (RP-SPE), normal-phase mode (NP-SPE), or ion exchange mode (IE-SPE). Each SPE mode can be conducted in various formats that are commercially available. These formats include the commonly used cartridge in addition to small packed columns, pipette tips,

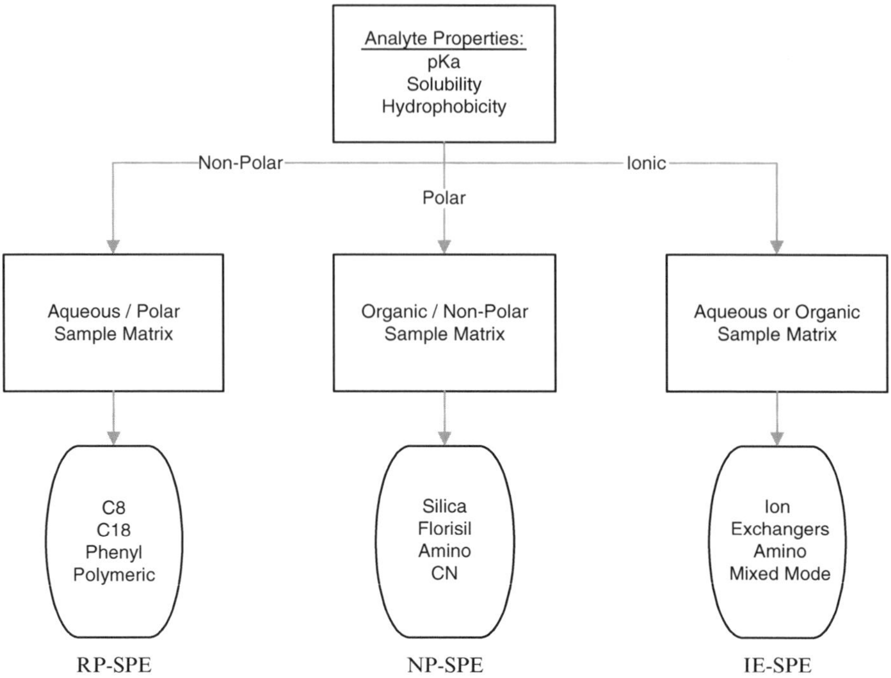

Fig. 4.4 General guidelines for solid phase extraction (SPE) adsorbent selection for reverse-phase SPE (RP-SPE), normal-phase SPE (NP-SPE) and ion-exchange SPE (IE-SPE)

disks, and the miniaturized versions of microextraction in packed syringe (MEPS) and SPME, which will be discussed in Sect. 4.5. These concepts and the theory behind SPE extractions have been described in detail in a book chapter by Poole (2002) and in a book by Thurman and Mills (1998) and therefore only general concepts will be included in this chapter.

Choosing the SPE mode of operation is the first step in method development and it is determined based on the characteristics of the analyte and the matrix; in particular, polarity, pK_a, and solubility as summarized in Fig. 4.4. In NP-SPE, the mechanism of retention is based on polar interactions: dipole–dipole, hydrogen bonding, π–π interactions, and induced dipole–dipole interactions (Thurman and Mills 1998). These types of interactions can be classified as low to moderate. The most common inorganic adsorbents for normal-phase SPE are silica gel, alumina, magnesium silicate (Florisil), and diatomaceous earth. These adsorbents are efficient in extracting moderate to high polarity compounds from non-polar matrices. Other phases (analogous to chromatography) are available for normal-phase operation: amino (NH_2) and cyano (CN). For RP-SPE, the partitioning occurs via hydrophobic interactions (i.e., van der Waals or dispersion forces). This is analogous to the LLE mechanism, now with the organic phase attached to a silica particle (as in a reversed-phase chromatography column). However, it should be considered that some of the polymeric phases (e.g., divinylbenzene (DVB)) are also capable of π–π interactions with aromatic compounds. Reversed-phase adsorbents might be more familiar based on the stationary phases commonly used for HPLC analysis: C8, C18, Phenyl, CN, and mixed polymeric phases. These are most efficient for the extraction of low to moderate polarity or hydrophobic analytes from aqueous or buffered matrices, making them ideal for pharmaceutical applications. In fact, the HPLC method employed for the chromatographic analysis of an extract and the main band elution conditions can give clues to the appropriate starting conditions for SPE method development. Ionization can be a very efficient way to selectively enrich the analyte of interest or minimize unwanted impurities/excipients from an extract. Therefore, ion exchange adsorbents should be explored for ionizable analytes/matrices. The ion exchange phases consist of weak exchangers (e.g., RCO_2^-, RNH_3^+) or strong exchangers (e.g., RSO_3^-, NR_4^+). The retention behavior and selectivity can be altered by carefully selecting the pH of the matrix solvent and the elution solvent. As shown, the amino functionality can be used in both the normal-phase and the ion-exchange mode. Note that, as with other extraction modes, a successful SPE procedure can be developed only when there is a difference, even if it is small, in the interaction between the analyte and the sample matrix with the adsorbent on which to base the extraction. Figure 4.4 presents these differences in terms of polarity and pK_a.

4.4.1 General Guidelines for Method Development and Execution of SPE

Once the SPE mode has been selected according to the desired goal of the extraction and the physical–chemical properties of the active, impurities and excipients; the method development stage generally includes three steps. These steps apply to all extraction modes and are summarized in the diagram presented in Fig. 4.5.

Step 1. Sample dispersion
The sample needs to be dissolved or dispersed in an appropriate solvent system. This solvent system not only needs to disperse and carry the sample through the SPE phase but also needs to be weak enough to allow the analytes of interest to have affinity for the adsorbent phase. As summarized by Majors and Slack (1997), common sample solvents in reversed-phase SPE are aqueous-based buffers with up to 10% organic. Increasing the organic composition of the sample solvent in RP-SPE will usually increase the analyte's affinity for the solvent and lower the retention and extraction efficiency of the SPE procedure. For ion exchange modes, similar solvents can be used. However, the ionic strength of any buffers used should be kept to a minimum. Solvents with low polarity (e.g., hexanes, chloroform) are the choice for NP-SPE.

Sample dissolution or dispersion is an important step, as it is with the sample solvent that one could reduce potential matrix interferences. For example, ionization of the matrix by choosing the appropriate sample solvent pH in RP-SPE can help selective extraction of an API. Or, ionization of the API can help in the selective extraction of a key low-level degradant.

Step 2. Choosing the adsorbent phase
Step 2 involves the selection of the adsorbent phase, both the type and the mass used. As discussed, the adsorbent type will be determined by the characteristics of

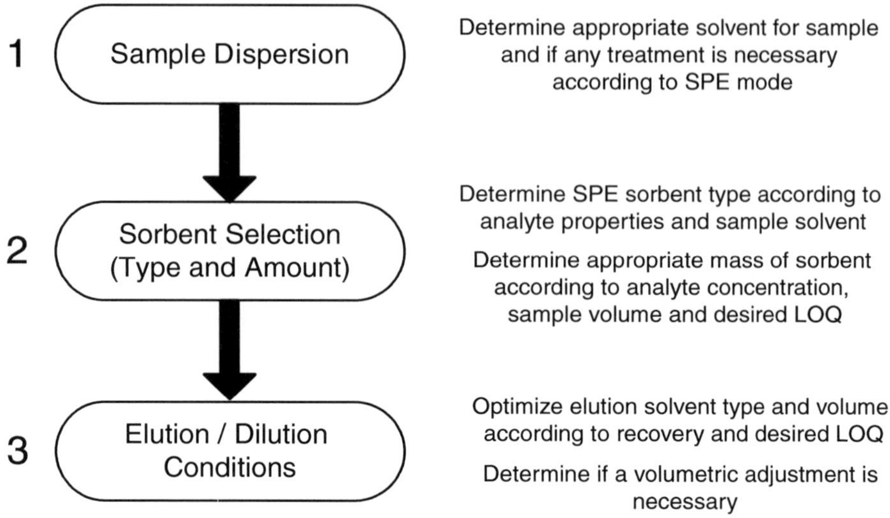

Fig. 4.5 General guidelines for SPE method development

the target analytes (polarity, for example) and their expected concentration. These general characteristics will determine the SPE mode and therefore the general choices for adsorbents. Figure 4.4 summarizes general guidelines for sorbent selection. It is important to keep in mind that the adsorbent phases used for SPE devices are usually of larger particle sizes (>40 μm) and more irregular in shape than stationary phases used in chromatography columns, in order to keep the costs at acceptable levels for disposable devices (Majors and Slack 1997). The irregularity of the packaging also leads to "less-well-packed" beds and therefore, the resulting efficiency is much lower ($N < 100$) than for chromatography.

For selection of the amount or mass of adsorbent to use, it is important to consider the volume of sample, analyte concentration (true or estimated), and how much pre-concentration is needed to achieve the desired sensitivity. This is analogous to "stacked injections" in chromatography. The concept of breakthrough volume is particularly important for quantitative analysis to avoid saturation of the adsorbent phase. This concept is discussed in detail by Poole (Poole 2002). The breakthrough volume is determined by a breakthrough curve (Fig. 4.6) and represents the point when analytes can no longer be retained by the adsorbent due to the sample volume at a given concentration exceeding the retention capacity of the adsorbent. Passing more sample through the cartridge will not yield additional retention and analytes will exit the SPE device at the same concentration as they enter. A breakthrough curve can be determined experimentally by analyzing the samples exiting the cartridge for the presence of the analytes or by calculating recoveries after eluting cartridges loaded with different sample volumes. Since generating a breakthrough curve could be time consuming, several methods exist for estimating or predicting breakthrough volumes based on parameters such as the octanol–water distribution constant (K_{ow}) as described by Hennion et al. (1998). Poole also offers a detailed discussion on the application of general theory of frontal chromatography to derive

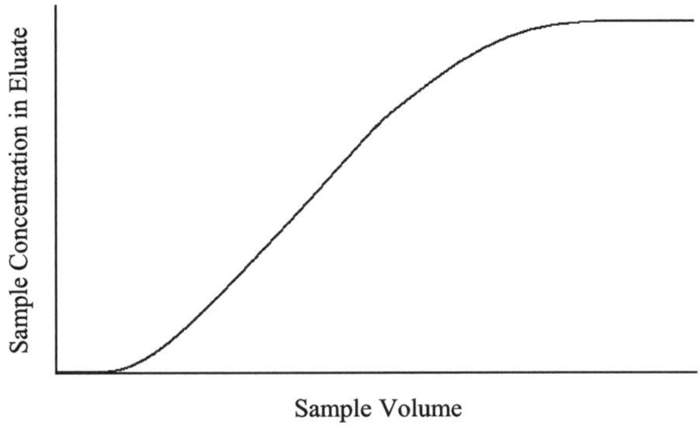

Fig. 4.6 Schematic diagram of a SPE breakthrough curve

a relationship between the breakthrough volume and the sorption properties of SPE devices (Poole 2002). However, for most pharmaceutical applications, there might be limited information on the analytes or impurities to derive these estimates and the best approach might be to go back to an empirical generation of the curve in the schematic shown in Fig. 4.6.

If the sorbent has little capacity for the analyte, some troubleshooting strategies may include (Thurman and Mills 1998): changing the adsorbent phase or amount, reducing the flow rate of sample introduction to allow more equilibration time, changing the form of the analyte through the dispersion solvent (e.g., ionization state), exploring the "salting out" effect or ultimately, trying a different SPE mode.

It is also important to note that in the analysis of dosage forms, SPE has been mostly applied for the cleanup of extracts prior to quantitative analysis. These applications will be discussed in more detail in Chap. 10. For these applications, retention of the analyte is not the main goal, but rather the goal is the retention of matrix components. Examples of these applications would be (1) ionization of the API to decrease its retention in a C18 SPE phase while retaining hydrophobic excipients or (2) using a Silica SPE phase to clean up an extract of a highly hydrophobic API.

Step 3. Elution

The type of elution solvent is determined by the SPE mode selected in the previous steps. It is also important to consider compatibility with the analytical system that will be used to analyze the final sample solution (i.e., normal-phase LC, reversed-phase LC, GC, spectroscopy, etc.). Again, chromatographic behavior is a good starting point in selecting a solvent or solvent system that will yield quantitative elution from an SPE phase. Great elution solvents for RP-SPE include ethyl acetate, acetonitrile, methanol, and water. These solvents are easily capable of disrupting van der Waals interactions between solutes and the hydrophobic adsorbents. For NP-SPE, some choices are hexanes, ethyl acetate, methylene chloride, and chloroform. In IE-SPE, the choices are strong acids (for cation exchange) or strong bases (anion exchange) keeping in mind that pH and ionic strength control are more critical.

The elution step is crucial in the sense that it will also determine the final concentration of an extract for quantitative applications. It is desirable to use stronger solvents to minimize the amount of solvent used and increase the concentration of the analytes, thereby decreasing the limit of quantitation of the analysis method.

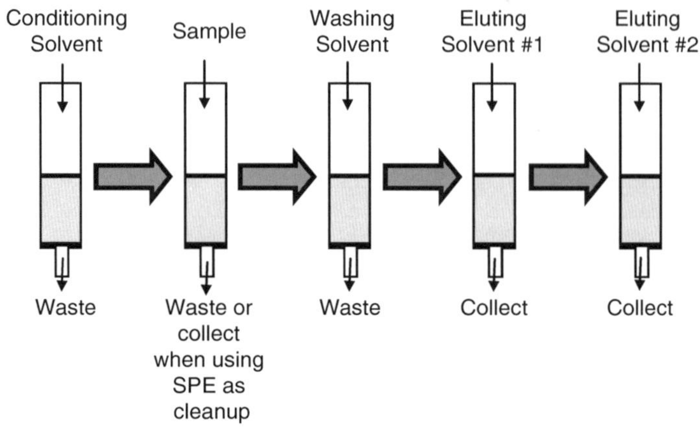

Fig. 4.7 Steps in a generic SPE procedure

In cases where larger amounts of solvents are needed to quantitatively elute the analytes of interest, a pre-concentration step by solvent evaporation might be required. During method validation, evaporation concerns should also be evaluated. It is common to have some solvent evaporation during elution, especially when using a vacuum manifold to perform SPE procedures and the elution solvents are volatile. In these cases, it is recommended to elute the cartridge into a volumetric flask and dilute the sample to a fixed volume to avoid method robustness issues.

The general steps to perform SPE in a disposable cartridge format are presented in Fig. 4.7. These steps also apply to SPE procedures in other formats. Prior to the procedure, the adsorbent phase is conditioned with a solvent or solvent mixture of similar polarity to that of the sample solvent. The same sample solvent could be used for this purpose, but it is recommended to include methanol or acetonitrile when using RP-SPE. The default is to use approximately 4× the adsorbent bed volume. This basically minimizes any interference due to the adsorbent itself and its exposure to laboratory conditions. It also increases efficiency by solvating the adsorbent phase (Majors and Slack 1997), although this is less critical for the newest generation of phases commercially available. Since this is just a conditioning step, the eluate is considered waste. The sample is then loaded into the cartridge at a predetermined flow rate. This eluate is considered waste when the analytes of interest are retained, but must be collected for further analysis in cases where SPE is utilized only for retention and removal of matrix interferences. If SPE is only used as a cleanup avenue, this eluate is collected, diluted if necessary, and analyzed by the preferred analytical technique. This would be the end of this SPE procedure. However, if the analytes are retained, the next step would be an optional washing step. This step will wash off additional interferences from the matrix. This solvent must be selected carefully, so that the retained analytes are not washed off to waste. As a general guideline, use 8× the bed volume with solvent of opposite polarity to the adsorbent phase for the wash. As an example for RP-SPE, it is common to use a water solution containing no more than 5% methanol. This step is key for desalting procedures as discussed in Chap. 10.

After the optional washing step, the retained analytes are eluted with the appropriate solvent. This eluate is collected, diluted if necessary and analyzed. This step can be repeated with different solvents if multiple analytes can be selectively desorbed from the SPE cartridge, similar to column chromatography. The volume of eluting solvent is usually determined by recovery of the analytes or by the final volume to achieve the desired limit of quantitation.

4.4.2 Pharmaceutical Applications

SPE has been used in several ways for pharmaceutical analyses. Most of the published pharmaceutical applications of SPE are related to pharmacokinetics and the analysis of pharmaceuticals in biological samples. A limited number of applications have been published related to direct analysis for dosage forms. Three major areas of these publications include the retention of analytes to increase their concentration for the development of more sensitive analytical methods, the retention of impurities to aid in their identification by MS or NMR, and the minimization of interferences from the matrix to help in its downstream analysis/detection. The latter application is the focus of the SPE discussion in Chap. 10.

Kenney et al. (1998) developed an SPE procedure for the quantitation of a proprietary drug (L-768673) in a microemulsion formulation. The analytical challenges for the development of a robust stability-indicating method included the low concentrations of the analytes (0.0125–0.1 mg/mL) and the emulsified matrix, with components of several polarities. A 6-mL capacity SPE cartridge containing 500 mg of a C18 phase was used to retain the highly hydrophobic API. The cartridge was conditioned using 6 mL of acetonitrile followed by 6 mL of water. One milliliter of the emulsion was charged, immediately followed by 2 mL of water to prevent overloading of the column. This charge was done drop by drop and repeated 4 more times to increase the concentration of analytes retained. The cartridge was then washed with water until the waste eluate was clear. Analytes were then eluted with 4.5 mL of acetonitrile into a volumetric flask and diluted to 5.0 mL for analysis by HPLC with UV detection. The method was validated, with accurate quantitation down to 0.1 μg/mL. It also demonstrated to be stability indicating as demonstrated by forced degradation samples.

The analysis of vitamins has also been improved by the use of SPE. These methods are usually time consuming and due to the differences in polarity among essential vitamins, methods can usually analyze only a few species at a time. Moreno and Salvado (2000) utilized SPE to develop a sample preparation procedure that allows the analysis of both fat-soluble (A, E, and D_3) and water-soluble (B_6, B_1, PP, B_2, and B_{12}) vitamins from the same multi-vitamin sample solution. A 3-mL C18 cartridge was conditioned with methanol followed by water. A diluted sample was loaded onto the cartridge. The fat-soluble vitamins were retained in the cartridge, while the water-soluble vitamins were unretained and collected into 10-mL volumetric flasks. The cartridge was then washed with water and a water/methanol mixture prior to the elution of the fat-soluble vitamins with one column volume of methanol and one column volume of chloroform. Due to the nature of the extracts, these were chromatographed separately. The method was validated with acceptable reproducibility, although recovery for D_3 was only 78%. It is important to note that it is acceptable to have lower recoveries, as long as the method meets other validation criteria (i.e., reproducibility, sensitivity, accuracy, etc.).

Rebbeck et al. (2006) published the use of a cation exchange cartridge (SCX) for the analysis of antimicrobial preservatives in oxytetracycline injectable suspension. In this application, the oxytetracycline API is considered an interference to the quantitation of the smaller components methylparaben and propylparaben. Since oxytetracycline is cationic at low pH (the preservatives are neutral), a cation exchange phase was selected for the procedure. A sample of the suspension (300 mg) was dissolved in 25 mL of 0.1 N HCl followed by 25 mL of MeOH. Ten milliliter of the sample was loaded into the SCX cartridge and the eluate (the preservatives) collected into a volumetric flask. The cartridge was washed with 2–4 mL MeOH. This wash was combined with the eluate to yield the final solution for analysis after diluting to volume. The procedure was successful in eliminating the API interference and validated with %RSD of <1% (same analyst) and LOQ of 1.3 mg/mL for methylparaben and 0.15 mg/mL for propylparaben.

In terms of identification of impurities and degradants, SPE has been used to increase the concentration of analytes, so that it is feasible to obtain MS or NMR structural data. Huidobro et al. (2007) present this particular application with the use of SPE to pre-concentrate a degradation product of alprazolam in tablets. Although this degradant was generated at 20% of the initial alprazolam content, its concentration was still a minor component of the sample and not enough to obtain structural information. The authors employed a polymeric phase (Waters Oasis HLB) due to its similarity with the chromatographic stationary phase, and were able to use conditions analogous to the chromatographic method for elution. The sample solution was prepared by dissolving the solid sample in dimethyl sulfoxide, centrifuging, and concentrating the supernatant further by vacuum. This sample was loaded onto the SPE cartridge. The cartridge was then washed with 10 mM ammonium bicarbonate (Solvent A). The elution was carried out using two solvent systems (analogous to a Mobile Phase A and B in chromatography): 10 mM ammonium bicarbonate (Solvent A), and acetonitrile (Solvent B). First, 15 volumes of 70:30 A:B (v/v) were used to elute the alprazolam API to waste. Second, smaller acetonitrile fractions were collected and the impurity was isolated almost completely in fraction 17. This 17th fraction was further concentrated by vacuum and reconstituted in an appropriate solvent for MS/MS and NMR analysis. With this concentration scheme, the degradant was successfully identified as triazolaminoquinoleine.

Online SPE has also been used to isolate impurities in order to automate the loading and elution steps. For this purpose, short pre-columns are involved and all steps are automated by means of switching valves. This mode has been successful in the preparation of samples for LC/NMR as reported by Pan et al. (2006) for the isolation and identification of a photodegradant of a proprietary compound TCH346. Larsen et al. (2009) also reported the use of an SPE interface for the identification of a degradant caused by the reaction of 5-aminosalicylic acid with one of the excipients (citric acid) in an enema formulation. One of the advantages of using SPE for this purpose is that the solvent exchange (for deuterated solvents necessary for NMR analysis) is much simpler, as the isolated material retained in the SPE column is directly eluted with the deuterated solvent without the need of evaporation to dryness and reconstitution.

Landis proposed a novel use of IE-SPE to study degradation pathways (Landis 2007). The basic idea is that using cation exchange or anion exchange SPE can yield important information regarding the acid/base character of degradation products in a mixture. This will give preliminary structural information, especially with the aid of predictive software for degradation pathways and pK_a values. Several case studies involving forced degraded samples (acid, base, and oxidative) are presented. As an example, the oxidation of chlorpromazine with peracetic acid was predicted to yield four major degradation products. Only one of those was observed in the actual sample. The pK_a for all degradation products was predicted using ACD Labs software in order to select a pH for an IE-SPE procedure that would yield clues about the ionization behavior of the degradant of interest. Based on its elution or retention at a specified pH, the author was able to confirm or rule out the proposed structures.

4.5 Solid-Phase Microextraction Techniques

The need for speed and sensitivity has driven the development of many different modes of extraction utilizing very small volumes of sample and elution solvents. These so-called microextraction techniques are miniaturized versions of existing extraction modes. These have gained popularity for pharmaceutical analysis due to their ease of operation, potential for automation, high pre-concentration power, and low solvent consumption. In many cases, these techniques allow scientists to go straight from sample to analysis without any other treatment, transfers, or manipulation. The majority of these microextraction techniques are based on SPE dynamics and as such involve partitioning between an adsorbent phase and a liquid medium. The other few are based on miniaturization of LLE procedures and were discussed in Sect. 4.3. Some popular microextraction modes are listed in Fig. 4.8.

Although this chapter will not include exhaustive details on the theoretical aspects of all of these techniques, they have been extensively described in other references throughout this text. Most pharmaceutical applications of these techniques are related to the bioanalytical field for the analysis of actives and metabolites in PK samples and there are very few applications published for the support of synthesis and/or formulation development. Kataoka (2005) mentions this gap, although recognizes the potential of all these techniques in clinical or pharmaceutical sciences.

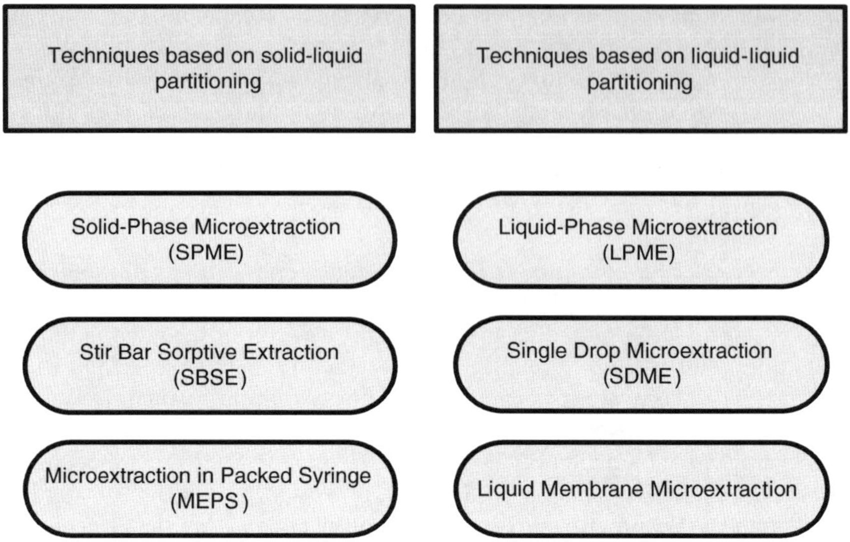

Fig. 4.8 Selected microextraction modes

If the SPE procedure is strictly miniaturized, the result is MEPS. This technique was recently reviewed by Abdel-Rehim (2010) for bioanalytical applications. In MEPS, approximately 1 mg of the solid-phase adsorbent material is packed as a plug inside a small volume syringe (100–250 μL). In this arrangement, the syringe acts as the SPE cartridge and also provides easier automation by connecting it directly to a GC or HPLC autosampler system. The miniaturization then provides for smaller sample volumes needed and therefore smaller elution volumes, having the potential for higher preconcentration factors. Since it is essentially SPE on a smaller scale, all the principles of method development and troubleshooting presented in Sect. 4.4 apply to this technique as well. Syringes are usually packed right in the laboratory with any of the adsorbents previously described, for operation in the RP, NP, or IE modes. Packed syringes are now also commercially available through SGE (2010).

As mentioned, the majority of MEPS pharmaceutical applications published are in the pharmacokinetics field (Abdel-Rehim 2010). To our knowledge, this technique has not been readily used for sample preparation in pharmaceutical sciences applications, but it has plenty of potential for this use as it is based on SPE principles.

SPME is one of the most popular microextraction techniques and was introduced in 1990 by Pawliszyn and coworkers (Arthur and Pawliszyn 1990). This technique can be described in simplistic terms as "chromatography inside-out." It uses a polymer-coated silica fiber as the extraction device and a modified syringe for its handling and housing (Supelco 2010). This coated fiber is exposed to the headspace or immersed directly into a sample. The headspace vs. direct immersion mode is selected based on the analyte volatility or matrix composition. Analytes are adsorbed onto

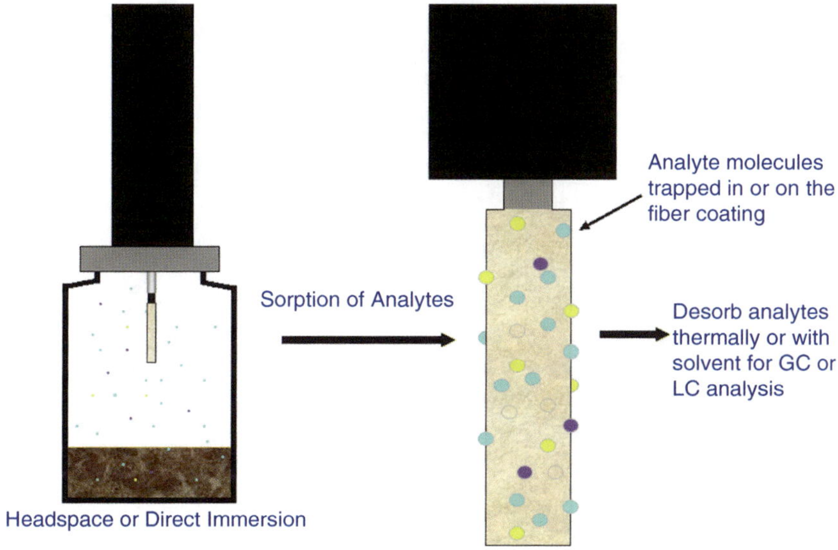

Fig. 4.9 Schematic of a SPME procedure

the fiber and subsequently desorbed thermally (GC injector) or by washing it with a solvent mixture (LC). A schematic of the extraction procedure is presented in Fig. 4.9. The fiber coatings resemble chromatographic stationary phases, involving polymers such as polydimethylsiloxane (PDMS), carbowax (CW), polyethylene glycol (PEG), carboxen (CAR), divinylbenzene (DVB), and polyacrylate (PA). Shortly after the introduction of SPME, it was being conducted by using short pieces of GC capillary columns (namely in-tube SPME, open tubular trapping, or capillary microextraction). This mode became extremely popular due to its ease of automation.

The theory behind SPME has been discussed in a book by the inventor of the technique himself (Pawliszyn 1997) and therefore only the basics will be offered in this chapter. Analytes will start partitioning onto the fiber immediately after contact and the extraction will reach an equilibrium state where the amount extracted is relatively constant. The equilibrium conditions can be described by (4.5):

$$n = \frac{K_{\mathrm{fs}} V_{\mathrm{f}} V_{\mathrm{s}} C_0}{K_{\mathrm{fs}} V_{\mathrm{f}} + V_{\mathrm{s}}}, \tag{4.5}$$

where

n = amount extracted by the coating,
K_{fs} = fiber-sample distribution constant,
V_{f} = fiber coating volume,
V_{s} = sample volume, and
C_0 = the initial concentration of analyte in the sample.

As discussed for SPE procedures, it is important to study saturation of the adsorbent phase as part of method development in order to assess linearity of the analyte's extraction with respect to sample concentration.

SPME method development can be summarized in a series of stages also discussed by Pawliszyn (1997). The first stage involves the decision on the extraction strategy that will be pursued. This stage involves the selection of the fiber coating, necessity for derivatizing agents to increase selectivity or sensitivity, selection of headspace vs. direct immersion based on volatility of the analytes and/or matrix and the agitation parameters. When selecting a fiber coating, the coating thickness is also a factor to consider. The selection of the fiber coating can be initially based on chromatographic behavior, although the matrix will have a big effect for direct immersion modes. The distribution constants could be predicted (Pawliszyn 1997) using Henry's constants, although in practice it might be easier to explore the coatings in an empirical manner. The PDMS coating has proven to be the most versatile, although the PA phase is very useful for polar analytes. Mixed phases can be useful for mixtures of analytes of different polarities. It is important to note that, in the author's experience, the fiber selection for pharmaceutical sciences applications can be difficult and somewhat limited, as some of these polymers swell or dissolve from the fused silica core in the presence of certain organic solvents commonly used in sample preparation of dosage forms.

When selecting an extraction mode, it is important to consider the volatility of the analytes and interferences. Headspace analysis can usually yield cleaner extractions as the matrix interferences are usually non-volatile. However, the analytes need to partition adequately into the gas phase to be able to achieve the required sensitivity. The selection of this extraction mode goes hand in hand with the selection of the downstream analytical technique, namely GC or HPLC.

Agitation is a key optimization parameter in SPME as this will allow the analytes to reach equilibrium state faster with the fiber coating, decreasing extraction time. Several agitation modes have been explored such as vortexing, moving the sample container, sonication, flowing sample, and magnetic stir bars. The latter are the most common agitation mechanism used in SPME.

The second stage in method development involves the optimization of the desorption conditions. For GC applications, this is accomplished by thermal desorption of the analytes right inside the injection port. A narrow insert is used to "focus" the desorbed analytes and increase sensitivity. The temperature must be selected, so that all analytes are desorbed and introduced into the chromatographic system. The maximum temperature allowed for the specific fiber coating is usually a good starting point. Injections are usually performed in a splitless mode for these low-level analyses, so the sampling time before an injection purge needs to be optimized. For LC analysis, the desorption step involves the exposure of the fiber to a solvent system for which the analytes have greater affinity. In this case, optimization of desorption is similar to an elution in SPE. This desorption step can be done on line with the aid of a sample loop or off-line into a sample vial from which a small aliquot can be analyzed later.

The third development or optimization stage of an SPME procedure involves all parameters related to the extraction dynamics. These include the selection of a sample volume, sample dissolving solvent, extraction time and the need for temperature control, or the use of the "salting out" effect. All of these can be simultaneously evaluated using a thorough design of experiments (DoE). One of the most important factors is the selection of the sample dissolving solvent. This is key, as it will dictate the "availability" of the analyte to be extracted and the number of potential interferences from the matrix that could be detected downstream. The pH of this sample solvent should be selected to maximize interaction of the analytes with the fiber while minimizing the interactions with the undesired matrix components as dictated by acid–base properties. The addition of salts to this solvent can increase analyte's partition onto the fiber, particularly for neutral molecules. The extraction time is selected to ensure that equilibrium conditions have been achieved under a given set of parameters. It is possible to conduct analysis under non-equilibrium conditions, but the method precision might be compromised. Increasing the extraction temperature usually reduces the extraction time.

In terms of quantitation, it is important to evaluate (and validate) the most appropriate mode: external standards, standard additions or more sophisticated internal standards such as ^{13}C labeled analogs. Method precision is affected by all the parameters discussed during optimization, but additional factors should also be considered, such as potential for fiber carryover, complexity of the matrix, age/condition of the fiber, and adsorption of analytes to sample containers among others.

For pharmaceutical development, SPME has been mostly used for the analysis of active drugs and their metabolites in biological matrices for pharmacokinetic assessment. Some of these applications have been reviewed by Kataoka (2005) and Kataoka and Lord (2002). In terms of formulation and synthesis support, SPME has been successful for the determination of residual solvents in APIs as reviewed by B'Hymer (2003). In these applications, headspace sampling of volatiles is employed after dispersing the samples in a suitable solvent such as DMF or DMSO. SPME has quickly gained popularity over conventional GC headspace methods because of the increased sensitivity with minimal additional equipment. Due to the success demonstrated for volatile analysis, other applications have been published utilizing SPME for other volatile residues such as flavorings and leachables/extractables. Ligor and Buszewski (1999) described an SPME procedure for the analysis of menthol and menthone in peppermint tea, menthol candies, peppermint chewing gum, and gastric peppermint drops. All samples were extracted with methanol and diluted with water directly into a headspace vial where the SPME extraction was conducted for menthol and menthone and followed by GC/FID. Yeung et al. (2003) also described the determination of these analytes in a taste-masked tablet formulation that contained peppermint oil sprayed-dried onto a food grade encapsulant. The tablets were crushed and an aliquot extracted with water heated to 45°C. The SPME fiber was then exposed to the headspace to conduct the analyte extraction. Sides et al. (2001) described the use of the technique for the identification of an off-odor impurity released from a packaging component. The tremendous pre-concentration factors provided by SPME allowed the identification of the component as ethyl-2-mercaptoacetate. Akapo and McCrea (2008) described a simple SPME procedure for the analysis of 11 volatile potential leachables from preprinted foil used to overwrap LDPE vials used in aqueous pharmaceutical formulations.

The use of SPME for the analysis of preservatives has also been demonstrated by Lokhnauth and Snow (2005). In this application, parabens contained in topical products were extracted by direct immersion SPME and detected by IMS. The parabens were quantitated by internal standards and adjusting the ionic strength of the samples increased the sensitivity of the method. This method provided limits of quantitation lower than 10 ng/mL with a reproducibility of <8% RSDs.

SPME has also been successful in the determination of low levels of process-related impurities, including those with genotoxic potential. Colón and Richoll (2005) described the use of direct immersion SPME for the analysis of methyl and ethyl esters of methanesulfonic, benzenesulfonic, and *p*-toluenesulfonic acids in APIs. Samples were dissolved in aqueous buffered systems prior to extraction. The use of these buffered systems aids in the selective elimination of interferences, therefore achieving both effects with the same process: cleanup and pre-concentration. Frost et al. (2003) also developed a procedure for the analysis of several process-related impurities including benzyl chloride, triethylamine, chloroethyl methyl ether, *N*-methylpyrrolididone, and *N*-methylmorpholine. Although these are mostly related to API analysis, analogous methods could be devised for trace analysis in formulations, if they are properly dispersed prior to extraction. Note that one of the advantages of SPME is that samples do not have to be solutions in nature. If particles are present in the samples, the fibers can be washed with an aqueous buffer (analogous to the wash step for SPE) prior to desorption of the analytes.

Stir bar sorptive extraction (SBSE) is another microextraction mode with similarity to SPME in terms of its extraction mechanism. In SBSE, a modified stir bar contains the adsorbent onto which the analytes will be extracted. This technique was introduced in 1999 by Baltussen et al. (1999) and has been recently reviewed (Lancas et al. 2009; Prieto et al. 2010). The optimization process is very similar to that of SPME, but the extraction and agitation are combined into a single device. The adsorbent selection commercially available is more limited than for SPME and most applications use PDMS as the adsorbent phase. One of the advantages of SBSE is that there is much more adsorbent available to conduct the extraction, so it is possible to achieve higher sensitivity than in SPME. However, desorption of analytes from these devices cannot be conducted in routine equipment and thermal desorption units have to be adapted to GC instrumentation. Of course, analytes could also be desorbed off-line with appropriate solvents for GC or LC analysis.

As it is the case for other microextraction techniques, not many applications have been published using SBSE for pharmaceutical analysis. Most of the applications published are related to bioanalytical samples for drug metabolism or the analysis of pharmaceutical residues in environmental samples (Lancas et al. 2009; Prieto et al. 2010).

4.6 Summary

Liquid–liquid and solid-phase extraction techniques are classical sample preparation techniques that have been and continue to be used in the preparation of pharmaceutical dosage forms for analysis as discussed in this chapter. The miniaturization of these techniques, liquid-phase and solid-phase microextraction, have been widely used in many other fields but their use in the analysis of pharmaceutical dosage forms is limited despite their potential for these types of applications.

Acknowledgments The authors would like to acknowledge Joanne Lukaszewicz for the development of the LLE sample preparation procedure for Compound A and Sofia Davies, Christophe Chassaing and James Grayson for the development of the LLE sample preparation procedure and LC/MS/MS analysis method for Compound B.

References

Abdel-Rehim M (2010) Recent advances in microextraction by packed sorbent for bioanalysis. J Chromatogr A 1217:2569–2580

Akapo S, McCrea CM (2008) SPME-GC determination of potential volatile organic leachables in aqueous-based pharmaceutical formulations packed in overwrapped LDPE vials. J Pharm Biomed Anal 47:526–534

SGE Analytical Science (2010) http://www.sge.com/products/meps

Arthur C, Pawliszyn J (1990) Solid phase microextraction with thermal desorption using fused silica optical fibers. Anal Chem 62:2145–2148

B'Hymer C (2003) Residual solvent testing: a review of gas-chromatographic and alternative techniques. Pharm Res 20:337–344

Baltussen E, Sandra P, David F, Cramers C (1999) Stir bar sorptive extraction (SBSE), a novel extraction technique for aqueous samples: theory and principles. J Microcolumn Sep 11:737–747

Cantwell FF, Losier M (2002) Liquid-liquid extraction (Chapter 11). In: Pawliszyn J (ed) Sampling and sample preparation for field and laboratory. Elsevier, Amsterdam, The Netherlands

Colón I, Richoll SM (2005) Determination of methyl and ethyl esters of methanesulfonic, benzenesulfonic and p-toluenesulfonic acids in active pharmaceutical ingredients by solid-phase microextraction (SPME) coupled to GC/SIM-MS. J Pharm Biomed Anal 39:477–485

Daneshfar A, Khezeli T, Manafi MH (2009) Determination of anti-malaria agent chloroquine using single drop liquid-liquid microextraction. J Sep Sci 32:511–516

Frost RP, Hussain MS, Raghani AR (2003) Determination of pharmaceutical process impurities by solid phase microextraction gas chromatography. J Sep Sci 26:1097–1103

Hennion M-C, Cau-Dit-Coumes C, Pichon V (1998) Trace analysis of polar organic pollutants in aqueous samples. Tools for the rapid prediction and optimization of the solid-phase extraction parameters. J Chromatogr A 823:147–161

Holden AJ (1999) Solvent and membrane extraction in organic analysis (Chapter 2). In: Handley AJ (ed) Extraction methods in organic analysis. Sheffield Academic Press, Sheffield, England

Huidobro AL, Ruperez FJ, Barbas C (2007) Isolation, identification and determination of the major degradation product in alprazolam tablets during their stability assay. J Pharm Biomed Anal 44:404–413

Kataoka H (2005) Recent advances in solid-phase microextraction and related techniques for pharmaceutical and biomedical analysis. Curr Pharm Anal 1:65–84

Kataoka H (2010) Recent developments and applications of microextraction techniques in drug analysis. Anal Bioanal Chem 396:339–364

Kataoka H, Lord HL (2002) Sampling and sample preparation for clinical and pharmaceutical analysis. In: Pawliszyn J (ed) Sampling and sample preparation for field and laboratory, 1st edn. Elsevier Science BV, Amsterdam, The Netherlands

Kenney RR, Forsyth RJ, Jahansouz H (1998) Solid-phase extraction and liquid chromatographic quantitation of the antiarrhythmic drug L-768673 in a microemulsion formulation. J Pharm Biomed Anal 17:679–687

Kolb B, Ettre LS (2006) Static headspace-gas chromatography: theory and practice, 2nd edn. Wiley, Hoboken NJ

Lancas FM, Queiroz MEC, Grossi P, Olivares IRB (2009) Recent development and applications of stir bar sorptive extraction. J Sep Sci 32:813–824

Landis M (2007) The use of mixed-mode ion-exchange solid phase extraction to characterize pharmaceutical drug degradation. J Pharm Biomed Anal 44:1029–1039

Larsen J, Staerk D, Cornett C, Hansen SH, Jaroszewski JW (2009) Identification of reaction products between drug substances and excipients by HPLC-SPE-NMR: ester and amide formation between citric acid 5-aminosalicylic acid. J Pharm Biomed Anal 49:839–842

Lee J, Lee HK, Rasmussen KE, Pedersen-Bjergaard S (2008) Environmental and bioanalytical applications of hollow fiber membrane liquid-phase microextraction: a review. Anal Chim Acta 624:253–268

Ligor M, Buszewski B (1999) Determination of menthol and menthone in food and pharmaceutical products by solid-phase microextraction – gas chromatography. J Chromatogr A 847:161–169

Lokhnauth JK, Snow NH (2005) Determination of parabens in pharmaceutical formulations by solid-phase microextraction – ion mobility spectrometry. Anal Chem 77:5938–5946

Lord H, Pawliszyn J (2000) Microextraction of drugs. J Chromatogr A 902:17–63

Majors R (2002) Sample preparation perspectives: trends in sample preparation. LC GC 29(12):1098–1113

Majors RE (2009) Salting-out liquid-liquid extraction (SALLE). LC GC North America 27(7): 526–533

Majors R, Slack G (1997) Sample preparation (Chapter 4). In: Snyder LR, Kirkland JJ, Glajch JL (eds) Practical HPLC method development, 2nd edn. Wiley, New York

Moreno P, Salvado V (2000) Determination of eight water- and fat-soluble vitamins in multi-vitamin pharmaceutical formulations by high-performance liquid chromatography. J Chromatogr A 870:207–215

Nerin C, Salafranca J, Aznar M, Batlle R (2009) Critical review on recent developments in solventless techniques for extraction of analytes. Anal Bioanal Chem 393:809–833

Pan C, Liu F, Ji Q, Wang W, Drinkwater D, Vivilecchia R (2006) The use of LC/MS, GC/MS and LC/NMR hyphenated techniques to identify a drug degradation product in pharmaceutical development. J Pharm Biomed Anal 40:581–590

Pawliszyn J (1997) Solid phase microextraction: theory and practice, 1st edn. Wiley, New York

Pedersen-Bjergaard S, Rasmussen KE (2005) Bioanalysis of drugs by liquid-phase microextraction coupled to separation techniques. J Chromatogr B 817:3–12

Pedersen-Bjergaard S, Rasmussen KE (2008) Liquid-phase microextraction with porous hollow fibers, a miniaturized and highly flexible format for liquid-liquid extraction. J Chromatogr A 1184:132–142

Poole CF (2002) Principles and practice of solid-phase extraction. In: Pawliszyn J (ed) Sampling and sample preparation for field and laboratory, 1st edn. Elsevier Science BV, Amsterdam, The Netherlands

Prieto A, Basauri O, Rodil R, Usobiaga A, Fernandez LA, Etxebarria N, Zuloaga O (2010) Stir-bar sorptive extraction: a view on method optimisation, novel applications, limitations and potential solutions. J Chromatogr A 1217:2642–2666

Rebbeck C, Hammond R, Wong J, Nair L, Raghavan N (2006) Solid-phase extraction and HPLC analysis of methylparaben and propylparaben in a concentrated antibiotic suspension. Drug Dev Ind Pharm 32:1095–1102

USP Reference Tables, Description and Solubility – F (2010) United States Pharmacopeia, vol 33, reissue. USP-NF Online. http://www.uspnf.com

USP Reference Tables, Description and Solubility – T (2010) United States Pharmacopeia, vol 33, reissue. USP-NF Online. http://www.uspnf.com

Rossi DT, Miller KG (2003) Sample preparation methods for the analysis of pharmaceutical materials (Chapter 8). In: Ahuja S, Alsante KM (eds) Handbook of isolation and characterization of impurities in pharmaceuticals, vol 5. Academic Press/Elsevier, San Diego

Sadek PC (2002) The HPLC solvent guide, 2nd edn. Wiley-Interscience, New York
Sides SL, Polowy KB, Thornquest AD, Burinsky DJ (2001) Identification of a pharmaceutical packaging off-odor using solid-phase microextraction gas chromatography/mass spectrometry. J Pharm Biomed Anal 25:379–386
Slack GC, Snow NH, Kou D (2003) Extraction of volatile organic compounds from solids and liquids (Chapter 4). In: Mitra S (ed) Sample preparation techniques in analytical chemistry. Wiley, Hoboken, NJ
Snow NH, Slack GC (2002) Head-space analysis in modern gas chromatography. Trends Anal Chem 21(9–10):608–617
Supelco Analytical (2010) http://www.sigmaaldrich.com
Thurman EM, Mills MS (1998) Solid-phase extraction: principles and practice. Wiley, New York, NY
USP Monograph for Acetaminophen Suppositories (2010) United States Pharmacopeia, vol 33, reissue. USP-NF Online. http://www.uspnf.com
USP Monograph for Alclometasone Dipropionate Ointment (2010) United States Pharmacopeia, vol 33, reissue. USP-NF Online. http://www.uspnf.com
USP Monograph for Amantadine Hydrochloride Capsules (2010) United States Pharmacopeia, vol 33, reissue. USP-NF Online. http://www.uspnf.com
USP Monograph for Apomorphine Tablets (2010) United States Pharmacopeia, vol 33, reissue. USP-NF Online. http://www.uspnf.com
USP Monograph for Carbinoxamine Maleate Tablets (2010) United States Pharmacopeia, vol 33, reissue. USP-NF Online. http://www.uspnf.com
USP Monograph for Chlordizepoxide Hydrochloride and Clidinium Bromide Capsules (2010) United States Pharmacopeia, vol 33, reissue. USP-NF Online. http://www.uspnf.com
USP Monograph for Chlorothiazide Oral Suspension (2010) United States Pharmacopeia, vol 33, reissue. USP-NF Online. http://www.uspnf.com
USP Monograph for Chlorpromazine Hydrochloride Syrup (2010) United States Pharmacopeia, vol 33, reissue. USP-NF Online. http://www.uspnf.com
USP Monograph for Codeine Phosphate Tablets (2010) United States Pharmacopeia, vol 33, reissue. USP-NF Online. http://www.uspnf.com
USP Monograph for Codeine Sulfate Tablets (2010) United States Pharmacopeia, vol 33, reissue. USP-NF Online. http://www.uspnf.com
USP Monograph for Dexamethasone Phosphate Cream (2010) United States Pharmacopeia, vol 33, reissue. USP-NF Online. http://www.uspnf.com
USP Monograph for Dexchlorpheniramine Maleate Tablets (2010) United States Pharmacopeia, vol 33, reissue. USP-NF Online. http://www.uspnf.com
USP Monograph for Docusate Calcium Capsules (2010) United States Pharmacopeia, vol 33, reissue. USP-NF Online. http://www.uspnf.com
USP Monograph for Docusate Sodium Syrup (2010) United States Pharmacopeia, vol 33, reissue. USP-NF Online. http://www.uspnf.com
USP Monograph for Doxylamine Succinate Oral Solution (2010) United States Pharmacopeia, vol 33, reissue. USP-NF Online. http://www.uspnf.com
USP Monograph for Erythromycin Estolate and Sulfisoxazole Acetyl Oral Suspension (2010) United States Pharmacopeia, vol 33, reissue. USP-NF Online. http://www.uspnf.com
USP Monograph for Flurandrenolide Cream (2010) United States Pharmacopeia, vol 33, reissue. USP-NF Online. http://www.uspnf.com
USP Monograph for Flurandrenolide Ointment (2010) United States Pharmacopeia, vol 33, reissue. USP-NF Online. http://www.uspnf.com
USP Monograph for Guanethidine Monosulfate Tablets (2010) United States Pharmacopeia, vol 33, reissue. USP-NF Online. http://www.uspnf.com
USP Monograph for Haloperidol Oral Solution (2010) United States Pharmacopeia, vol 33, reissue. USP-NF Online. http://www.uspnf.com
USP Monograph for Hydrocodone Bitartrate Tablets (2010) United States Pharmacopeia, vol 33, reissue. USP-NF Online. http://www.uspnf.com
USP Monograph for Hyoscyamine Tablets (2010) United States Pharmacopeia, vol 33, reissue. USP-NF Online. http://www.uspnf.com

USP Monograph for Levocarnitine Oral Solution (2010) United States Pharmacopeia, vol 33, reissue. USP-NF Online. http://www.uspnf.com

USP Monograph for Mafenide Acetate Cream (2010) United States Pharmacopeia, vol 33, reissue. USP-NF Online. http://www.uspnf.com

USP Monograph for Mebendazole Oral Suspension (2010) United States Pharmacopeia, vol 33, reissue. USP-NF Online. http://www.uspnf.com

USP Monograph for Mesoridazine Besylate Oral Solution (2010) United States Pharmacopeia, vol 33, reissue. USP-NF Online. http://www.uspnf.com

USP Monograph for Mibolerone Oral Solution (2010) United States Pharmacopeia, vol 33, reissue. USP-NF Online. http://www.uspnf.com

USP Monograph for Morphine Sulfate Suppositories (2010) United States Pharmacopeia, vol 33, reissue. USP-NF Online. http://www.uspnf.com

USP Monograph for Neomycin Sulfate Ointment (2010) United States Pharmacopeia, vol 33, reissue. USP-NF Online. http://www.uspnf.com

USP Monograph for Norethindrone Acetate and Ethinyl Estradiol Tablets (2010) United States Pharmacopeia, vol 33, reissue. USP-NF Online. http://www.uspnf.com

USP Monograph for Norethindrone Acetate Tablets (2010) United States Pharmacopeia, vol 33, reissue. USP-NF Online. http://www.uspnf.com

USP Monograph for Phenmetrazine Hydrochloride Tablets (2010) United States Pharmacopeia, vol 33, reissue. USP-NF Online. http://www.uspnf.com

USP Monograph for Promazine Hydrochloride Syrup (2010) United States Pharmacopeia, vol 33, reissue. USP-NF Online. http://www.uspnf.com

USP Monograph for Simethicone Oral Suspension (2010) United States Pharmacopeia, vol 33, reissue. USP-NF Online. http://www.uspnf.com

USP Monograph for Thioridazine Hydrochloride Oral Solution (2010) United States Pharmacopeia, vol 33, reissue. USP-NF Online. http://www.uspnf.com

USP Monograph for Thioridazine Oral Suspension (2010) United States Pharmacopeia, vol 33, reissue. USP-NF Online. http://www.uspnf.com

USP Monograph for Tolnaftate Cream (2010) United States Pharmacopeia, vol 33, reissue. USP-NF Online. http://www.uspnf.com

USP Monograph for Tolnaftate Gel (2010) United States Pharmacopeia, vol 33, reissue. USP-NF Online. http://www.uspnf.com

USP Monograph for Undecylenic Acid Ointment (2010) United States Pharmacopeia, vol 33, reissue. USP-NF Online. http://www.uspnf.com

USP Monograph for Valproic Acid Oral Solution (2010) United States Pharmacopeia, vol 33, reissue. USP-NF Online. http://www.uspnf.com

Wells MJM (2003) Principles of extraction and the extraction of semivolatile organics from liquids (Chapter 2). In: Mitra S (ed) Sample preparation techniques in analytical chemistry. Wiley, Hoboken, NJ

Wille SMR, Lambert WEE (2007) Recent developments in extraction procedures relevant to analytical toxicology. Anal Bioanal Chem 388:1381–1391

Xu L, Basheer C, Lee HK (2007) Developments in single-drop microextraction. J Chromatogr A 1152:184–192

Yamini Y, Reimann CT, Vatanara A, Jönsson JÅ (2006) Extraction and preconcentration of salbutamol and terbutaline from aqueous samples using hollow fiber supported liquid membrane containing anionic carrier. J Chromatogr A 1124(1–2):57–67

Yeung Y-H, Lee T, Grant G, Ma M, Kwong E (2003) A SPME-GC procedure for monitoring peppermint flavor in tablets. J Pharm Biomed Anal 30:1469–1477

Chapter 5
Extraction Techniques Leveraging Elevated Temperature and Pressure

Daniel Brannegan, Carlos Lee, Jian Wang, and Larry Taylor

Abstract This chapter introduces, explains, and evaluates several important sample preparation techniques available for pharmaceutical applications. In-depth discussion of Pressurized Fluid Extraction (also known as Pressurized Liquid Extraction, or Accelerated Solvent Extraction), Microwave Assisted Extraction, and Supercritical Fluid Extraction is presented. The principles of each technique, including instrumentation, method development, key parameters, and advantages/limitations, are detailed. A number of pharmaceutical applications and case studies are described to illustrate each technique in practice. Pressurized Hot Water Extraction is also summarized as a member of this extraction group. While each technique is unique, with its own pros and cons with respect to pharmaceutical applications, none provide a universal solution to sample preparation. Failure to evaluate these techniques as part of an analytical toolbox, however, can lead to missing a very simple solution to difficult pharmaceutical extraction issues.

5.1 Introduction

The development of robust and efficient sample preparation and extraction procedures for pharmaceutical formulations has been, and continues to be, a challenge for analytical chemists. Conventional sample preparation and extraction approaches involving techniques such as sonication and mechanical shaking can, at times, be inadequate to efficiently and quantitatively extract APIs (active pharmaceutical ingredients) from dosage forms, especially solid oral dosage forms. Additionally, these

D. Brannegan (✉)
Analytical Development, Pfizer Global Research and Development, Pfizer Inc., 558 Eastern Point Road, Groton, CT 06340, USA
e-mail: Daniel.R.Brannegan@pfizer.com

B. Nickerson (ed.), *Sample Preparation of Pharmaceutical Dosage Forms: Challenges and Strategies for Sample Preparation and Extraction*, DOI 10.1007/978-1-4419-9631-2_5,

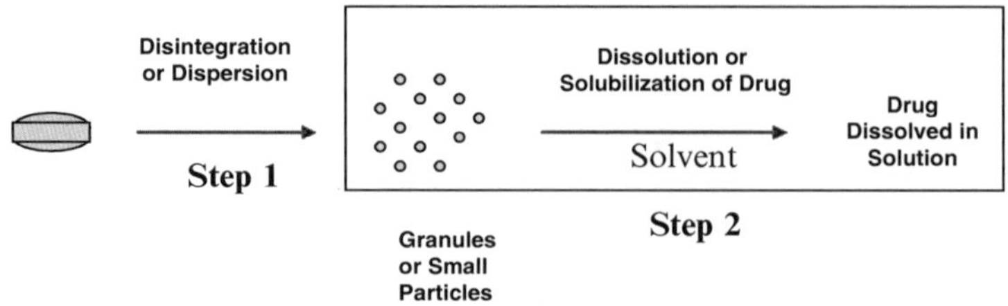

Fig. 5.1 Critical steps in the sample preparation and extraction process

conventional approaches can be time-consuming, cumbersome, and environmentally unfriendly – requiring the use of copious amounts of solvents. With the demand for increased productivity in the pharmaceutical industry, the need for the development of sample preparation and extraction techniques that reduce extraction time, reduce solvent consumption, and increase extraction efficiency is both critical and necessary. Over the last 15 years, several non-traditional sample preparation and extraction techniques have been developed that allow for the selective extraction of analytes from sample matrices, increase speed of extraction, and reduce solvent consumption. These techniques include pressurized fluid extraction (PFE), microwave assisted extraction (MAE), supercritical fluid extraction (SFE), and pressurized hot water extraction (PHWE), all of which operate at elevated temperatures and pressures. In this chapter, the principles of each of these four techniques will be discussed, including their advantages and limitations. Examples of applications and case studies for pharmaceutical dosage forms will also be presented.

5.1.1 Extraction Efficiency: The Key Variables

The key processes that impact the efficiency of sample preparation and extraction methods for pharmaceutical oral dosage forms are captured in Fig. 5.1. While steps 1 and 2 are applicable to solid and semi-solid dosage forms, only step 2 is applicable to suspension and powder formulations. These key processing steps, if not appropriately addressed, can lead to methods that are insufficiently robust – leading to low potency results during HPLC assay and/or lengthy sample preparation procedures. As shown in Fig. 5.1, the first critical step in the sample preparation and extraction process involves disintegration or dispersion of the solid oral dosage form to create small granules or particles. This increase in surface area is then followed by step 2, which involves dissolution of the API in the diluent. Factors that influence these two critical steps will impact the overall rate and extent of extraction of the API from the matrix. For example, most immediate release (IR) solid oral dosage formulations contain superdisintegrants, such as sodium starch glycolate or croscarmellose sodium, as part of the formulation matrix. Superdisintegrants facilitate rapid disintegration of tablets on exposure to aqueous solvents. Additionally, most sample preparation and extraction procedures utilize some type of agitation technique, such as shaking, sonication, homogenization, or stirring to help rapidly disperse the matrix. Rapid dispersion of the tablet leads to increased surface area of the matrix, which promotes faster extraction of the active from the tablet matrix. To facilitate rapid dissolution of the active, current sample preparation and extraction methods often use agitation techniques similar to those discussed above. Furthermore, solvent selection plays a critical role in facilitating rapid dispersion and dissolution. Often times, the physico-chemical properties of the active are used to determine whether aqueous, organic, or mixtures of the two can be used to extract the active from the pharmaceutical dosage formulation. The pH of the diluent is often also varied to enhance dissolution of the active.

5.1.2 Extractions at Elevated Temperature and Pressure

While the dispersion and dissolution strategies described above have helped to provide marginal increases in the rate and extent of extraction, they have not contributed much to help reduce solvent usage. One variable that has the potential to significantly increase the rate and extent of extraction of APIs from pharmaceutical dosage forms, and also decrease solvent usage, is temperature. The effect of temperature on the rate of dissolution and/or extraction efficiency is well understood. For example, the Stokes–Einstein equation (5.1) shows that increasing temperature increases diffusion of the analyte out of the matrix and into the solvent, thus increasing extraction rates. Furthermore, the Andrade equation (5.2) shows that increasing temperature leads to a decrease in solvent viscosity, thus allowing the sample diluent to better penetrate the pores of the sample matrix. Increasing temperature also leads to a reduction in the surface tension of the diluent (5.3) and increased solubility of the active from the matrix (5.4). Additionally, elevated temperature decreases the binding energy between analyte and sample matrix (5.5). The above effects of temperature lead to an overall faster extraction and improved recovery due to increased wettability of the matrix and reduced drug/excipient interaction, respectively.

– Diffusion (D)

$$D = kT / 6\eta\pi a \text{ (Stokes-Einstein equation).} \tag{5.1}$$

– Viscosity (η)

$$\ln\eta = A + B / T \text{(Andrade equation).} \tag{5.2}$$

– Surface Tension (σ)

$$\sigma = a - bT. \tag{5.3}$$

$$> T \rightarrow < \sigma \text{(solvent can better wet the matrix).}$$

– Solubility

$$> T \rightarrow > \text{Solubility}. \tag{5.4}$$

– Binding Energy

$$> T \rightarrow \text{Decreased interaction between analyte and matrix}. \tag{5.5}$$

Equations (5.1)–(5.5) suggest that significant gains in extraction efficiency can be realized by performing extractions at elevated temperatures – extraction times will be shorter, solvent usage lower, and extraction efficiency increased.

Pressure is another variable that impacts the extraction efficiency, especially for SFE and PHWE techniques. Density, diffusivity, and viscosity of supercritical/subcritical fluids are dependent on both temperature and pressure. As such the solvating power of the fluid can be tuned by changing both variables.

Over the last 15 years, a number of sample preparation and extraction techniques that operate at elevated temperatures and pressures have been developed and implemented. Four of these techniques, PFE, MAE, SFE, and PHWE, have shown great potential at increasing the speed and efficiency of extractions, and their use in this area has been on the rise. This chapter will focus on the principles of each of these four techniques, including their advantages and limitations. Several literature case studies will be discussed and evaluated, with a primary focus on pharmaceutical dosage formulations.

5.2 Pressurized Fluid Extraction/Accelerated Solvent Extraction

PFE also known as Pressurized Liquid Extraction (PLE), or Accelerated Solvent Extraction (ASE) – Dionex® trade name – is a semi-automated solvent extraction technique for solid and semi-solid samples. In PFE/ASE, which was first introduced in 1995 by Dionex® (Sunnyvale, CA), an extraction solvent is pumped from one or more reservoirs through a fritted stainless steel cell (1–100 mL cell volumes) containing the sample, usually in a finely ground state (Richter et al. 1995, 1996, 2001; Richter 1999). Following introduction of solvent to the cell, it is held for a specific amount of time at some elevated temperature (up to 200°C) and pressure (up to 21 MPa), and then fresh solvent, usually 40–60% of the cell volume, is added to the cell, discharging and filtering the previous volume into a collection vial. To force the remaining solvent out of the extraction cell and lines, ensuring that all solvent used is collected, compressed nitrogen gas is used (Richter et al. 2001). The extract is usually collected in 40- or 60-mL vials, sealed with Teflon-coated septa, under nitrogen. The temperature, pressure, and static times are all pre-determined by the scientist. Additionally, since any unextracted analyte remains within the extraction cell, a series of flush volume cycles (repeated extractions) can also be applied to the extraction

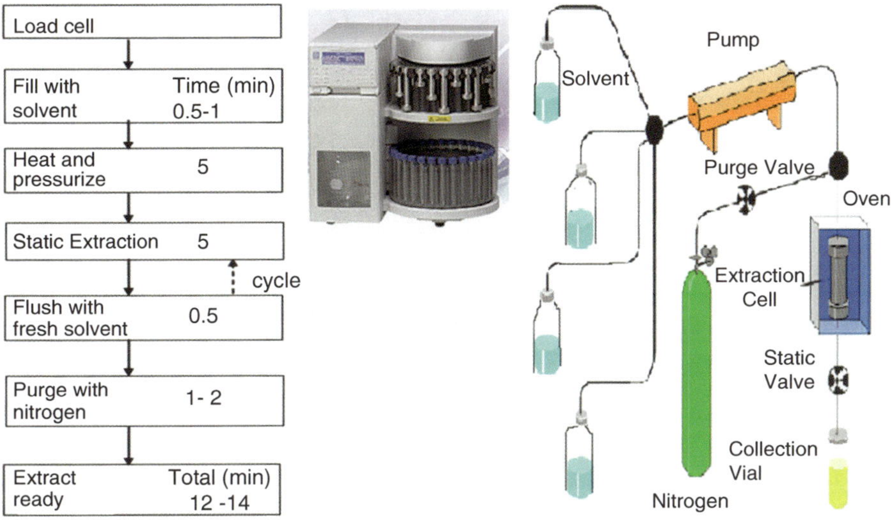

Fig. 5.2 Schematic of Dionex ASE 200 Extraction System (courtesy of Dionex, Sunnyvale, CA)

process to improve extraction efficiencies. Figure 5.2 shows a general schematic of the Dionex ASE 200 instrument (recently replaced by the ASE 350). With the unit shown, up to 24 samples can be processed sequentially and unattended in less than 15 min each. While the ASE 200 is equipped to handle extraction solvent volumes on the order of 5–60 mL, Dionex also has an ASE 300 system for larger volumes (up to 250 mL) (Richter et al. 2001). However, with the ASE 300 system, sample throughput is lower, as only 12 samples can be processed sequentially, unattended. A smaller ASE 100 (recently replaced by the ASE 150) system is also available for labs with modest sample throughput. The ASE 100 system, however, does not have the capability to handle multiple samples unattended. PFE systems capable of parallel processing are also now available. For example, both Applied Separation's PASE and Buchi's new Speed Extractor have the capability of extracting up to six samples in parallel. However, unlike Dionex's ASE 200, the PASE and Speed Extractor do not have an autosampler, so samples need to be manually loaded after every six samples. Buchi's Speed Extractor does have the capability to automatically seal the extraction vessels and comes with a Parallel Evaporator to facilitate rapid concentration of collected extracts. A schematic of the system is shown in Fig. 5.3.

5.2.1 Key PFE Parameters in Method Development

Method development in PFE is simple and the systems, especially the Dionex ASE 200 and 300, are user friendly and versatile. The key variables that have the greatest impact on extraction efficiency are sample preparation/introduction, solvent selection, temperature, pressure, and cycle time.

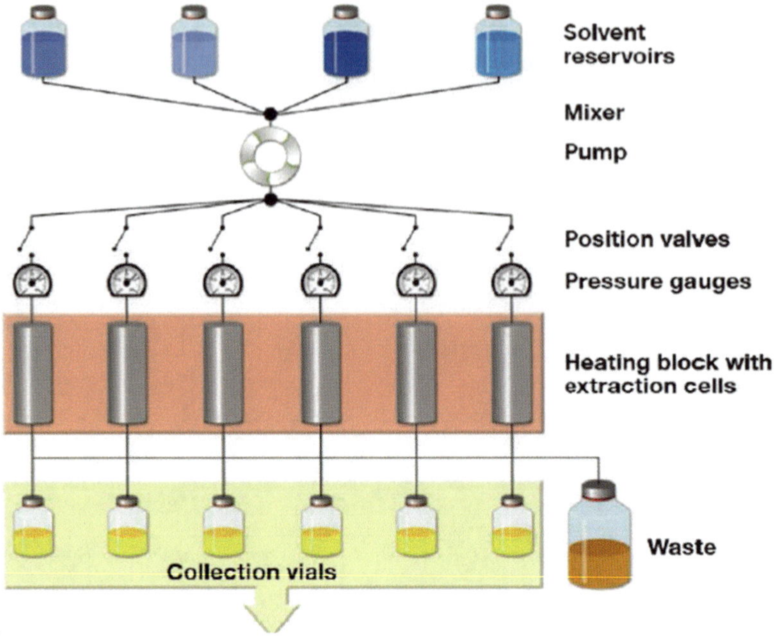

Fig. 5.3 Schematic of Buchi's Speed Extractor Unit (courtesy of Buchi, New Castle, DE)

5.2.2 Sample Introduction

As indicated previously, PFE is a technique for solid or semi-solid samples; therefore, liquid samples are typically not amenable to PFE. The technique works best on dry, finely dispersed solid samples, through which the solvent can easily and quickly traverse, effectively penetrating the matrix, and interact with the analyte. Semi-solid, wet, and/or sticky samples are often mixed with drying agents such as sodium sulfate, diatomaceous earth, or Ottawa sand/hydromatrix to help dry and disperse the sample prior to extraction. To minimize the potential for oxidative degradation of the analyte in the cell during extraction, and to help reduce solvent usage, residual space in the cell is often reduced by filling the cell completely with additional dispersing agent. The presence of additional dispersing agent in the cell also helps to prevent the sample from floating up to the top of the cell, thereby reducing contact and interaction with the extraction solvent. Extractions on intact pharmaceutical solid oral dosage forms are rarely successful due to the lack of penetration of the tablet matrix by the solvent (Bjorklund et al. 1998). However, with some creative efforts, success may be attained (Hoang et al. 2002). In most cases, tablets need to be ground prior to extraction by PFE and the ground material quantitatively transferred to the extraction cell.

5.2.3 Extraction Solvent

In addition to extraction temperature, the nature of the extraction solvent often has a significant impact on a successful extraction via PFE. The extraction solvent should have a high solubility for the analyte, but leave the other sample matrix components intact. A wide variety of solvents can be used in PFE, including organic solvents such as acetonitrile, methanol, and those with lower vapor pressures such as hexane, methylene chloride, or acetone. Aqueous solvents are also amenable to PFE, as long as they do not contain high levels of strong mineral acids such as hydrochloric, nitric, or sulfuric acid. Single solvents, mixtures of solvents, or multi-solvents can be used to selectively extract the analyte(s) from the sample matrix. If multi-solvents are used to extract the analyte from the single sample, the individual extracts must be collected into separate collection vials due to current limitations with most PFE systems.

5.2.4 Extraction Temperature

As indicated previously, solvent selection and extraction temperature are the critical variables in PFE. The impact of elevated temperature on extraction efficiencies has already been discussed in this chapter. Most extractions are performed at temperatures between 40 and 150°C, with 100°C the typical starting point for method development. In the 40–150°C temperature range, significant increases in extraction efficiencies are often observed, with minimal analyte degradation. The fact that minimal degradation is observed for most analytes at elevated temperatures is most likely due to the short exposure times – typically 10–20 min. Analytes that are known to be thermally unstable are often extracted at temperatures closer to ambient, typically 40–70°C, with reduced static times.

5.2.5 Extraction Pressure

The role of pressure in the extraction process is twofold: (1) to maintain the solvents in their liquid states while being heated and (2) to rapidly fill and flush the extraction cells. The standard operating pressure in most PFE extraction is 1,500 psi, which is well above the threshold necessary to keep most solvents in their liquid state, when being heated above their boiling points.

5.2.6 General Approach to Method Development: Extraction of Tablets

In most cases, an analyst looking at the suitability of PFE as a tablet extraction technique already has a manual method and is looking at increasing extraction efficiency, reducing solvent usage and extraction time, and/or semi-automation of the sample preparation and extraction process. As such, the approach to method development is usually not significantly challenging. In terms of solvent selection, analysts should start with the solvent system used in the manual extraction process. If a sequential solvent extraction approach was used, for example aqueous followed by organic, the analyst should attempt to start with a pre-mixed solvent system, consisting of the final ratios/compositions of the solvents in the manual sample preparation process. After solvent selection, method development should be undertaken initially with ground tablets rather than intact tablets. Transfer the pulverized (pulverized within filter paper) tablet to an appropriately sized cell, previously fitted with a suitable filtration disc (cellulose or glass). Fold the filter paper and add to the cell. Fill the cell with a dispersion agent such as Ottawa sand, cap and transfer to the carousel. Following solvent selection, extraction should be performed with the following parameters: pressure, 1,500 psi (10.3 MPa); temperature, 100°C; heat time, 5 min; flush volume, 60% of cell volume; purge time, 60 s; static cycle, 1. The size of the extraction cell should be selected based on sample size – small cell for small samples and large cells for large samples. If incomplete extraction is obtained under the initial method development conditions, increase the number of static cycles to at least three and analyze the collected extracts for the presence of analyte. If additional analyte is obtained in the second and third extracts, consider changing the following parameters (one at a time) and repeat the extraction on freshly prepared samples: (1) increase temperature in 20°C increments; (2) increase static cycle to 2 or 3; (3) increase static time (5-min increments); (4) change extraction solvent.

5.2.7 Pharmaceutical Applications/Case Studies

The use of PFE as a sample preparation and extraction tool in the pharmaceutical industry is relatively new, and on the rise. Pharmaceutical applications include the extraction of natural products from plants; the extraction of parent compounds and metabolites in tissues such as liver and kidney; and the extraction of APIs from various formulations, including animal feed, transdermal patches, emulsions, and tablets. Bjorklund et al. first used the technique in 1998 for extraction of Felodipine from tablets (Bjorklund et al. 1998). Hoang et al. (2002) later used ASE to extract the active ingredient, Montelukast Sodium, from oral chewable tablets and Abend et al. (2003) applied the technique to the extraction of Ivermectin in a meat-based chewable formulation. In 2004, Blanchard et al. successfully utilized ASE to extract and concentrate low-level degradation products from a tablet formulation (Blanchard et al. 2004) and in 2007 Lee highlighted the utility of the technique as a troubleshooting tool for low potency results in solid oral dosage forms (Lee 2007). In the following case studies, the potential and limitations of PFE as a sample preparation and extraction tool will be discussed.

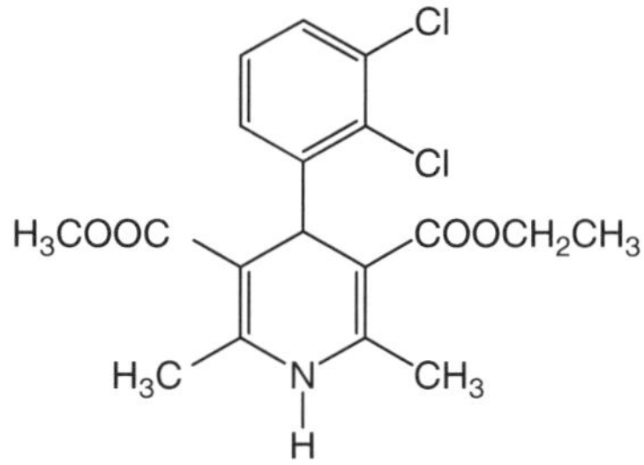

Fig. 5.4 Felodipine

Table 5.1 ASE parameters for the extraction of Felodipine tablets

Static temperature	Pressure (psi)	Number cycles	Static time (min)	Pre-heat time (min)	Flush volume	Solvent	Cell volume
50°C	1,500	1	15	5	100%	ACN	11 mL

Table 5.2 Extraction of Felodipine tablets by ASE in comparison with its manual method

Technique	Tablet form (10 mgA)	Solvent	Temperature (°C)	Recovery (%)/ %RSD	Preparation time (min)
Manual (Sonication)	Crushed	75/25 ACN/ MeOH	RT	101.7/0.6 ($n=5$)	35
ASE	Crushed	ACN	50	98.8/4 ($n=10$)	20
ASE	Intact	ACN	50	35	20
			100	60	20

RT = Room temperature, *mgA* = mg Active

5.2.7.1 Case Study 1: Extraction of Felodipine from Tablets

Bjorklund et al. (1998) were the first to evaluate the application of PFE as a sample preparation and extraction tool for tablet formulations. In their studies, a Dionex ASE 200 system was used to successfully extract the active ingredient, Felodipine (Fig. 5.4) from its tablet formulation. Employing the extraction conditions shown in Table 5.1, complete extraction of the active tablet formulation was achieved in 20 min, compared to the 35 min required by the manual method (Table 5.2). Unlike the manual method, which required use of a binary mixture consisting of 75/25 acetonitrile (ACN)/methanol (MeOH), extraction via ASE only required the use of 100% acetonitrile ACN. As indicated in Table 5.2, attempts to perform extraction on intact tablets by ASE resulted in very poor recovery. This was due to the inability to agitate and sufficiently disperse the tablet matrix within the extraction cell, thus limiting mass transfer of the analyte from the tablet to the extraction

Table 5.3 Initial ASE 200 extraction conditions

Static temperature (°C)	Pressure (kPa)	Number cycles	Static time (min)	Flush volume (%)	Cell volume (mL)
40	10,000	1	5	60	22

Table 5.4 Initial extraction of Montelukast Sodium from ground and intact IR tablets

Tablet form (5 mg tablets)	Solvent	Temperature (°C)	Recovery (%)
Ground	50/50 MeOH/H_2O	40	50
Ground	75/25 MeOH/H_2O	40	90
Ground	100% MeOH	40	87
Intact	50/50 MeOH/H_2O	40	45
Intact	75/25 MeOH/H_2O	40	12
Intact	100% MeOH	40	15

solvent. Increasing the extraction temperature led to an increase in recovery of the active from the intact tablet, but still significantly short of the complete recovery obtained by crushed, well-dispersed tablets.

5.2.7.2 Case Study 2: Extraction of an Asthma Drug from Oral Chewable Tablets

In a 2002 published study, Hoang et al. evaluated the application of PFE as a sample preparation and extraction tool for the extraction of the asthma drug, Montelukast Sodium, from an intact oral chewable tablet formulation (Hoang et al. 2002). Initial attempts to extract the active from both intact and ground IR tablets, under the conditions shown in Table 5.3, provided significantly lower recoveries for the intact tablets (Table 5.4). The extraction solvents employed consisted of 50, 75, and 100% of methanol in water. The highest recovery of Montelukast Sodium from intact tablets was observed with 50% MeOH/water extraction solvent (Table 5.4). The authors noted that when higher levels of MeOH were used (75 and 100%), the IR tablets did not disintegrate in the extraction cells. Higher levels of MeOH, however, facilitated increased recovery of the active from the ground tablets, as shown in Table 5.4.

Based on the above information, the authors decided to use a two-step approach to extract the active from the intact tablets. In the first extraction step, water was used as the extraction diluent to disintegrate the tablet, while 100% MeOH was used in step 2 to solubilize and extract the active from the disintegrated tablet. Because of limitations with the design of the ASE 200 system, two separate methods had to be run on the sample cell and the resulting extracts collected into separate collection vials. The design of the ASE 200 system does not allow for the collection of the extracts into the same collection vial, when two separate methods are employed. Following an assessment of the impact of temperature and number of cycles on the two-step extraction process, the method shown in Table 5.5 was deployed to

Table 5.5 Final ASE 200 extraction conditions for the extraction of Montelukast Sodium from its 5 mg tablet formulation

	Static temperature (°C)	Pressure (psi)	Number cycles	Static time (min)	Flush volume (%)	Solvent	Cell volume (mL)
Method 1: Disintegration	40	1,450	2	2	60	Water	22
Method 2: Dissolution	70	1,450	3	3	60	100% MeOH	22

Table 5.6 Extraction of Montelukast Sodium from 5 mg tablets via Accelerated Solvent Extraction (ASE) and mechanical shaking

Technique	Tablet form (5 mg Tablets)	Temperature	Recovery (%)/%RSD
Mechanical shaking	Intact	Ambient	97.6/0.9 ($n=10$)
ASE	Intact	40°C (Water) followed by 70°C (MeOH)	98.2/1.3 ($n=10$)

successfully extract the active from intact tablets. As shown in Table 5.6, results compared favorably with the manual extraction method, which consisted of transferring one tablet to a volumetric flask and allowing it to completely disintegrate in 50 mL of water. The flask was then filled to 80% of its volume with 100% MeOH and shaken for 60 min on a mechanical shaker, followed by dilution to volume with methanol. Using the two-step approach, the authors were able to completely extract the active from intact tablets in about half the time as compared to the manual method. Additionally, a significant reduction in solvent usage was achieved. Combined with the fact that the ASE 200 system employed in the study is capable of extracting up to 24 samples unattended, the resulting ASE method also provided an advantage in the area of throughput.

5.2.7.3 Case Study 3: Extraction of the Active Pharmaceutical Ingredient from a Spray Dried Dispersion Tablet Formulation

In a 2007 published study, Lee et al. evaluated the suitability of ASE as an extraction tool for the extraction of Compound A, from a 15% Spray Dried Dispersion (SDD) IR tablet formulation (Lee et al. 2007). The authors had hoped to utilize the elevated temperature advantage of ASE to reduce the sample preparation and extraction time from the 5.5 h required as per the manual method. Additionally, the authors had hoped that the semi-automated aspect of the ASE 200 would provide a significant advantage over the cumbersome process involved for the manual method. However, in spite of the unique potential advantages of ASE, including elevated temperature, capability for repeated extractions on the same sample, and flexible solvent selections, the authors reported significant challenges during attempts to extract Compound A from its 50-mg SDD IR tablet formulation. The authors looked at intact vs. crushed tablets, static times, elevated temperature, cell volume, and the use of hydromatrix to help increase surface area, none of which provided complete extraction of Compound A from the SDD formulation. Attempts to extract Compound A from the 50-mg SDD tablet formulation with an optimized ASE method led to approximately 91.1% recovery of the active (Table 5.7). The primary factor contributing to the observed low recovery results was an apparent gelling effect of the hydroxypropyl cellulose (HPC) polymer present in the formulation. The authors indicated that HPC, which is well known to gel in the presence of certain solvents, entraps the API within the tablet matrix, decreasing contact between the analyte and solvent, thus making complete extraction very difficult. Additionally, the fact that the sample matrix is in a fixed position within the cell, unable to be agitated, further contributed to the resulting low recovery of the active. A comparison to other sample preparation and extraction techniques such as MAE and the Sotax (formerly Caliper Life Sciences) Tablet Processing Workstation (TPW), which are capable of agitating samples during the extraction process, provided good support to the authors' low recovery explanation. Attempts to extract Compound A from the same tablet formulation by MAE and the TPW resulted in quantitative extraction of the active. Both MAE and the TPW are capable of agitating the sample solution during the extraction process, thus minimizing the gelling capabilities of the polymer.

5.3 Microwave Assisted Extraction

MAE is a partially automated sample preparation and extraction technique in which extraction solvents are rapidly heated to temperatures 2–3 times higher than their atmospheric boiling points (Renoe 1994; Eskilsson and Bjorklund 2000; Domini et al. 2006). For example, as shown in Table 5.8, solvents such as acetonitrile and methanol, with atmospheric boiling points of 81.6°C and ~64.7°C have closed vessel boiling points of 194 and 151°C, respectively, at 175 psi (Renoe 1994). This rapid, direct heating of the solvent medium is unique to MAE and leads to faster extraction times, higher recoveries, and reduced solvent usage. Additionally, unlike in PFE, MAE allows for the sample to be agitated or stirred during the heating process, creating a more homogeneous solution and increasing solvent/solute interaction. The net result is increased extraction efficiency.

Current MAE technology allows the operator to control the wattage, temperature, and length of time that go into the extraction process. Temperature fluctuations are typically within ±2° and throughput is high, as samples are usually processed in parallel. Although only polar solvents are microwave absorbers, this drawback of MAE is not generally an issue for most oral dosage forms, since most pharmaceutically relevant extraction solvents such as acetonitrile, methanol, and water are polar in nature and excellent microwave absorbers. As with PFE, only a handful of

Table 5.7 Optimized method for the extraction of compound A by Accelerated Solvent Extraction (ASE)

Technique	Temperature (°C)	Tablet form (50 mgA)	Static time (min)	Flush volume (%)	# Cycles	Solvent	Recovery (%)	Prepration time (min)
Manual (Sonication and mechanical shaking)	RT	Intact	N/A	N/A	N/A	Water followed by acetonitrile (ACN)	100.0	330
ASE	70	Crushed (and dispersed in hydromatrix)	3	60	3	80/20 ACN/H_2O	91.1	70

Table 5.8 Solvent boiling points and closed vessel temperatures

Solvent	Boiling point (°C)	Closed vessel temperature (°C) at 175 psi
Dichloromethane	39.8	140
Acetone	56.2	164
Methanol	64.7	151
Hexane	68.7	
Ethanol	78.3	164
Cyclohexane	80.7	
Acetonitrile	81.6	194
2-Propanol	82.4	145
Petroleum ether	35–52	
Acetone–hexane (1:1)	52+	156
Acetone–cyclohexane (73:30)	52+	160
Acetone–petroleum ether (1:1)	39+	147

Reprinted from Renoe (1994) with permission from International Scientific Communications Inc.

publications surfaced in the literature in the late 1990s utilizing MAE as a sample preparation and extraction tool for pharmaceutical dosage forms. However, the number of publications since then has been on the rise. In 1996, Bouhsain et al. reported on the use of MAE for analysis of paracetamol in various pharmaceutical formulations (Bouhsain et al. 1996). Eskilsson et al. (1999) were some of the first to use MAE as a technique for extraction of the active ingredient and degradation products from tablets. Labbozzetta et al. (2005) later used MAE for extraction and LC determination of the active ingredient in naproxen-based suppositories. An excellent review of the use of MAE, including details on the principles behind microwave heating, was provided by Eskilsson and Bjorklund (2000) and by Lopez-Avilla (2000). Similarly, an excellent comparative review of elevated temperature extraction techniques, including MAE, was done by Camel (2001). A more recent review was completed by Domini et al. (2006).

5.3.1 *Instrumentation*

MAE is often performed under atmospheric conditions (open vessels) or under controlled pressure (closed vessels). The latter approach is the more predominant one for extraction of oral dosage forms, as it allows for heating samples above their boiling points (Renoe 1994). The general schematic of a closed vessel microwave instrument is shown in Fig. 5.5. The system consists of an oven or microwave cavity, with extraction vessels on a turntable or rotor, a magnetron/microwave generator, and various devices for monitoring temperature and pressure. Most systems also come with a stirring plate to allow for sample agitation (Fig. 5.6). A number of safety devices are also available, such as solvent rupture vent tubes in the event of vessel

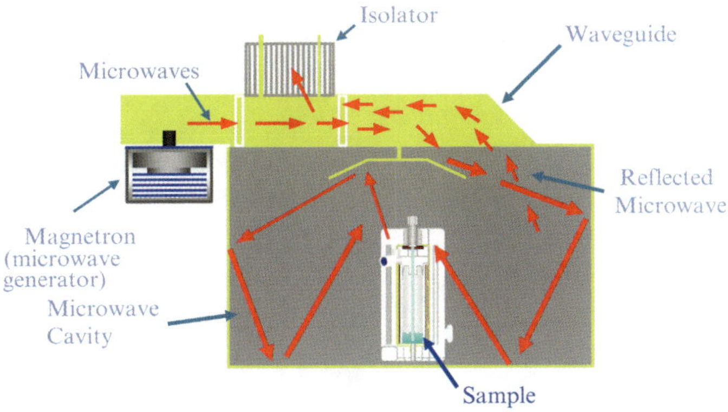

Fig. 5.5 Schematic of MAE instrument – multiple mode design (courtesy of CEM, Matthews, NC)

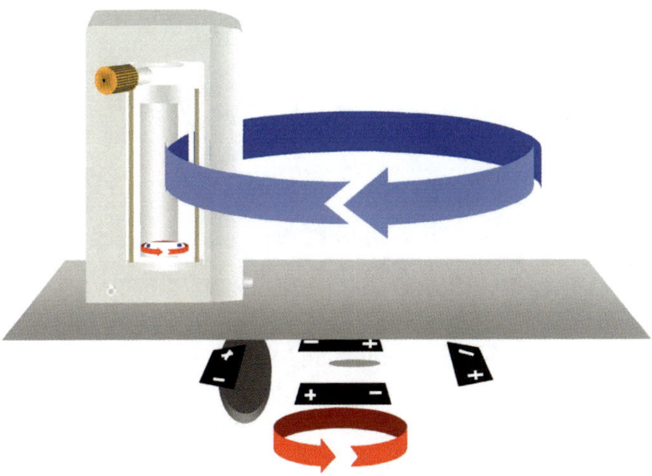

Fig. 5.6 Schematic of the CEM MARS extractor cavity floor, with rotating magnetic plate. Note: The magnetic plate contains four 2,800 gauss magnets, which is capable of stirring all vessels on the turntable (courtesy of CEM, Matthews, NC)

membrane failure/rupture due to increased pressure in the vessel and solvent vapor detector for monitoring the buildup of solvent vapors in the microwave cavity. Once solvent is detected in the instrument cavity, microwave irradiation stops, as power is removed from the magnetron. To facilitate direct heating of solvents, microwave vessels are often lined with microwave transparent material such as perfluoroalkoxy (PFA) or Teflon®. However, vessel body and caps are often made with polyether-imide (PEI). Temperature and pressure control is done via use of a control vessel. The control vessel is often modified to allow for connection of fiber-optic temperature probes and pressure sensing tubes. Like the vessels, the fiber-optic probe and pressure sensing tubes are also microwave transparent. Some microwave systems come

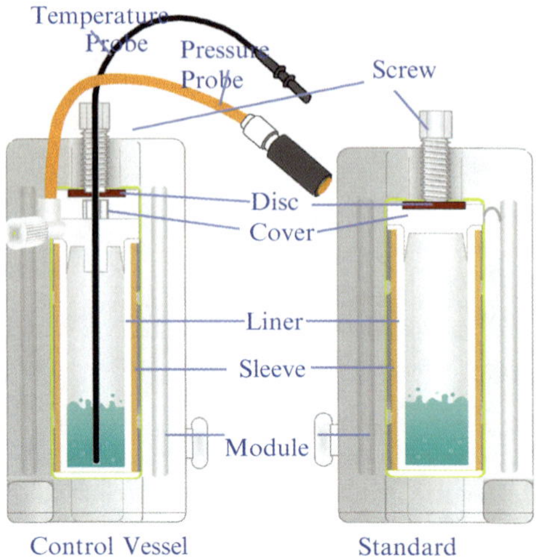

Fig. 5.7 Schematics of the control and standard vessels (courtesy of CEM, Matthews, NC)

with infrared temperature sensors that are capable of monitoring the temperature of each vessel in the system as the turntable rotates. Schematics of control and standard extraction vessels are shown in Fig. 5.7.

5.3.2 Key MAE Parameters/Method Development

Method development in MAE is simple and very user friendly. The key variables that have the greatest impact on extraction efficiency are sample pretreatment, solvent composition, temperature, and extraction time.

5.3.3 Sample Pretreatment

Depending on the dosage form and solvent composition, MAE can be successfully performed on intact tablets. IR tablets in particular are designed to rapidly disintegrate/disperse when exposed to aqueous media and therefore sample pretreatment may not be necessary to facilitate efficient dispersion and extraction of the active from the formulation. However, controlled release (CR) formulations are usually designed to disperse slowly and may therefore require appropriate sample pretreatment to increase surface area prior to MAE extraction. For CR formulations, intact tablets may need to be cut in half, quartered, or preferably crushed/homogenized to increase surface area and expose the inner core of the tablets. The above increases wettability of the matrix and increases solute–solvent interaction, thereby facilitating faster extraction.

5.3.4 Solvent Selection

Solvent selection is a critical variable for effective MAE extraction. When selecting solvents, analysts should consider not only the solubility of the analyte for the solvent and its compatibility with the analytical method but also the microwave absorbing properties of the solvent. In general, if extraction solvents are not capable of absorbing microwave energy, heating will not occur and extraction will be ineffective and time-consuming. Additionally, consideration must also be given to its interaction with the matrix. Most of the polar solvents used in conventional sample preparation and extraction procedures for pharmaceutical oral dosage forms, such as methanol, acetonitrile, water/buffers, and mixtures of the above, are all excellent microwave absorbers and are therefore excellent solvents for MAE extractions. Recent MAE developments have allowed for the use of non-microwave absorbing solvents for the extraction of actives from pharmaceutical dosage forms. These solvents can be heated indirectly through the use of polymeric bars (CEM's Carboflon and Milestone's Weflon bars), which absorb microwave energy and transfer the heat to the surrounding medium (Eskilsson and Bjorklund 2000; Lopez-Avilla 2000).

5.3.5 Extraction Temperature

Extraction temperature is probably the most critical and one of the most investigated variables in MAE. In closed vessel MAE, extraction temperatures of solvents often reach well above their atmospheric boiling points. In Table 5.8, the closed vessel boiling points of a number of non-polar and polar solvents are compared to that of their corresponding atmospheric pressure boiling points. For some solvents, such as dichloromethane, methanol, and acetonitrile, the closed vessel boiling points are 2–3 times that observed at atmospheric pressure. This increased temperature facilitates improved extraction of APIs from pharmaceutical dosage forms for the many reasons discussed previously, such as increased diffusion, solubility, and reduced surface tension.

5.3.6 Other Variables

Two other parameters that are often key to improved extraction efficiency in MAE are extraction time and agitation. In most cases, extraction times are on the order of 5–15 min, but can be longer due to time required for vessels to cool down when higher temperatures are employed. Rapid heating rates are often recommended, especially for thermally labile compounds, to help minimize degradation that is likely to occur due to increased exposure to elevated temperatures. Longer extraction times, while often positive for compounds that are thermally stable, can also have a negative impact on extraction recoveries. For example, Eskilsson et al. (1999) observed a reduction in recoveries at extraction times exceeding 60 min due to increased dissolution of the polymer matrix at longer extraction times. This increased dissolution of the polymer in the matrix caused an increase in the viscosity of the solvent, encapsulating the analyte within the matrix (Eskilsson et al. 1999).

Agitation of the sample within the sample vessel is also key for facile extraction and increased recoveries of actives from pharmaceutical dosage forms. Several microwave units come with built-in stirring plates at the bottom of the carousel (Fig. 5.6), allowing for efficient stirring of the samples during the heating process. Agitation by stirring allows for increased analyte–solvent interaction and also aids in tablet dispersion. This key variable was shown to play a critical role by Lee et al. (2007) during their investigation into the use of MAE for an SDD tablet formulation. Low recovery of the API was observed following extraction by ASE, where agitation was not possible, while a significant increase in recovery was observed by MAE and the Sotax (formerly Caliper Life Sciences) TPW. The authors suggested that the increased recovery observed by MAE and the TPW was primarily due to the ability to agitate the samples during the heating and extraction process (Lee et al. 2007).

5.3.7 Pharmaceutical Applications/Case Studies

Early sample preparation applications of MAE were focused on the analysis of environmental samples, including air, water, soil sediments, and sludge (Renoe 1994; Zuloaga et al. 1998; Eskilsson and Bjorklund 2000; Lopez-Avilla 2000; Camel 2001; Domini et al. 2006; Sanchez-Prado et al. 2010). As with PFE, the use of MAE as a sample preparation and extraction tool in the pharmaceutical industry is relatively new, but on the increase. Pharmaceutical applications include the extraction of natural products from plants (Pan et al. 2002); the extraction of parent compounds and metabolites in tissues such as liver and kidney (Eskilsson and Bjorklund 2000); and the extraction of APIs from various formulations (Eskilsson et al. 1999; Labbozzetta et al. 2005, 2008; Hoang et al. 2007; Lee et al. 2007). In the following case studies, the potential and limitations of MAE as a sample preparation and extraction tool for pharmaceutical dosage forms will be discussed.

5.3.7.1 Case Study 1: Extraction of Felodipine from Tablets by MAE

In a 1999 study, Eskilsson et al. reported on the use of MAE for the development of a robust method for the extraction of Felodipine (Fig. 5.4) and its degradation products from a tablet matrix (Eskilsson et al. 1999). The study was done to compare the effectiveness of MAE to that obtained previously with two other elevated temperature extraction techniques, SFE (Howard et al. 1994) and PFE/ASE

Table 5.9 Recoveries of Felodipine from tablets using MAE under various extraction solvent and temperatures

		Temperature (°C)			
		30	40	60	80
Solvent	Extraction time (min)	Recovery (%)/(%RSD)	Recovery (%)/(%RSD)	Recovery (%)/(%RSD)	Recovery (%)/(%RSD)
Methanol (MeOH)	10	91/(6)	90/(3)	50/(41)	55/(14)
	20	96/(2)	97/(2)	74/(7)	66/(10)
	40	99/(3)	97/(1)	81/(16)	81/(7)
	60	93/(2)	93/(2)	80/(9)	88/(4)
Acetonitrile (ACN)	10	65/(43)	29/(6)	36/(5)	102/(3)
	20	55/(59)	44/(3)	66/(35)	96/(1)
	40	64/(45)	68/(36)	83/(19)	88/(10)
	60	83/(30)	75/(5)	80/(14)	94/(1)
ACN/MeOH (2:1)	10	20/(9)	28/(7)	37/(13)	75/(8)
	20	60/(27)	59/(29)	70/(30)	96/(1)
	40	90/(18)	86/(23)	81/(4)	93/(3)
	60	91/(6)	90/(7)	88/(3)	95/(2)

Reprinted from Eskilsson et al. (1999), with permission from Elsevier

(Bjorklund et al. 1998). In their studies, a CEM MSP 1000 MAE unit (CEM, Matthews, NC), capable of processing up to 12 samples simultaneously, was used for the extraction of Felodipine and its degradation products from the tablet formulation. Variables such as solvent type, temperature, extraction time, and solvent volume were evaluated by the authors and the results were compared to those obtained by the manual ultrasonication method. The results from the evaluation of the impact of solvent type, extraction time, and temperature on the extraction of Felodipine from its solid oral dosage form are highlighted in Table 5.9. As indicated in Table 5.9, near complete recovery of the active was obtained within 40 min using 100% MeOH at temperatures between 30 and 40°C. Recoveries were, however, poor at temperatures above 40°C even when longer extraction times were applied. Incomplete disintegration of the tablet matrix at elevated temperatures was suggested as a reason for the lower recoveries observed when using MeOH at temperatures above 40°C. At these higher temperatures, the tablet matrix did not disintegrate into small particles, but rather into relatively large viscous residues, possibly trapping the active inside resulting in long diffusion times out of the bulk and into the solvent. At temperatures below 40°C, the tablet disintegrated into small granules allowing for the rapid release of the active out of the matrix and into the solvent. In the case of ACN, near complete recovery of the active was obtained at 80°C within 20 min of extraction time. At temperatures below 80°C, recoveries were poor even when extraction times were increased to 60 min. The inability of ACN, unlike MeOH, to dissolve and/or effectively fracture the outer layer of the tablet, especially at temperatures below 80°C, was provided as the primary reason for the lower observed recoveries at temperatures below 80°C. At 80°C, while the outer layer did not dissolve in the solvent, cracking was observed and effective release of the active was possible.

Table 5.10 Recoveries of Felodipine from tablets using MAE with 5% methanol in acetonitrile under various temperatures

	Temperature (°C)		
	40	60	80
Extraction time (min)	Recovery (%)/ (%RSD)	Recovery (%)/ (%RSD)	Recovery (%)/ (%RSD)
0.5	95/(3)		
1.5	96/(5)	98/(1)	
3	78/(27)	99/(2)	100/(3)
5	98/(3)	101/(1)	99/(1)
10	99/(1)	99/(1)	100/(2)
15			99/(1)

Reprinted from Eskilsson et al. (1999), with permission from Elsevier

The authors also noted that ACN, unlike MeOH, resulted in the swelling of the tablet core, causing the tablet to almost double in size. This swelling of the tablet core allowed for cracking of the outer layer at temperatures below 80°C. Mixtures of ACN and MeOH (2:1) did not have a significant impact on extraction efficiencies in general, except for at 80°C and extraction times of 20 min. A more detailed investigation by the authors showed that increased recoveries of Felodipine could be obtained using ACN/MeOH mixtures in the range of 5–10% MeOH. Consistently higher recoveries were obtained using 95/5 ACN/MeOH mixtures in shorter extraction times and at elevated temperatures (between 40 and 80°C) (Table 5.10). The final optimized MAE method was performed at 80°C with an extraction time of 5 min, plus 2 min ice-bath cooling, and utilized 95/5 ACN/MeOH as the diluent. Under these conditions, Felodipine and its degradants could be quantitatively extracted from tablet formulations (99.0% recovery for Felodipine) with good precision (RSD = 1.5%). When compared to the manual extraction procedure and other non-traditional elevated temperature techniques, such as PFE (Bjorklund et al. 1998) and SFE (Howard et al. 1994), MAE provided excellent results in reduced time and with higher throughput (Table 5.11). Additionally, unlike with PFE and SFE, up to 12 tablets could be extracted simultaneously using the MSP 1000 CEM microwave system employed in the study. With PFE and SFE, samples have to be analyzed one at a time.

5.3.7.2 Case Study 2: Extraction of Montelukast Sodium from Various Oral Dosage Forms

In a 2007 study, Hoang et al. reported on the successful application of MAE for the extraction of the asthma drug, Montelukast Sodium (Singulair®), from various oral formulations (Hoang et al. 2007). Extractions were performed using an Ethos EX microwave lab station from Milestones Inc. (Shelton, CT). With the Ethos system, up to 24 samples could be processed simultaneously, one of which is the control

Table 5.11 Comparison of extraction efficiencies between microwave assisted extraction (MAE), PFE, SFE and the manual method for the extraction of Felodipine from tablets

Technique	Recovery (%)/ (%RSD)	Temperature (°C)	Total preparation time (per tablet) (min)	Diluent
Manual	101.7/(0.6)	Ambient	35	75/25 ACN/MeOH
MAE	99.0/(1.5)	80	7	95/5 ACN/MeOH
PFE	98/(4)	50	20	Acetonitrile (ACN)
SFE	98.6/(1.2)		80	CO_2 (with 8% methanol (MeOH) modifier)

Table 5.12 Recovery data from single unit dose extraction of Montelukast Sodium (Singulair®) by microwave assisted extraction (MAE)

	% Label claim ($n=3$)	
Sample ID	Control (manual procedure)	MAE
10 mg Film coated tablets	101.7 (Sulfoxide adduct: 0.33)	100.7 (Sulfoxide adduct: 0.33)
4 mg Oral granules	102.3 (Sulfoxide adduct: 0.46)	101.9 (Sulfoxide adduct: 0.42)
4 mg Chewable tablets	97.6 (Sulfoxide adduct: 0.36)	98.6 (Sulfoxide adduct: 0.43)

vessel for monitoring temperature and pressure. Following initial method development work, the optimized MAE method consisted of an extraction temperature of 50°C for 5 min under stirring at 400 rpm. The power was kept at 300 W and 75/25 (v/v) MeOH/water was used as the extraction solvent, consistent with that used for the validated manual method. Under the above-mentioned conditions, Montelukast Sodium and its primary degradation product (sulfoxide adduct) could be quantitatively extracted from its various oral formulations (Table 5.12). The results compared very well to that obtained by the manual extraction procedure, but throughput was higher (up to 23 samples could be processed simultaneously with the Ethos system), and extraction times were significantly shorter (up to 60 min required by the manual procedure). The MAE method was also comparable to that obtained by PFE, where recoveries approached 98.2% for the chewable tablet formulation, following a sequential extraction approach involving disintegration in water at 40°C using two cycles at 2 min and 1×10^4 kPa, followed by extraction in methanol at 70°C and three cycles at 3 min each (Hoang et al. 2002). However, in PFE, samples were extracted one at a time, so throughput was lower when compared to MAE. The impact of stirring/agitation was also investigated by the authors. Results showed that complete recovery of the active and degradation products could be obtained in the absence of stirring for the oral granules formulation. Less than 80% recovery was obtained for the other two formulations in the absence of stirring. The above findings show that agitation by stirring is key to MAE, especially when dealing with intact solid oral dosage forms.

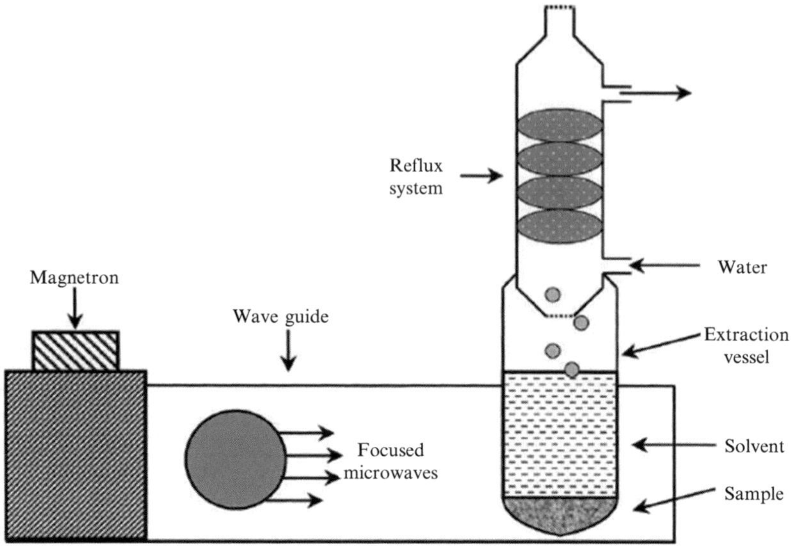

Fig. 5.8 Schematic diagram of FMAE instrument (reprinted from Labbozzetta et al. (2005), with permission from Elsevier)

5.3.7.3 Case Study 3: Extraction of the Active Pharmaceutical Ingredient in Naproxen-Based Suppositories by Open Vessel MAE

Labbozzetta et al. (2005) were the first to investigate the application of open vessel MAE for the extraction of APIs from suppositories. In their 2005 study, an open vessel or focused microwave assisted extraction (FMAE) unit was used to extract naproxen from suppositories. In FMAE, the extraction sample is placed in an open vessel and focused microwave radiation is used to irradiate the sample. In FMAE, the extraction solvent is refluxed at atmospheric pressure until extraction is completed. A CEM Star System 2 FMAE unit (CEM, Matthews, NC) equipped with a 2,450 MHz magnetron was used in their study. A schematic of the unit used is shown in Fig. 5.8. Extractions were performed at 70°C with a 10 min extraction time (including a 7 min linear ramp to reach 70°C), using a buffered diluent consisting of 50/50 MeOH/0.1 M sodium hydrogen carbonate (pH 8.7). Attempts to use solvents such as 100% MeOH, which was the diluent selected in the manual extraction procedure, or ethanol, proved futile and provided incomplete recovery of naproxen from suppositories. When compared to the manual extraction procedure, which involved ultrasonic dispersion of the suppositories in 500 mL of MeOH for 40 min, followed by cooling for 1 h at 5°C, recoveries by FMAE proved to be very comparable, but more efficient and utilized less solvent (Table 5.13). Additionally, results obtained via FMAE showed significantly higher precision than those obtained by the manual procedure and no further sample cleanup was required when using FMAE. In a later study, Labbozzetta et al. (2008) highlighted the effectiveness of

Table 5.13 Recovery of naproxen from suppositories

Mg/suppository (% label claim)	
Manual procedure	FMAE
527.3 (105.5)	542.2 (108.2)
440.4 (88.1)	511.0 (102.2)
600.3 (120.1)	544.7 (109.0)
523.6 (104.7)	516.7 (103.3)
576.3 (115.3)	540.5 (108.1)
481.5 (96.3)	546.5 (109.3)
593.8 (118.8)	538.2 (107.6)
585.1 (117.0)	539.6 (107.9)
600.8 (120.2)	523.7 (106.5)
503.0 (100.6)	547.0 (109.4)
Average = *543.2*	*Average* = *534.9*
%RSD = *10.4*	*%RSD* = *2.4*

Reprinted from Labbozzetta et al. (2005), with permission from Elsevier

FMAE for the extraction of active ingredients from pharmaceutical cream formulations. In their 2008 study, Ketoprofen was efficiently extracted from its topical cream formulation in the presence of preservatives.

5.4 Supercritical Fluid Extraction

SFE is an extraction technique that is characterized by, and takes advantage of, the use of an extraction solvent in its supercritical state. In SFE, the extraction fluid is a substance, which is pressurized, and then may or may not be mixed with an organic modifier/additive to enhance extraction. This pressurized fluid is pumped into a high-temperature environment, which makes the fluid either supercritical or near supercritical, and allowed to flow into the extraction vessel containing the sample matrix. The fluid interacts with the sample matrix under zero flow (static) or under a set flow rate (dynamic), or a combination thereof. Based on fluid composition and density, and the solubility of analytes of interest, the extraction will take place, carrying analytes to a trap. Just prior to the trap, a restrictor allows the pressurized fluid to decompress at ambient pressure, and if set up properly, the trap will catch analytes of interest. While topically simple, the principles and practice of SFE have significantly more complexity than most other extraction techniques, beginning with the unique properties of supercritical fluids themselves.

The supercritical state of a substance is defined as having temperature and pressure beyond that substance's critical point on a phase diagram as shown in Fig. 5.9. The critical point is the intersection of the substance's critical temperature, the highest temperature that a gas can phase transition to a liquid with an increase in pressure, and its critical pressure, the highest pressure that a liquid can phase transition to a gas with an increase in temperature. Supercritical fluids (SF), and substances

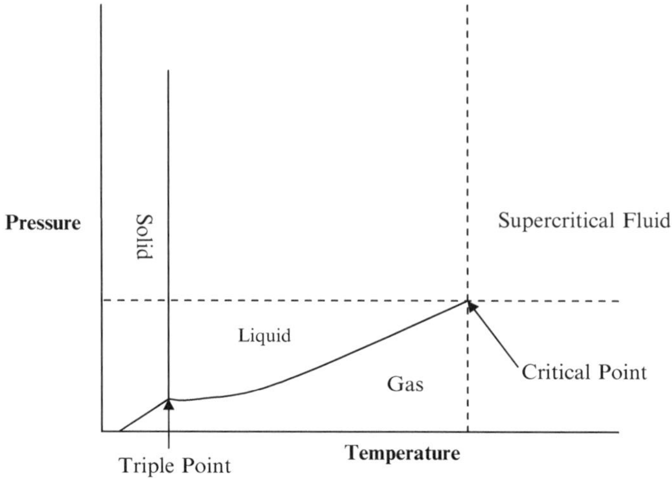

Fig. 5.9 Phase diagram extended beyond the critical point of a substance

approaching their critical point, possess a mix of both liquid and gas-like properties, and are typically characterized by having lower viscosities than liquids but with higher diffusion coefficients. The density of these fluids can be changed or tuned based on pressure and temperature settings. These characteristics are what SFE takes advantage of, and are the basis of this extraction mechanism.

5.4.1 Instrumentation and Principles of Operation

Figure 5.10 illustrates the typical SFE instrument setup. In most cases, the bulk fluid used in SFE is CO_2 in a high pressure aluminum cylinder with a full-length dip tube. The dip tube enables liquid CO_2 to be introduced into the pump heads for pressurization, a procedure that is more efficient that attempting to pressurize gas phase CO_2 into a liquid state. Alternatives to aluminum cylinders are bulk delivery setups, which also introduce liquid CO_2 to the pump heads. Pre-mix cylinders, cylinders containing both CO_2 and methanol, are not preferred as they limit analyst flexibility and are less reproducible, as the ratio of CO_2 to methanol can change over the lifetime of the cylinder, in comparison to an in-line modifier pump.

SFE employs high-pressure pumps to pressurize liquid CO_2. These pumps are typically reciprocating piston pumps, but single or tandem syringe pumps can also be used. As compared to HPLC, SFE pumps and their software are more sophisticated as a target pressure is desired and the fluids in question are compressible. SFE pump heads are chilled to ensure that liquid CO_2 reaches the pump head. At a higher temperature, a combination of liquid and gas CO_2 could produce cavitation, which is undesirable. Typically, in line with the flow (i.e., post-SF pump) a modifier pump, usually an HPLC-style pump, is present for the introduction of modifier.

The flow of fluid then enters an oven where elevated temperatures convert the substance into a supercritical or near supercritical state. Subcritical fluids (fluids

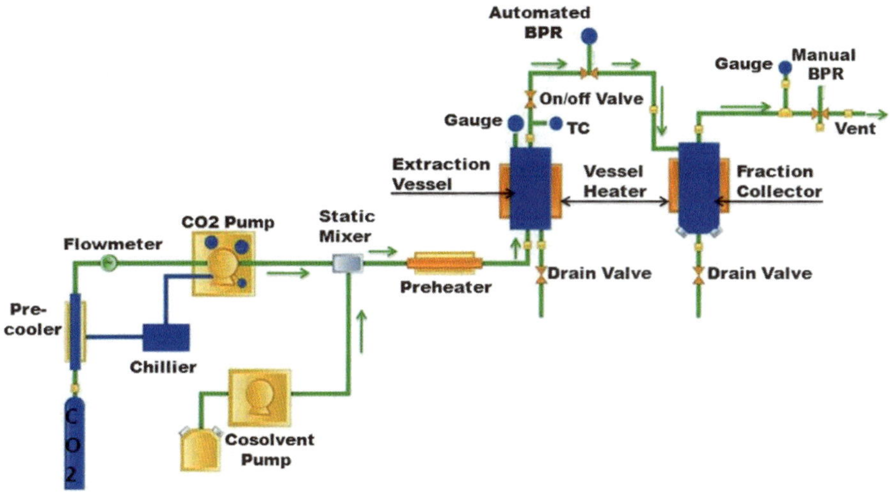

Fig. 5.10 Typical setup of an SFE instrumentation (courtesy of Waters Corporation, Milford, MA)

that are close to but not above a critical point due to lower temperatures or pressures, or because the addition of modifier raises critical point) still embody most of the attributes of classical supercritical fluids but may exist as a two-phase system. In this high temperature environment, the fluid interacts with the sample in the extraction vessel. The extraction vessel is typically a stainless steel body, as it must be capable of withstanding up to 10,000 psi during extractions. The body of the vessel contains two end fittings that screw or lock into place over a seal. Each end fitting contains a frit to keep solid material from moving outside of the vessel.

Post extraction, a back pressure regulator, or restrictor, maintains pressure while the fluid decompresses from the system. Variable restrictors are the norm for most systems, but fixed restrictors can be used. Fixed restrictors are not optimal as they create a pressure dependence on the flow rate, which limits analyst flexibility. The fluid decompresses through the restrictor to ambient pressure, and the extracts of interest are introduced to the trap. The rapid decompression of CO_2, and associated Joule Thompson effect, makes heating of the restrictor and/or solid phase trap essential to avoid clogging of the line, especially when modifiers are used. Solid phase, liquid phase, and combination of solid/liquid phase traps can be used for trapping in SFE. Some instruments limit this choice based on design, but the importance of trap type and conditions cannot be overemphasized. A recent review (Turner et al 2002) summarizes much of the work that has been done on the subject.

5.4.2 Key SFE Parameters in Method Development

SFE method development is not straightforward, and the key parameters that need to be assessed are multiple. However, if done correctly, SFE can provide an extraction method that is significantly more selective and environmentally friendly than other extraction techniques. Manipulation of the extraction fluid, by changing pressure and temperature, or by adding in an organic modifier or modifier/additive combination, is the primary parameter for optimization. More traditional parameters such as fluid flow rate and extraction time also need to be optimized. Once extraction is complete, a proper trapping mechanism must obviously be in place, thus trapping optimization is inherently linked to the extraction development.

5.4.3 Extraction Fluid

CO_2 is the preferred material, both practically and environmentally, used in SFE. In terms of supply, SFE grade CO_2 is widely available in high purity form from various vendors or one can purchase industrial grade CO_2 and purify it. Additionally, as compared to other fluids, the critical point of CO_2 is readily achievable with contemporary instrumentation. Although CO_2 has some polar character – it contains two polar bonds, has a quadrupole moment, and acts as a weak Lewis acid and Lewis base, the overall non-polar nature of CO_2 does limit the use of pure CO_2 as an extraction solvent, especially for polar pharmaceuticals. The practice of online mixing of pressurized CO_2 with organic modifier (i.e., methanol), however, is readily preferred to the use of more polar fluids. On paper, use of Freon-113 ($CHClF_2$) and fluoroform (CHF_3), as supercritical fluids, overcomes polarity problems seen with CO_2; however, those gains come at the cost of higher price, lower availability, environmental problems, and typically still involve the use of organic modifier to enhance extractions. While other fluids appear to be a more attractive choice due to their more polar nature, limitations of safety (N_2O), price and availability (Propane), and accessible critical point (H_2O) make modified CO_2 the logical and most widely used choice of supercritical fluid.

In terms of being green, or environmentally friendly, SFE with CO_2 is unrivaled in comparison to other extraction techniques. CO_2 can be vented to the atmosphere, thus eliminating expensive disposal of large amounts of organic solvents. Even with organic modified CO_2 extractions, the disposal amounts are small compared to traditional liquid extractions. Typically, extraction waste is trapped in a collection vessel to avoid aerosolization along with any residual modifier left in the vessel, and the CO_2 is simply vented to a hood. CO_2 is nontoxic and nonflammable, making it very worker friendly. Lastly, high purity tanks of CO_2 are typically prepared using the by-products of other chemical reactions; thus, they are not deemed to be a specific contribution to rising atmospheric CO_2 levels.

5.4.4 Pressure and Temperature

Pressure and temperature of the extraction fluid are directly linked to the solvating power of that fluid, and therefore are the main, and most interesting, parameters to optimize in SFE. The explanation of pressure and temperature relationship of the extraction fluid is made easier by replacing the parameter of pressure with that of density (Taylor 1996), as this simplifies the fact that at low pressures the solvating power of CO_2 decreases with increasing temperature, yet at higher pressures the solvating power will increase with an increase in temperature (Brogle 1982). Thus, by replacing density for pressure, it can be summarized that the solvating power of an SF will increase with density at a set temperature, or, that the solvating power of an SF will increase with temperature at a set density.

Pressure and temperature are also used to tune the fluid's diffusivity and viscosity. As described, SF possess characteristics between that of a liquid and a gas, with liquid-like solvating power but with gas-like diffusivity, which, as with density, can be tuned with changes in pressure and temperature. This relationship is simplified as follows: at a fixed density, with an increase in temperature, there will be an increase in the fluid's diffusivity and a decrease in its viscosity. With an increasing density at a fixed temperature, there will be a decrease in the fluid's diffusivity and an increase in the fluid's viscosity (Taylor 1996; Kamat et al. 1993).

The more traditional extraction parameters of fluid flow rate and extraction time also need to be optimized. Special attention needs to be paid to fluid flow rate with respect to trapping efficiency, especially where liquid phase traps are used. Extraction time optimization should be mapped out in development where multiple samples are collected at various time points. For diffusion limited extractions, extraction time may be the most important variable to monitor for exhaustive extractions.

5.4.5 Method Development

For successful method development in SFE, fluid type, pressure, temperature, and flow rate must be optimized in conjunction with proper setting of the trap conditions and sample introduction. Procedures for proper restriction and trapping are compound and method specific. Time must be taken to ensure that appropriate trapping measures are in place throughout method development. Use of DOE type matrixes for SFE method development, which include both extraction and trapping, would be a valuable and time saving tool for an analyst.

The first step of SFE method development should be extraction of a pure analyte from an inert matrix. In practice, a solution of analyte is spiked onto an inert matrix or directly into the extraction vessel. Spiking solvent is then allowed to evaporate, such that organic solvents do not affect extraction results. If a highly volatile analyte is under evaluation, care must be taken in the spiking phase in order to inhibit loss of analyte before the extraction. Initial method development extractions should take place with pure CO_2. Due to the polarity of most pharmaceuticals, and to overcome any matrix effects, extractions with small amounts of organic modifier should also be performed.

A simple method of extraction optimization for SFE would be to simply increase organic modifier to increase recovery. This is not a preferred approach as it ignores the power of the other SFE parameters that are adjustable, and it may lead to potential trapping issues. Therefore, to increase extraction efficiency, it is recommended that the following are evaluated in method development: (1) increase CO_2 density by increasing pressure, (2) increase fluid flow rate, (3) increase extraction time, and (4) increase extraction temperature. Compound-specific issues may limit method development. For example, a thermally labile compound should not use a high temperature in the extraction. It is recommended to evaluate these criteria prior to extraction from the real matrix such that a baseline of effects is available, as changing any of these parameters can affect both extraction and trapping efficiency. If difficulties with extractions are encountered at this point, method development should expand to include adding modifier directly to the extraction vessel or performing static and dynamic extractions.

After extraction of a pure analyte from an inert matrix, extractions of the analyte from the matrix in question should be performed. If possible, extraction of pure analyte spiked onto placebo matrix should be executed such that any issues with desorption from active sites on the matrix that come about from a manufacturing process can be identified. If recoveries are not quantitative, both extraction and trapping parameters should be re-addressed.

In the case of real-world samples, analytes may have a much different interaction with the matrix as compared to spiked samples, and greater solvating power may be necessary to overcome analyte matrix interactions. As mentioned previously, key SFE parameters (pressure, flow rate, temperature, and organic modifier content) should be re-evaluated. DOE experiments can facilitate fast method development and optimization. In some cases, specifically with some controlled release formulations, extractions may be diffusion limited. In this regard, the most effective avenue is to increase extraction time, or to include static extraction steps in conjunction with the dynamic step.

5.4.6 Advantages and Limitations of SFE

The advantages of SFE are inherently linked to the use of a supercritical fluid as the extraction solvent. The use of CO_2 gives SFE an environmental advantage over extraction techniques that utilize larger amounts of organic solvents. The ability to tune fluid density with changes in pressure potentially makes SFE a very selective technique. Other extraction techniques that extract without discrimination many times lead to dirtier samples or the need to perform additional sample cleanup steps, as compared to SFE.

As with the advantages of SFE, the limitations of the technique are also directly linked to the extraction fluid. Primarily, supercritical CO_2 is very non-polar, limiting its advantages in dealing with polar or ionic compounds. SFE with CO_2 can be very selective, but this may come at a drop off in effectiveness for breaking some analyte matrix effects. Additionally, the tunable nature of CO_2 leads to a lack of universal type methods. Specifically, SFE methods for one analyte in one matrix may not work for the same analyte in another matrix, thus starting the process of method development over again. In terms of pharmaceuticals, these limitations may seem great with the polar nature of most active ingredients, and with the complexity of many of today's dosage forms, however, the variety of drug formulations that are available are the exact reason that SFE should always be considered, as there are certain active ingredients in certain formulations that would match up perfectly with this technique.

5.4.7 Pharmaceutical Applications/Case Studies

Many pharmaceutical applications using SFE were published in the 1990's. However, due to the lack of robust commercial SFE instrumentation at analytical scale, most recent SFE work in the literature has focused on food (Comim et al. 2010; Egydio et al. 2010), environmental matrices (Ramsey et al. 2010; Wang et al. 2009; Yarita 2008), and natural products (Grosso et al. 2010; Pereira and Meireles 2010). The overlap of natural product work and examples in other fields with analytes similar to typical pharmaceuticals indicates that SFE is still a promising technique for pharmaceutical analysis.

Case studies are presented below to provide an overview of the capability of SFE in terms of sample preparation for pharmaceutical applications. Examples of active ingredient extraction, impurity extraction, and inverse SFE are provided.

5.4.7.1 Application 1: Extraction of Active Ingredients from Pharmaceutical Dosage Forms

Early work with SFE on solid dosage forms illustrated extraction of the active ingredient ibuprofen from immediate release tablets (Khundker et al. 1993). It was noted, even in this early work, that the same extraction conditions for different matrixes caused recovery issues. Evaluation of SFE with more complex dosage forms can be seen in the recovery of felodipine from controlled release tablets (Howard et al. 1994). In this case, method development was optimized on spiked samples, but conditions had to be adjusted when tablet extraction was attempted. Methanol modifier and repeated static and dynamic steps were needed to optimize the extraction and trapping of the analyte for quantitative recovery. Examples of polar (Eckard et al. 1998) and ionic active ingredients (Eckard and Taylor 1997, 1999) extracted via SFE have been evaluated.

The use of experimental design to map out method development for extraction of three pharmaceutical-type molecules, methimazole, phenazopyridine, and propranolol, from solid dosage forms (Bahramifar et al. 2005) is an example of a modern approach to SFE. In this work, method development was carried out with samples spiked onto glass beads to optimize recoveries. Extractions of active ingredients from pharmaceutical dosage forms showed lower recoveries than were seen from

Table 5.14 Extraction of active spiked onto inert sand

SFE parameters[a]	% Recovery run 1	% Recovery run 2	% Recovery run 3	Average recovery	%RSD
10% MeOH+0.7% TFA, 10 min	94.51	93.08	89.46	92.35	2.82
25% MeOH+0.7% TFA, 10 min	92.36	99.66	108.85	100.29	8.24
25% MeOH+0.7% TFA, 20 min	95.42	91.37	95.20	94.00	2.42
25% MeOH+0.7% TFA, 30 min	96.90	95.60	92.80	95.10	2.20

[a]Extraction pressure 350 atm, extraction temperature 40°C, fluid flow rate 2.0 mL/min

glass beads. While the same conditions were used in each case, along with matrix effects, it was noted that the two analytes existed as hydrochloride salts in the tablets, which was different than their state in method development. The use of basic modifiers provided quantitative extractions for these hydrochloride salts.

SFE of active ingredient from solid dosage forms was evaluated (Brannegan et al. 2005) with a focus on improvement of difficult or lengthy existing sample preparations/extractions. The first part of this work dealt with a controlled release matrix tablet, which releases drug by diffusion out of the matrix or erosion of the tablet. The current sample preparation consisted of weighing the tablet, allowing it to disperse in methanol, mixing with dissolving solvent, centrifuging for 45 min, and dilution. SFE, with the following parameters, was evaluated to optimize extraction of spiked samples off of inert sand: extraction pressure 350 atm, extraction temperature 40°C, flow rate 2.0 mL/min, fluid composition of CO_2 modified with 10–25% methanol with 0.7–1.0% (Trifluoroacetic acid) TFA, extraction time of 30–120 min, with collection of a solid phase trap followed by a liquid trap of methanol. Table 5.14 illustrates the recoveries attained off of inert sand.

Further optimization was required to get quantitative recovery of this active component from crushed tablets. Table 5.15 illustrates, as seen in the extractions from the inert sand, that while recoveries over 90% are attainable, a higher than expected %RSD is present. Development work also indicated that a nonhomogeneous sample, i.e., a crushed tablet that is not evenly distributed into an appropriate matrix, in this case, inert sand, would provide low recovery values. This application required modifier with additive for extraction from the tablet matrix, which may decrease trapping of analyte by aerosol formation during decompression. While not quantitative, recoveries and %RSDs may be able to be improved with further extraction and trapping development.

The second part of the work utilized SFE in a more complex matrix system, a swellable core tablet (Thombre et al. 2004) with a different active ingredient. Current sample preparation methodology for this swellable core tablet includes quartering each tablet with a razor, transferring to a 100-mL volumetric flask, stirring solutions overnight, centrifuging, and further dilution.

Initial SFE experiments showed quantitative recovery of the active from spiked sand. Extractions from tablets indicated that sample preparation was required,

Table 5.15 Extraction of active ingredient from crushed tablets

SFE parameters[a]	% Recovery run 1	% Recovery run 2	% Recovery run 3	Average recovery	%RSD
100 mg 25% MeOH + no TFA, 30 min	58.76	64.23	61.21	61.40	4.46
100 mg 25% MeOH + 0.7% TFA, 15 min	74.48	77.75	78.71	76.98	2.88
100 mg 25% MeOH + 0.7% TFA, 20 min	91.80	89.54	94.35	91.89	2.88
100 mg 25% MeOH + 0.7% TFA, 30 min	88.26	96.96	92.27	91.41	4.76
100 mg 25% MeOH + 0.7% TFA, 60 min	93.42	95.16	90.23	92.94	2.69
50 mg 25% MeOH + 0.7% TFA, 30 min	94.31	97.41	93.72	95.15	2.08
50 mg 25% MeOH + 0.7% TFA, 60 min	92.15	95.35	91.25	92.92	2.32

[a]50 or 100 mg of crushed tablet are mixed with 6 g of sand for extraction. Extraction pressure 350 atm, extraction temperature 40°C, fluid flow rate 2.0 mL/min

crushing of tablets to put into the extraction vessel, and longer extraction times were necessary. It was also noted that higher temperatures were needed. Due to the nature of the dosage form, there were multiple small changes in method development that eventually lead to recoveries over 95%, including very specific requirements for vessel size/sand amount/sample amount.

The use of SFE for extraction of active ingredients from pharmaceutical dosage forms is a user friendly and green alternative to most liquid extraction techniques. Unfortunately, SFE is typically only evaluated in situations where conventional sample preparations have been unsuccessful or have been deemed nonideal. In these situations, as detailed above, SFE can be attempted, but may not be the logical choice. As with assessments of use with analyte type, SFE can be the optimal choice for certain dosage forms, and should be evaluated routinely, not simply as a last resort.

5.4.7.2 Application 2: SFE Extraction of Impurities from Solid Dosage Forms

While most pharmaceutical examples of SFE are focused on extraction of the main active ingredient from a dosage form, the extraction of impurities in these dosage forms is also possible. Much like environmental or natural product work, extraction of impurities from dosage forms takes advantage of SFE's selectivity, but also must pay special attention to trapping and transfer of analytes for appropriate analysis. Applications such as this are of high interest to pharmaceutical testing labs, as typically analysis of both potency and purity are assessed in testing protocols.

A good example of the use of SFE in working with impurities in tablets (Bochkareva et al. 2006) evaluated benzodiazepine compounds, diazapam, phenazepam, nitrazepam, and clonazepam, for moderately volatile impurities. Tablet preparations were ground into fine powders, put into 0.5-mL extraction vessels, and extracted with 100% supercritical fluids (no modifiers) at 40°C, 250 atm, and 1.5 mL/min. In an effort to minimize loss of impurities post extraction, the team used a special cartridge for tapping of the extracts. Analytes trapped on the cartridge were desorbed at high temperature and analyzed by GC/MS. Both CO_2 and N_2O were used as SFs in this work, and as no modifiers were used, it is likely that N_2O was evaluated to extract more polar analytes.

Extractions were performed in a series of 5 min extractions. Results showed that 75–100% of most impurity's total amount was extracted in the first 5 min extraction. Thirty-five percent of all impurities showed full extraction in the first 5 min. The extractions reported two USP regulated impurities in the diazepam tablets and one in the clonazepam tablets. Extractions were optimized to provide a potential method for determination of semi-volatile impurities in pharmaceutical tablets of the benzodiazepine series through the use of SFE with GC/MS detection. This type of SFE work may gain traction in analysis of trace impurities in dosage forms or impurities of exotic dosage forms due to the high selectivity and extraction efficiency of SFE.

5.4.7.3 Application 3: Inverse SFE

Inverse SFE is a technique that finds use with pharmaceutical formulations such as creams, ointments, and suppositories that contain a more polar active ingredient. As compared to traditional SFE, where the main active ingredient is extracted and trapped for analysis, in inverse SFE, the inactive ingredients are removed and the more polar active ingredient remains in the extraction vessel for transfer for analysis.

While only a few pharmaceutical-type references are available in the literature, this niche technique is highlighted as pharmaceutical companies put more time and effort into product enhancements and exploring different formulations and delivery mechanisms with existing active ingredients. An early evaluation of inverse SFE (Messer and Taylor 1994) used Zovirax ointment. This work was continued (Moore and Taylor 1994) by further evaluating inverse SFE with cream and ointment formulations. This work attempted extraction of the active ingredient, polymyxin B sulfate from Neosporin Cream, and Neosproin Ointment. The cream formulation contained methyl paraben, emulsifying wax, mineral oil, polyoxyethylene polyoxypropylene compound, propylene glycol, purified water, and white petroleum. The ointment formulation was a white petroleum base. Once the extraction vessel was optimized to retain the unextracted active in the vessel while exhaustively extracting the matrix, exceptional results were attained for both the cream and ointment extractions.

Evaluation of inverse SFE with suppositories and the active ingredient of acetaminophen (Almodovar et al. 1998) further shows a number of dosage forms that this technique is amenable to. This work used very mild SFE conditions to minimize fast melting of the suppository as well as minimizing risk of loss of the active via entrapment during extraction or mechanical transfer through the frits. Results of this work showed quantitative recovery of acetaminophen, comparable to the USP method, yet was able to attain such results with significant time savings and without the use of larger amounts of organic solvents.

These examples all exhibit the main criteria for a successful inverse SFE experiment. First, the main active ingredient must be insoluble in the SF. This requirement is typically met when 100% CO_2 or CO_2 with small amounts of organic modifier are used as the extraction solvent, and polar pharmaceutical active ingredients are present. Secondly, the matrix, or formulation, must be readily soluble in the SF. This criterion is typically met when dealing with creams, ointments, and suppositories. Additional concerns with transferring the left over active ingredient from an extraction vessel, as well as evaluation of drug loading and the limit of quantitation of the method analysis, are of significance in attempting inverse SFE.

A unique application of inverse SFE is provided in the attempt to remove common reactive impurities from several pharmaceutical excipient powders (Ashraf-Khorassani et al. 2006). Pharmaceutical solid dosage forms are typically made up of a number of highly purified inactive ingredients, or excipients. While international pharmacopeia guidelines are in place to ensure that the impurities in excipients are below a specified level, in many cases, these impurities are still present. The intimate contact of active ingredients with certain impurities in these excipients can have negative impact on the dosage form, manifesting as a significant decrease of active content, the forming of undesired drug-related impurities, or a combination of both. In the inverse SFE experiments evaluated, results showed that, while some polymeric excipients could be changed (swelled) by the extraction procedure, the extraction of surface-bound formic acid and formaldehyde was possible. The low solubility of hydrogen peroxide in CO_2 and the long extraction times to remove embedded impurities obviously indicate the limits of this technique; however, the promise of niche use of SFE in various pharmaceutical laboratories is interesting.

5.5 Pressurized Hot Water Extraction

PHWE (Ong et al. 2000; Teo et al. 2010) or superheated water extraction (SWE) (Hawthorne et al. 1994; Smith 2002) is another PLE tool and a true green extraction method. This technique utilizes the changes of physical chemical properties of water at high temperature. While water at room temperature is too polar to extract any non-polar compound, the polarity of water can be reduced by increasing the temperature. At higher temperature, the viscosity, dielectric constant, and surface tension of water are reduced, with additional pressure to keep water in the liquid phase, the polarity of water is similar to that of methanol at higher temperature. The decreased polarity and improved mass transfer of water at high temperature result in improved extraction of less polar compounds. The instrument setup for PHWE is similar to PLE/ASE and SFE instrumentation. The critical parameters for this extraction method are temperature, pressure, and flow rate. In some cases, organic modifier or surfactant was also added to increase the extraction efficiency. The major applications of this technique have been for the extraction of components or contaminants from food, environmental, and natural product samples. A pharmaceutical application of SHWE was reported by Richter et al. who used this technique to prepare nifedipine tablets for assay and content uniformity analysis (Richter et al. 2006). Detailed technical aspects and applications of this method can be found in a recent review article by Teo et al. (2010).

5.6 Conclusions

In comparison with traditional sample preparation (e.g., mechanical shaking), PLE, MAE, SFE and PHWE methods can provide reduced solvent consumption, shorter extraction time, and ease of method development. PLE, SFE, and PHWE use similar general technology, except the extraction media (organic solvents or subcritical/supercritical fluids). In terms of extraction efficiency, PLE, MAE, and SFE are comparable if the methods are optimized for a given drug product. Since PLE and MAE techniques use higher extraction temperature and a wide range of solvents, the drug–matrix interaction can be minimized and they are less matrix dependent compared to SFE. In addition, PLE and MAE methods are easier to optimize than SFE methods as they use fewer parameters to control the extraction. SFE has the best selectivity among the extraction techniques. SFE is also a suitable technique to extract thermolabile compounds as it performs at moderate temperature. In summary, there is no superior technique in a general term, all the extraction techniques have their advantages and shortfalls. Choosing the right technique for a drug product application depends on the individual compound and its matrix. One should carefully evaluate the compound and matrix properties and select the most suitable technique(s) based on the principles and case studies of these techniques discussed in this chapter.

References

Abend AM, Chung L, McCollum DG, Wuelfing WP (2003) Development and validation of an automated extraction method (Accelerated Solvent Extraction®) and a reverse-phased HPLC analysis method for assay of ivermectin in a meat-based chewable formulation. J Pharm Biomed Anal 31:1177–1183

Almodovar RA, Rodriguez RA, Rosario O (1998) Inverse supercritical extraction of acetaminophen from suppositories. J Pharm Biomed Anal 17:89–93

Ashraf-Khorassani M, Taylor LT, Waterman KC, Narayan P, Brannegan DR, Reid GL (2006) Purification of pharmaceutical excipients with supercritical fluid extraction. Pharm Dev Technol 10:507–516

Bahramifar N, Yamini Y, Shamsipur M (2005) Investigation on the supercritical carbon dioxide extraction of some polar drugs from spiked matrices and tablets. J Supercrit Fluids 35:205–211

Bjorklund E, Jaremo M, Mathiasson L, Karlsson L, Strode JT III, Eriksson J, Torstensson A (1998) Determination of felodipine in tablets using accelerated solvent extraction. J Liq Chromatogr Relat Technol 21(4):535–549

Blanchard A, Lee C, Nickerson B, Lohr LL, Jensen AJ, Alsante KM, Sharp TR, Santafianos DP, Morris R, Synder KD (2004) Identification of low-level degradants from low dose tablets. J Pharm Biomed Anal 36(2):265–275

Bochkareva NL, Glaskov IN, Revelsky IA (2006) Combination of supercritical fluid extraction and gas chromatography-mass spectrometry: determination of impurities extracted from tablet preparations of the benzodiazepine series. J Anal Chem 61:1082–1089

Bouhsain Z, Garrigues S, Morales-Rubio A, de Guardia M (1996) Flow injection spectrophotometric determination of paracetamol in pharmaceuticals by means of online microwave-assisted hydrolysis and reaction with 8-hydroxyquinoline (8-quinolinol). Anal Chim Acta 330(1):59–69

Brannegan DR, Wang J, Taylor, LT, Ashraf-Knorassani M (2005) Internal Pfizer SFE study report

Brogle H (1982) CO_2 as a solvent: its properties and applications. Chem Ind 12:385–390

Camel V (2001) Recent extraction techniques for solid matrices – supercritical fluid extraction, pressurized fluid extraction and microwave-assisted extraction: their potential and pitfalls. Analyst 126:1182–1193

Comim SR, Madella RK, Oliveira JV, Ferreira SRS (2010) Supercritical fluid extraction from dried banana peel (*Musa* spp., genomic group AAB): extraction yield, mathematical modeling, economical analysis and phase equilibria. J Supercrit Fluids 54(1):30–37

Domini CE, Hristozov D, Almagro B, Roman IP, Prats S, Canals A (2006) Chapter 2: Sample preparation for chromatographic analysis of environmental samples. In: Nollet LML (ed) Chromatographic analysis of the environment, 3rd edn. CRC Press, Boca Raton

Eckard PR, Taylor LT (1997) Feasibility of ion-pair/supercritical fluid extraction of an ionic compound-pseudoephedrine hydrochloride. J Pharm Biomed Anal 15:613–619

Eckard PR, Taylor LT (1999) Modifier and additive effects in the supercritical fluid extraction of pseudoephedrine hydrochloride from spiked-sand and suphedrine tablets. J High Resolut Chromatogr 22:469–474

Eckard PR, Strode JTB, Taylor LT, Howard AL, Ip D (1998) Investigation of primary and secondary modifiers for the subcritical extraction of lovastatin from MEVACOR tablets with carbon dioxide. J Chromatogr Sci 36:139–145

Egydio JA, Moraes AM, Rosa PTV (2010) Supercritical fluid extraction of lycopene from tomato juice and characterization of its antioxidation activity. J Supercrit Fluids 54(2):159–164

Eskilsson CS, Bjorklund E (2000) Analytical-scale microwave assisted extraction. J Chromatogr A 902:227–250

Eskilsson CS, Bjorklund E, Mathiason L, Karlsson L, Torstensson A (1999) Microwave-assisted extraction of felodipine tablets. J Chromatogr A 840:59–70

Grosso C, Figueiredo AC, Burillo J, Mainar AM, Urieta JS, Barroso JG, Coelho JA, Palavra AMF (2010) Composition and antioxidant activity of *Thymus vulgaris* volatiles: comparison between supercritical fluid extraction and hydrodistillation. J Sep Sci 33(14):2211–2218

Hawthorne SB, Yang Y, Miller DJ (1994) Extraction of organic pollutants from environmental solids with sub- and supercritical water. Anal Chem 66:2912–2920

Hoang TH, Farkas R, Wells C, McClintock S, Di Maso M (2002) Application of pressurized liquid extraction technology to pharmaceutical solid dosage form analysis. J Chromatogr A 968:257–261

Hoang TH, Sharma R, Susanto D, Di Maso M, Kwong E (2007) Microwave-assisted extraction of active pharmaceutical ingredient from solid dosage forms. J Chromatogr A 1156:149–153

Howard AL, Sahh MC, Ip DP, Brooks MA, Strode JT III, Taylor LT (1994) Use of supercritical fluid extraction for sample preparation of sustained-release felodipine tablets. J Pharm Sci 83(11):1537–1542

Kamat SV, Iwaskewycz B, Beckman EJ, Russell AJ (1993) Biocatalytic synthesis of acrylates in supercritical fluids: tuning enzyme activity by changing pressure. Proc Natl Acad Sci USA 90:2940–2943

Khundker S, Dean JR, Hitchen SM (1993) Extraction of ibuprofen by supercritical carbon dioxide. Anal Proc 30:472–475

Labbozzetta S, Valvo L, Bertocchi P, Manna L (2005) Focused microwave assisted extraction and LC determination of active pharmaceutical in naproxen-based suppositions. J Pharm Biomed Anal 39:463–468

Labbozzetta S, Valvo L, Bertocchi P, Alimonti S, Gaudiano MC, Manna L (2008) Focused microwave-assisted extraction and LC determination of ketoprofen in the presence of preservatives in a pharmaceutical cream formulation. Chromatographia 69(3–4):365–368

Lee C (2007) Troubleshooting low potency results in solid oral dosage forms: Is it the method or the formulation? Am Pharm Rev 10(6):142–147
Lee C, Gallo J, Arikpo W, Bobin V (2007) Comparison of extraction techniques for spray dried dispersion tablet formulation. J Pharm Biomed Anal 45:565–571
Lopez-Avilla V (2000) Chapter 3: Applications of microwave-assisted extraction in environmental analysis. In: Barcelo D (ed) Sample handling and trace analysis of pollutants: techniques, applications and quality assurance. Elsevier Science B, Amsterdam, pp 15–153
Messer DC, Taylor LT (1994) Inverse analytical supercritical fluid extraction of Zovirax ointment. Anal Chem 66:1591–1592
Moore WN, Taylor LT (1994) Analytical inverse supercritical fluid extraction of polar pharmaceutical compounds from cream and ointment matrices. J Pharm Biomed Anal 10:1227–1232
Ong ES, Woo SO, Yong YL (2000) Pressurized liquid extraction of berberine and aristolochic acids in medicinal plants. J Chromatogr A 22:57–64
Pan X, Niu G, Liu H (2002) Comparison of microwave-assisted extraction and conventional extraction techniques for the extraction of tanshinones from Salvia miltiorrhiza bunge. Biochem Eng J 12:71–77
Pereira CG, Meireles MAA (2010) Supercritical fluid extraction of bioactive compounds: fundamentals, applications and economic perspectives. Food Bioprocess Technol 3(3):340–372
Ramsey ED, Rees AT, Wei G, Liu JY, Wu XH (2010) Direct aqueous supercritical fluid extraction coupled on-line with liquid chromatography-tandem mass spectrometry for the analysis of polyether ionophore antibiotics in water. J Chromatogr A 1217(20):3348–3356
Renoe BW (1994) Microwave assisted extraction. Am Lab 26(12):34–40
Richter BE (1999) The extraction of analytes from solid samples using accelerated solvent extraction. LC GC 17(6S):S22–S28
Richter BE, Ezzell JL, Felix WD, Roberts KA, Later DW (1995) An accelerated solvent extraction system for the rapid preparation of environmental organic compounds in soil. Am Lab 27(4):24–28
Richter BE, Jones BA, Ezzell JL, Porter NL, Avdalovic A, Pohl C (1996) Accelerated solvent extraction: a technique for sample preparation. Anal Chem 68(6):1033–1039
Richter BE, Rothe N, Ezzell JL, Francis ES, Covino L, Toth J (2001) Expanding the capabilities of automated sample prep. Am Lab 33(6):56–60
Richter P, Ines Toral M, Toledo C (2006) Subcritical water extraction and determination of nifedipine in pharmaceutical formulations. J AOAC Int 89(2):365–368
Sanchez-Prado L, Garcia-Jares C, Llompart M (2010) Microwave-assisted extraction: application to the determination of emerging pollutants in solid samples. J Chromatogr A 1217: 2390–2414
Smith RM (2002) Extractions with superheated water. J Chromatogr A 975:31–46
Taylor LT (1996) Chapter 2: Properties of supercritical fluids. In: Supercritical fluid extraction. Wiley, New York
Teo CC, Tan SN, Yong JWH, Hew CS, Ong S (2010) Pressurized hot water extraction (PHWE). J Chromatogr A 1217:2484–2494
Thombre AG, Appel LE, Chidlaw MB, Daugherity PD, Dumont F, Evans LAF, Sutton SC (2004) Osmotic drug delivery using swellable-core technology. J Control Release 94:75–89
Turner C, Eskilsson CS, Bjorklund E (2002) Collection in analytical-scale supercritical fluid extraction. J Chromatogr A 947:1–22
Wang W, Yang H, Hu J, Guo C (2009) Extraction of metal ions with non-fluorous bipyridine derivatives as chelating ligands in supercritical carbon dioxide. J Supercrit Fluids 51:181–187
Yarita T (2008) Development of environmental analysis methods using supercritical fluid extraction and supercritical fluid chromatography. Chromatography 29:19–23
Zuloaga O, Etxebarria N, Fernandez LA, Madariaga JM (1998) Comparison of accelerated solvent extraction with microwave-assisted extraction and Soxhlet for the extraction of chlorinated biphenyls in soil samples. Trends Anal Chem 17(10):642–647

Section C
Sample Preparation Method Development and Validation for Various Dosage Form Types

Chapter 6
Addressing Drug–Excipient Interactions

Carlos Lee

Abstract Drug–excipient interactions consist of physical and chemical interactions. On the one hand, chemical interactions lead to degradation products and are a formulation or stability concern. On the other hand, physical interactions can pose sample preparation and extraction challenges by hindering complete drug extraction. Types of physical drug–excipient interactions and their potential impact on sample preparation will be discussed. Strategies to address these interactions and ensure complete extraction of the drug will also be covered. Selected case studies will also be considered and discussed in detail.

6.1 Introduction

Drug–excipient interactions can have tremendous impact on analytical methods for pharmaceutical dosage forms. These types of interactions, which can be chemical or physical in nature, can pose sample preparation challenges by hindering complete drug extraction. They often reflect themselves in terms of low recovery or potency during drug product assays. Drug–excipient interactions can also lead one to question the robustness of the analytical method and potentially stress the analyst/formulator relationship (method vs. formulation debate). This chapter will focus on physical drug–excipient interactions and their potential impact on sample preparation. Strategies to address these interactions and ensure complete extraction of the drug will be covered, and selected case studies will be evaluated.

C. Lee (✉)
Research Analytical, Pfizer Global Research and Development, Pfizer Inc.,
558 Eastern Point Road, Groton, CT 06340, USA
e-mail: Carlos.W.Lee@Pfizer.com

B. Nickerson (ed.), *Sample Preparation of Pharmaceutical Dosage Forms: Challenges and Strategies for Sample Preparation and Extraction*, DOI 10.1007/978-1-4419-9631-2_6,

6.2 Excipients and Their Role in the Pharmaceutical Industry

Excipients play a critical role in the pharmaceutical industry. They aid in drug product manufacturability, administration, formulation stability, safety, and esthetics (Jackson et al. 2000; Akers 2002; Rowe et al. 2009). Microcrystalline cellulose (MCC) and dibasic calcium phosphate (DCP), for example, are very critical to achieving workable compactibility in solid oral dosage forms. Sodium starch glycolate, on the contrary, is a superdisintegrant, facilitating rapid dispersion of tablet formulations once in the presence of water. Buffers, antioxidants, and chelating agents help to stabilize otherwise unstable formulations, helping to extend their use periods under various storage conditions. Polymers, such as hydroxypropyl methylcellulose (HPMC), are critical to spray dried dispersion formulations and serves to enhance the solubility and bioavailability of poorly water soluble active pharmaceutical ingredients (APIs). The above types of solubility enhancing formulations are becoming more and more critical, as the number of poorly soluble drug substances is on the increase, approaching approximately 10% for currently marketed drugs.

6.3 Drug–Excipient Interactions: Chemical vs. Physical

Excipients are often thought to be inert components of drug product formulations. The above is far from the truth, as these important pharmaceutical components can be highly reactive, interacting chemically or physically with APIs to produce both negative and positive effects. Although in some cases excipient–drug interactions help to increase drug stability, solubility, and/or bioavailability, they more often than not lead to degradation of the API and thus impact the quality and safety of the drug product formulation. Additionally, excipient–drug interactions can lead to a decrease in drug solubility and hence activity/bioavailability.

6.3.1 Chemical Drug–Excipient Interactions

There are essentially two primary types of drug–excipient interactions, chemical and physical. Drug–excipient interactions that are chemical in nature are not the focus of this chapter; however, these type of interactions often present themselves in two forms: those involving direct interaction between the API and the excipient, and those involving reactions between excipient-related impurities and the API. Both types of drug–excipient interactions facilitate the formation of degradation products – compromising the stability and potential safety of the dosage form. An excellent review on chemical solid-state drug–excipient interactions was provided by Byrn et al. (2001). Similarly, see Akers for a review on chemical drug–excipient interactions in liquid-based formulations (Akers 2002).

6.3.2 Physical Interactions: Adsorption

In addition to chemically reacting with APIs, some excipients can adsorb APIs onto their surfaces, increasing the surface area of the APIs and optimizing drug dissolution/solubilization and absorption. For example, the weak acid, Dicumarol, shows increased dissolution/absorption in the presence of excipients such as MgO and $Mg(OH)_2$ (Jackson et al. 2000). Chelating of the drug with magnesium was provided as a possible explanation for the increased absorption of the drug. The above example and many others highlight a positive consequence of physical interactions. However, if attractive forces are high, desorption becomes unfavorable and absorption decreases. The net effect is a decrease in the extent and rate of dissolution/solubilization. For example, a significant reduction in the antimicrobial activity of cytylpyridium chloride (CPC) was observed in tablet-based lozenges containing the popular lubricant, magnesium stearate (Richards et al. 1996). The adsorption of CPC onto the surface of magnesium stearate was postulated as the main cause of the reduced microbial activity of CPC in the tablet-based lozenges formulations. Talc, another drug product lubricant, did not have inhibitory effects on CPC. Likewise, Senderoff et al. showed that the k-opoid agonist analgesic, CI-977, was capable of adsorbing onto MCC, leading to incomplete drug release from capsules (Senderoff et al. 1992). Electrostatic interaction between the positively charged API and negatively charged surface of MCC was the predominant adsorptive force behind the reduction in dissolution characteristics of the API in the formulation. Similar strong electrostatic adsorption was observed for the analgesic, oxymorphone, in the presence of disintegrants such as cross-linked carboxymethylcellulose and sodium starch glycolate (Chien et al. 1981). Maximum binding was observed at pH 6–7 resulting in reduced drug dissolution. Under these conditions, the positively charged drug and negatively charged surface of the disintegrants facilitated good electrostatic interaction, and thus reduced dissolution.

6.3.3 Physical Interactions: Excipient–Drug Entrapment

Adsorption of drug molecules onto excipient components is probably the most common type of physical drug–excipient interaction. A less common, but related physical type interaction, which also has the potential to lead to low potency results during drug product assays, is excipient–drug entrapment. Excipient–drug entrapment involves entrapment of drug molecules within the excipient matrix where, like in the case of electrostatic type drug–excipient interaction, some of the drug molecules are no longer available to dissolve in the diluent. This type of phenomenon is especially common in tablet matrices containing excipients such as HPMC and polyethylene oxide (PEO), which are capable of swelling and gelling when in contact with certain solvents. The hydrophilic excipient, HPMC, for example is known to swell and gel in the presence of water and other polar solvents. If the rate of gelling is faster than the rate of dissolution, then the drug can become entrapped within the excipient matrix, leading to low potency results during drug product assays (Williams et al. 2001). Care must be taken to devise appropriate sample preparation and extraction procedures for formulations containing HPMC and PEO, especially in the presence of lipophilic drugs. Extraction issues due to the gelling properties of excipients such as HPMC will become more and more common as the use of solubility enhancing formulations such as spray-dried formulations increases, and as their use in control release formulations continues to increase.

6.4 Impact of Physical Excipient–Drug Interactions and Excipient–Drug Entrapment on Analytical Methods

As discussed previously, drug–excipient interactions that are physical in nature (adsorption-ionic interactions/excipient–drug entrapment) can have both positive and negative effects on drug dissolution/absorption. In addition to negatively impacting the performance of the pharmaceutical dosage form, drug–excipient physical interactions can also have significant impact on the development of analytical methodology for assay/potency evaluation of the dosage form. The impact is prevalent when adsorptive forces or entrapment due to gelling of the excipient is high and desorption or release is unfavorable. This leads to low recovery/potency during drug product assays, because of the inability of the diluent to effectively compromise the strong drug–excipient adsorptive/entrapment forces. To ensure the development of robust and effective sample preparation and extraction procedures for drug product assays, analysts must take drug–excipient type interactions into account during the method development phase. Failure to account for drug–excipient interactions will impact drug product release and stability testing – leading to failed batches, time-consuming investigations, and unnecessary method rework and revalidation. Several case studies highlighting the impact of drug–excipient interactions on sample preparation and extraction methods will be discussed in the following sections. Strategies to minimize and/or resolve drug–excipient interactions will also be discussed in detail.

6.5 Case Studies on the Impact of Physical Drug–Excipient Interactions on Analytical HPLC Methods

6.5.1 Case Study 1: Adsorption of Scopolamine HBr onto MCC and Ac-Di-Sol

As discussed previously, adsorption of APIs onto pharmaceutical excipients reduces the amount available for dissolution and diffusion in the diluent, leading to low potency/recovery during drug product assays. A good example of this phenomenon was reported by Pramar and Gupta for the drug molecule, Scopolamine HBr, shown in Fig. 6.1 (Pramar and Gupta 1991). Attempts to extract 0.4 mg Active (mgA) tablet

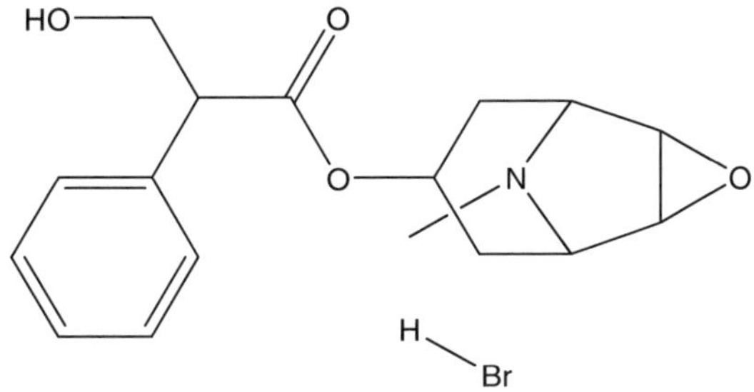

Fig. 6.1 Structure of Scopolamine HBr

formulations of the drug product under aqueous conditions (water) resulted in a recovery of only 57.8% of the API. Subsequent extraction studies involving binary mixtures of the drug with MCC and with sodium carboxymethylcellulose (Ac-Di-Sol) showed that about 75% of the drug was adsorbed onto Ac-Di-Sol and ~32% onto MCC (Pramar and Gupta 1991). Extraction of the 0.4 mgA tablets using 1 N HCl, on the contrary, resulted in quantitative recovery of the active. Similar results were obtained when the acidic diluent was used to extract the active from binary powder mixtures of Scopolamine HBr and MCC or Ac-Di-Sol. The results discussed above indicate that the interaction between Scopolamine HBr and the excipients, MCC and Ac-Di-Sol, is electrostatic in nature, with the positively charged drug interacting strongly with the negatively charged surfaces of MCC and Ac-Di-Sol. The resulting strong adsorption is not capable of being sufficiently overcome by the solvation properties of water, in spite of the high solubility of the drug in water. Under acidic conditions, however, the basic amine, Scopoalamine, exists in its cationic form, while the excipients become deionized. The resulting nonpolar excipients now have much less affinity for the ionized drug molecules, making the drug molecules available to be solvated by the diluent. One would expect that the use of high pH diluents would have a similar effect on extraction efficiencies. The high pH environment would deionize the basic amine, converting it to its free base form, while keeping the surfaces of the excipients negatively charged. The above would again facilitate the desorption process. However, if an entirely high pH aqueous solvent is used, potencies may still be low, as the free base of the drug may have limited solubility in water. To overcome this, extractions at high pH would need to be done in a mixture of aqueous and organic diluents.

6.5.2 *Case Study 2: Adsorption of CI-977 onto Common Tablet Excipients*

Another good example highlighting the impact of adsorption type drug–excipient interaction on drug product assay methods was reported by Senderoff et al. (1992). The k-opoid, CI-977 (Fig. 6.2), intended for use as an analgesic for pain was shown to adsorb strongly onto MCC and two common superdisintegrants, Ac-Di-Sol and

Fig. 6.2 Structure of CI-977

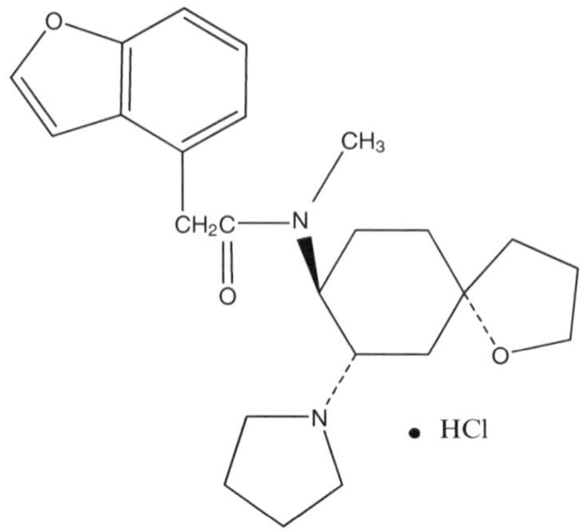

Table 6.1 Summary of recovery data from evaluation of CI-977 drug–excipient binary mixtures

	Fraction of CI-977 recovered				
Excipient	Water (pH 5.8)	Phosphate (pH 7.0)	Citrate (pH 5.0)	HCl (pH 1.1)	0.9% NaCl (pH 5.9)
MCC	0.34	0.99	1.00	1.01	0.99
Dicalcium phosphate dihydrate	1.02	–	–	–	–
Croscarmellose (Ac-Di-Sol)	0.22	0.63	0.68	0.74	0.72
Sodium starch glycolate (Explotab)	0.60	0.94	0.95	1.00	0.95
Pregelatinized starch	1.00	–	–	–	–
Corn starch	0.99	–	–	–	–

Reprinted from Senderoff et al. (1992), with permission from Elsevier

sodium starch glycolate (Explotab). In their reported work, the authors evaluated binary mixtures consisting of 1 mL of a 1.0 μg/mL stock solution of CI-977 and approximately 50 mg each of the excipients, MCC, dicalcium phosphate dihydrate, croscarmellose sodium, sodium starch glycolate, corn starch, and pregelatinized starch. Samples were evaluated in various diluents including water (pH 5.8), 0.05 M phosphate buffer (pH 7.0), 0.05 M citrate buffer (pH 5.0), 0.9% NaCl, and 0.1 N HCl. The samples were shaken for 1 h at ambient temperature using a rotary shaker and centrifuged prior to being assayed. The assay results are shown in Table 6.1.

As shown in Table 6.1, incomplete recovery of CI-977 was observed in the presence of MCC, Ac-Di-Sol, and Explotab, with water as the diluent. In the other diluents, however, essentially complete recovery was obtained from MCC and Explotab, with increased recovery from Ac-Di-Sol. The above results demonstrated that CI-977 is capable of adsorbing to MCC, Explotab, and Ac-Di-Sol. Although very soluble in water (>200 mg/mL), this diluent (water) was unable to overcome the strong electrostatic attractive forces between the cationic drug and the negatively charged surfaces of the individual excipients. However, decreasing the pH of the

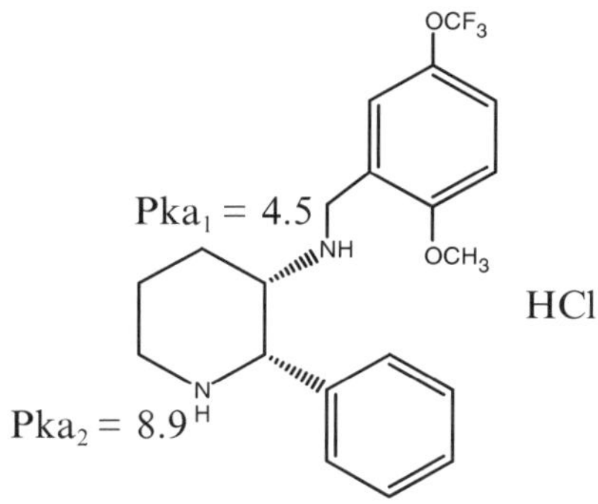

Fig. 6.3 Structure of CP-122,721

diluent, as in the case of HCl, is able to facilitate the desorption process and provide complete extraction of the drug from MCC, Explotab, and to a smaller extent, Ac-Di-Sol. As in the case of Scopolamine HBr (Case Study 1) under acidic conditions, the low pH environment allowed for protonation of the negatively charged surfaces of the excipients and the resulting generation of a nonpolar surface, while keeping the drug in its ionized form. The net result is quenching of the electrostatic attractive force between drug and excipient. As Table 6.1 shows, increasing the ionic strength of the diluent also aids in the desorption process. By increasing the ionic strength of the medium, as in the case of NaCl, a competitive environment is set up whereby the negatively charged surface of the excipients compete with the abundant, smaller, and more labile negatively charged chloride ions for the positively charged drug molecules. Likewise, the positively charged drug molecules compete with the abundant, smaller, and more labile sodium ions for the negatively charge surface of the excipients. The net result is a reduction in drug–excipient interaction and concomitant increase in drug dissolution.

6.5.3 Case Study 3: It's Not Always the Method

As indicated in the two case studies above, electrostatic drug–excipient interactions can significantly impact extraction of API from solid oral dosage forms. This phenomenon is particularly common with ionizable APIs, especially with weakly basic drugs – a fact that is confirmed by the relatively large number of publications involving basic APIs and certain tablet excipients (Chien et al. 1981; Hollenbeck et al. 1983; Okada et al. 1987; Hollenbeck 1988; Pramar and Gupta 1991; Ghannam et al. 1992; Senderoff et al. 1992; Richards et al. 1996; Al-Nimry et al. 1997; Steele et al. 2003; Cory et al. 2004). Of note is the publication by Cory et al. involving the basic amine, CP-122,721, shown in Fig. 6.3 (Cory et al. 2004). Low potency (93–97% label claim) results on prototype tablets during formulation development work led to an extensive analytical investigation to determine whether the low drug recovery was related to the analytical assay method (e.g., insufficient recovery of drug) or to the formulation process. With a drug solubility of ~60 mg/mL in water and >50 mg/mL in MeOH, the authors first decided to perform a solvent screen using 10 mgA prototype tablets to determine whether the dissolving solvent selected in the original method (50/50 ACN/0.01 N HCl) was the most appropriate diluent. The diluents used in the solvent screen included 0.01 N HCl, water, 50/50 ACN/0.01 N HCl, and 50/50 ACN/water. In spite of the high aqueous solubility of the drug, none of the solvents in the solvent screen was capable of providing 100% recovery of the active from the prototype tablet formulation after approximately 2-h of shaking on a reciprocal shaker. The highest recovery obtained was ~93%, which was observed for 0.01 N HCl, 50/50 ACN/0.01 N HCl, and 50/50 water/ACN diluents. Only ~80% recovery was obtained in water, although the drug is highly soluble in this diluent. On the basis of the solvent screen result, the diluent system consisting of 50/50 ACN/0.01 N HCl was selected for all subsequent investigative work.

Concerned about the potential for drug–excipient interaction, the authors next conducted extraction studies on binary mixtures of the API and three of the primary tablet excipients – MCC, croscarmellose sodium (Ac-Di-Sol), and mannitol. With water as the diluent, essentially complete extraction of the drug was obtained from MCC and mannitol, while only 86% recovery was obtained from the API/Ac-Di-Sol binary mixture, suggesting that the drug is potentially being adsorbed onto Ac-Di-Sol. As a result of the above findings, the authors then looked at ways to minimize the interaction and increase the dissolution of API; the various parameters investigated were ionic strength and pH.

In the case of the API-Ac-Di-Sol binary mixtures, an increase in the recovery of the API was observed with increasing ionic strength – with maximum recovery occurring at an ionic strength of 0.1 (NaCl). The results suggested that the interaction between the drug and Ac-Di-Sol was electrostatic in nature, between the positively charged basic amine and the negatively charged AC-Di-Sol. The competitive environment presented by the addition of aqueous NaCl to the matrix helped to decrease the adsorption allowing more free drug available to be solubilized in the diluent.

Evaluation of the impact of pH on the CP-122,721/Ac-Di-Sol interaction provided some interesting but expected results. At a low ionic strength ($\mu=0.001$), the recovery of the API went from approximately 95% at pH 3.0 to approximately 80% at pH 8.0. Alternatively, at an ionic strength of 0.2, 100% recovery of the drug from the API/AC-Di-Sol binary mixture was obtained from pH 1 through pH 9, followed by a drop in recovery through pH 11. The results from the pH study provided additional support for an electrostatic type interaction between CP-122,721 and Ac-Di-Sol. When the impact of ionic strength is negligible ($\mu=0.001$), drug–excipient interaction is driven by the pH of the local environment. At pH less than 4, the carboxylic groups of Ac-Di-Sol exist in their neutral, protonated form and are therefore unable to interact with the positively charged drug molecules. Under these conditions, the highest recovery is obtained. As the pH increases, the surfaces of the excipient become increasingly ionized/negatively charged, while the drug remains positively charged. As a result drug–excipient interaction is at its highest from pH 4 through 9. The low concentration of competitive ions from NaCl is unable to have any impact on the interaction. If the ionic strength of the solution is kept high

(μ=0.2), no significant drop in recovery is observed through pH 9. This is because the high ionic strength is allowing desorption of the positively charged drug from the surface of the Ac-Di-Sol, because of the competition with Na^+ ions. At pH 9 and above, however, we see a significant drop in recovery. This is nicely explained by the deionization/deprotonation of the basic drug to its corresponding free base form, which is not very soluble in water. The pK_a of the drug (8.9) suggests that deprotonation to the free base should occur around pH 9.

The numerous investigative experiments described earlier suggested to the authors that the low recovery results on the prototype tablets was unlikely a result of the analytical method and more likely due to the manufacturing process, with possible drug lost to the manufacturing equipment during the early period of the run. Testing of this hypothesis showed that segregation was in fact occurring during the tablet manufacture, leading to the production of subpotent tablets during the initial period of the run and super-potent tablets toward the end of the run.

6.5.4 *Case Study 4: Recovery of a Lipophilic Drug from HPMC Matrix Tablets – Excipient–Drug Entrapment*

In a 2001 study by Williams et al., the recovery of the anxiety disorder drug, alprazolam, from a solid oral dosage form containing the hydrophilic polymer, HMPC, was evaluated (Williams et al. 2001). Two sample preparation procedures/diluents were evaluated in their study: Sample Preparation I – Acetonitrile (ACN), which has a high solubility for the drug, was used as the diluent; and Sample Preparation II – A binary diluent system consisting of the addition of ~17% by volume of hot water, followed by the addition of ~33% by volume of cold water, then diluting to volume with ACN. Samples of the drug were wet spiked (in 100% ACN) and dry spiked onto various excipient/placebo blends and extracted using either Sample Preparation 1 or Sample Preparation 2. Extraction was also performed on two different 10 mg tablet formulations using Sample Preparation 2, the results of which are shown in Table 6.2.

As the results in Table 6.2 indicate, with the exception of samples A and I, extractions utilizing Sample Preparation 1, with 100% of the diluent, ACN, facilitated low recovery of the API from the matrix. In spite of the high solubility of the API in ACN, this diluent was not capable of completely extracting the API from the matrix. This is because the polar nature of ACN facilitated swelling and gelling of the HPMC within the matrix, leading to entrapment of the API within the HPMC excipient. Entrapment of the API within the gel layers of HPMC did not occur in the wet spike samples, where the API was already dissolved in the diluent, because in these samples the dissolution rate of the API was faster than the gelling rate of the HPMC and as such drug recovery was complete. However, when the powdered API is mixed with the powdered placebo blend, the gelation rate of the polymer was faster than the dissolution rate of the API, entrapping some of the API within the gel layers.

Table 6.2 Extraction of alprazolam from HPMC based formulations

Sample	Wet or dry spike	Solvent/ temperature	Sample prep procedure	Recovery (%alprazolam)
A	Wet	ACN/22°C	Placebo powder blend[a]: add ACN, and API stock solution, mix, and adjust to volume	100.5
B	Dry	ACN/22°C	Placebo powder blend[a]: add ACN, sonicate for 30 min, add powdered API, sonicate for 30 min, mix, and adjust to volume	86.8
C	Dry	ACN/22°C	Add powdered API, add ACN, sonicate for 30 min, add placebo powder blend,[a] mix, and adjust to volume	95.1
D	Dry	ACN/22°C	Placebo powder blend[a]: add powdered API, add ACN, sonicate for 30 min, mix, and adjust to volume	91.6
E	Dry	ACN/5°C	Placebo powder blend[a]: add powdered API, add cold ACN, sonicate for 30 min, stir at 25°C for 12 h, mix, and adjust to volume	96.0
F	Dry	ACN/5°C	Add powdered API, add cold ACN, sonicate for 30 min, add placebo powder blend[a]: stir at 25°C for 12 h, mix, and adjust to volume	95.7
G	Dry	ACN/5°C	Placebo powder blend[a]: add cold ACN, sonicate for 30 min, add powdered API, sonicate for 30 min, stir at 25°C for 12 h, mix, and adjust to volume	95.9
H	Dry	ACN/−10°C	Placebo powder blend[a]: add powdered API, add cold ACN, sonicate for 30 min, stir at 25°C for 12 h, mix, and adjust to volume	94.9
I	Wet	ACN/−10°C	Placebo powder blend[a]: add API stock solution, add cold ACN (−10°C), sonicate for 30 min, stir overnight at 25°C, mix, and adjust to volume	99.1
J	Dry	ACN/−20°C	Placebo powder blend[a]: add powdered API, add cold ACN (−20°C), stir at −17°C for 2 h, mix, and adjust to volume	90.6
K	Dry	Water/ACN	Blend placebo powder blend[a] and powdered API and transfer to a 1 L volumetric flask. Add 170 mL hot water (~90°C), stir to disperse blend, add 330 mL of cold water (~5°C). Stir in an ice bath (~2°C) for 3 h to dissolve the HPMC gel. Add ~450 mL ACN and stir for 4 h. Adjust to volume with ACN to give a 50/50 mixture of water/ACN	100.5

(continued)

Table 6.2 (continued)

Sample	Wet or dry spike	Solvent/ temperature	Sample prep procedure	Recovery (%alprazolam)
L	Dry	Water/ACN	Blend placebo powder blend[b] and powdered API and transfer to a 1 L volumetric flask. Add 170 mL hot water (~90°C), stir to disperse blend, add 330 mL of cold water (~5°C). Stir in an ice bath (~2°C) for 3 h to dissolve the HPMC gel. Add ~450 mL ACN and stir for 4 h. Adjust to volume with ACN to give a 50/50 mixture of water/ACN	101.6
Q	10 mg tablet[c]	Water/ACN	Add tablet transfer to volumetric flask. Add hot water (~90°C), stir to disperse blend, add cold water (~5°C). Stir in an ice bath (~2°C) for 3 h to dissolve the HPMC gel. Add sufficient ACN to bring to solution to approximately 50/50 ACN/water and stir for 4 h. Adjust to volume with ACN to give a 50/50 mixture of water/ACN	101.9% label
R	10 mg tablet[d]	Water/ACN	Add tablet transfer to volumetric flask. Add hot water (~90°C), stir to disperse blend, add cold water (~5°C). Stir in an ice bath (~2°C) for 3 h to dissolve the HPMC gel. Add sufficient ACN to bring to solution to approximately 50/50 ACN/water and stir for 4 h. Adjust to volume with ACN to give a 50/50 mixture of water/ACN	98.8% label

Sample prep concentrations ~ 0.05 mg/mL, except for sample A = 0.02 mg/mL

Reprinted from Williams et al. (2001), with permission from Springer Science + Business Media

[a]Placebo powder contained 23% wt/wt of HPMC K4M premium (high molecular weight distribution polymer), 20% wt/wt of MCC, 56% wt/wt of dicalcium phosphate dehydrate, 0.5% wt/wt of silicon dioxide, and 0.5% wt/wt of magnesium stearate

[b]Placebo powder contained 40% wt/wt of HPMC K100LV (low molecular weight distribution polymer), 20% wt/wt of MCC, 39% wt/wt of Dicalcium Phosphate dehydrate, 0.5% wt/wt of silicon dioxide, and 0.5% wt/wt of magnesium stearate

[c]Ten milligram tablet (400 mg tablet weight). Thirty-seven percent wt/wt of HPMC K4M (high molecular weight distribution polymer), 20% wt/wt of MCC, 42% wt/wt lactose, 0.5% wt/wt of silicon dioxide, and 0.5% wt/wt of magnesium stearate

[d]Ten milligram tablet (400 mg tablet weight). 45% wt/wt of HPMC K100LV (low molecular weight distribution polymer), 20% wt/wt of MCC, 34% wt/wt lactose, 0.5% wt/wt of silicon dioxide, and 0.5% wt/wt of magnesium stearate

The net result is incomplete extraction of the API from the matrix. In extraction studies involving Sample Preparation 2, complete extraction of the API was obtained. Complete solubilization of the polymer via use of the hot, followed by cold water process negated the swelling and gelling capabilities of the HPMC polymer. As such, when the organic solvent, ACN, was added to the solution, the API was available to be solubilized by it. The above provides additional key information supporting the fact that it is critical to dissolve the HPMC polymer before dissolving the API in the formulation. The above prevents entrapment of the API within the gel layers of the polymer, allowing the drug to be free for dissolution by the organic co-solvent. The same was observed for the 10 mg tablets, where complete extraction of the API was observed for formulations involving both low and high molecular weight distribution polymers.

6.6 Conclusions

The potential impact of electrostatic drug–excipient interactions and/or excipient–drug entrapment on extraction of API from solid oral dosage forms cannot be overstated. It is therefore critical that these type of drug–excipient interactions be clearly understood by analysts and factors such as pH, ionic strength, and the need for polymer solubilization to minimize gelling considered during the sample preparation and extraction method development process. This will ensure the development of robust, effective, and efficient sample preparation and extractions methods for oral dosage forms.

References

Akers MJ (2002) Excipient-drug interactions in parenteral formulations. J Pharm Sci 91(11):2283–2300

Al-Nimry SS, Assaf SM, Jalal IM, Najib NM (1997) Adsorption of ketotifen onto some pharmaceutical excipients. Int J Pharm 149:115–121

Byrn SR, Xu W, Newman AW (2001) Chemical reactivity in solid-state pharmaceuticals: formulation implications. Adv Drug Deliv Rev 48:115–136

Chien YW, Nostrand PV, Hurwitz AR, Shami EG (1981) Drug-disintegrant interactions: binding of oxymorphone derivatives. J Pharm Sci 70(6):709–710

Cory W, Field K, Wu-Linhares D (2004) Is it the method or the process – separating the causes of low recovery. Drug Dev Ind Pharm 30(8):891–899

Ghannam MM, Gautam J, Restaino FA (1992) Devazipide (MK-329) adsorption on microcrystalline cellulose. Drug Dev Ind Pharm 18:973–996

Hollenbeck RG (1988) Bioavailability of phenylpropanolamine HCl from tablet dosage forms containing croscarmellose sodium. Int J Pharm 47:89–93

Hollenbeck RG, Miitrevej KT, Fan AC (1983) Estimation of the extent of drug-excipient interactions involving croscarmellulose sodium. J Pharm Sci 72:325–327

Jackson K, Young D, Pant S (2000) Drug-excipient interactions and their affect on absorption. Pharm Sci Tech Today 3(10):336–345

Okada S, Nakahara H, Isaka H (1987) Adsorption of drugs on microcrystalline cellulose suspended in aqueous solutions. Chem Pharm Bull 35:761–768

Pramar Y, Gupta VD (1991) Quantitation of scopalamine hydrobromide when adsorbed onto microcrystalline cellulose and sodium carboxymethylcellulose in tablets. Drug Dev Ind Pharm 17(17):2401–2407

Richards RME, Xing JZ, Mackay KMB (1996) Excipient interaction with cytylpyridinium chloride activity in tablet based lozenges. Pharm Res 13(8):1258–1264
Rowe RC, Sheskey PJ, Owen SC (2009) Handbook of pharmaceutical excipients. Pharmaceutical Press, London
Senderoff RI, Mahjour M, Radebaugh GW (1992) Characterization of adsorption behavior by solid dosage form excipients in formulation development. Int J Pharm 83:65–72
Steele DF, Edge S, Tobyn M, Moreton R, Staniforth JN (2003) Adsorption of an amine drug onto microcrystalline cellulose and silicified microcrystalline cellulose samples. Drug Dev Ind Pharm 29:475–487
Williams RO III, Sykora MA, Mahaguna V (2001) Method to recover lipophilic drug from hydroxypropyl methylcellulose matrix tablets. AAPS Pharm Sci Tech 2(2):1–9

Chapter 7
Sample Preparation for Solid Oral Dosage Forms

Beverly Nickerson and Garry Scrivens

Abstract Development of extraction and sample preparation methods for solid oral dosage forms for potency and purity analysis can be challenging. Complete extraction of drug and impurities is required using reasonable extraction and sample preparation conditions, and the final prepared sample must be compatible with the analysis method. A systematic approach for the development of extraction and sample preparation methods for potency and purity analysis of solid oral dosage forms is presented. Key steps of the process include the selection of an appropriate diluent to allow complete extraction and solubilization of the analytes of interest and the selection of an appropriate mechanism to disperse the dosage form to facilitate extraction of the analytes. Each step of the method development process is discussed and potential problem areas are highlighted.

7.1 Developing a Sample Preparation Method for Solid Oral Dosage Forms

Approximately two-thirds of all prescriptions are dispensed as solid oral dosage forms and half of these are compressed tablets. Solid oral dosage forms include capsules, tablets, orally disintegrating tablets and films. A number of challenges exist in the extraction and sample preparation of solid oral dosage forms for potency and purity analysis as well as isolation and identification of impurities and degradation products. Complete extraction of drug/impurities is required using reasonable

B. Nickerson (✉)
Analytical Development, Pfizer Global Research and Development,
Pfizer Inc., 558 Eastern Point Road, Groton, CT 06340, USA
e-mail: beverly.nickerson@pfizer.com

B. Nickerson (ed.), *Sample Preparation of Pharmaceutical Dosage Forms: Challenges and Strategies for Sample Preparation and Extraction*, DOI 10.1007/978-1-4419-9631-2_7,

extraction and sample preparation conditions (i.e., length of time, solvents, etc.). The final prepared sample must be compatible with the analysis method (e.g., HPLC). In addition, the extraction and sample preparation method needs to be rugged and robust to allow testing by different analysts at different facilities over an extended period of time. During method development, both for the sample preparation and the analysis method, there may be time and resource constraints.

During the early stages of drug development, there is usually an attempt to minimize resources and efforts in developing an extraction and sample preparation method as the chances of the drug candidate progressing to later stages of development are low. This method needs to be "fit-for-purpose" in that it gives accurate and reliable results when used by the developing laboratory, but the method may not necessarily be robust to changes in the API or dosage form and may not be transferable to other sites. Initial sample preparation method development is typically based on the API method and available information, such as drug pK_a, API solubility in organic solvents and buffers, and evaluating the potential of drug–excipient interactions based on the structure and properties of the API and the excipients expected to be used in the formulation.

During the late stages of development, more effort is spent on developing and ensuring a robust sample preparation method due to the likelihood of the compound being filed and therefore more willingness to invest time and resources into the method development and optimization. In addition, a robust and rugged method is needed for transfer and long-term use at contract laboratories and manufacturing sites.

As shown in Fig. 7.1, the sample preparation method development strategy presented in this chapter for solid oral dosage forms is based on two key processes: (1) dosage form disintegration/dispersion; and (2) dissolution or solubilization of the drug. The sample preparation method should disperse the solid oral dosage form to allow efficient extraction of the drug and it must effectively and completely solubilize the drug. It is also important to understand the relationship between the dispersion step and the extraction/solubilization step in developing the sample preparation method. This relationship can be a key factor for certain types of dosage forms such as nondisintegrating dosage forms (extended-release or delayed-release dosage forms). The dispersion step is typically less critical or not an issue for nonsolid oral dosage forms such as powders, solutions, and suspensions. Although various means can be used to disperse the dosage form, solubilization and recovery of the active ingredient is limited by its solubility in the diluent. Therefore, selection of the appropriate diluent is critical in the development of the sample preparation method. For nonsolid oral dosage forms (i.e., suspensions, lyophiles), the first process (i.e., disintegration/dispersion of the dosage form) is not a factor, and sample preparation focuses on solubilization of the drug.

A systematic sample preparation method development strategy is outlined in Fig. 7.2. This development strategy can be used to initially develop a sample preparation method or to optimize one. In addition, the amount of time and effort spent at various steps can be adjusted based on the purpose of the method (e.g., "fit for purpose" or "good enough" method vs. final optimized commercial method) and stage of development of the compound. Although all steps of the method development

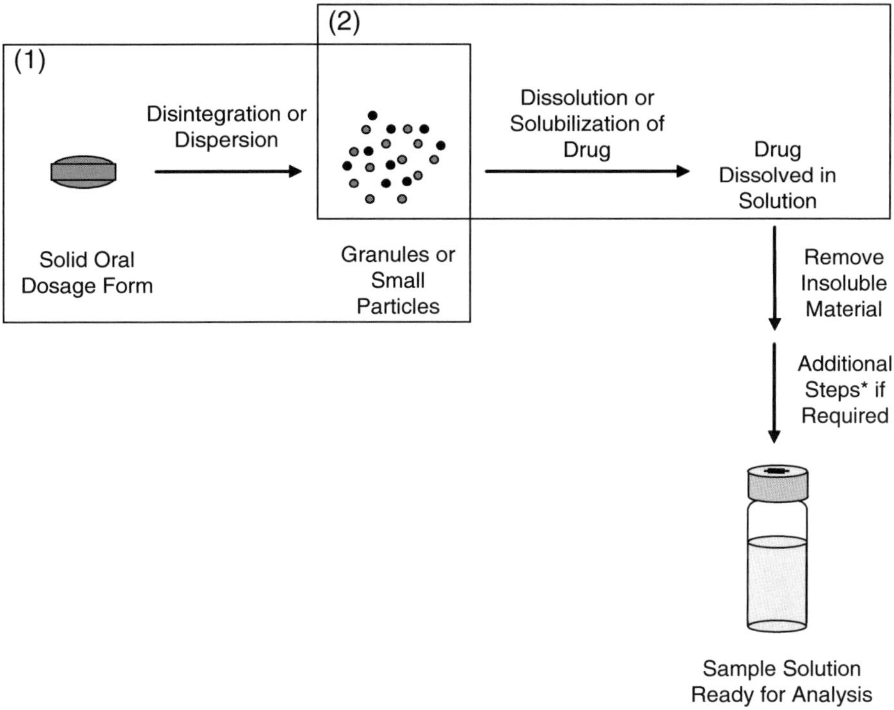

Fig. 7.1 Key processes in solid oral dosage form sample preparation and extraction: (1) disintegration, or dispersion of the dosage form; (2) dissolution or solubilization of the drug. *Additional steps may include sample derivatization, sample concentration, or sample clean up to remove interferences

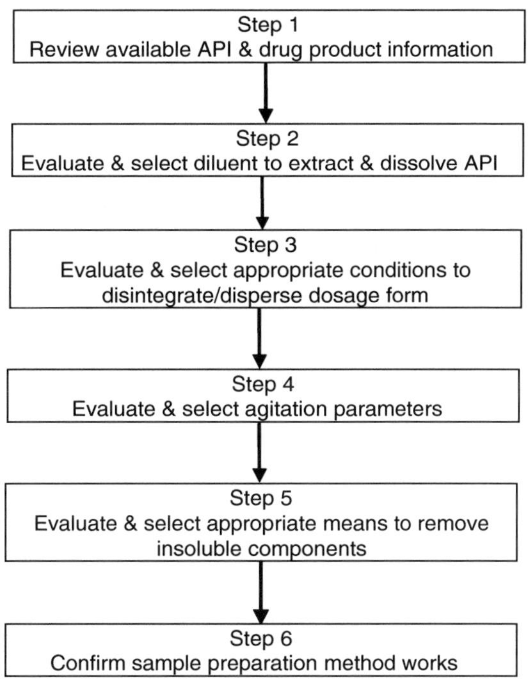

Fig. 7.2 Overview of sample preparation method development for solid oral dosage form

process are important, a critical step of the method development strategy is to select an appropriate diluent that will allow complete extraction and solubilization of the analytes of interest. Another key step of the method development process is ensuring adequate dispersion of the dosage form to minimize sample preparation time. Each of the steps in the method development strategy is described in detail later. Steps performed after extraction of the active ingredient from the dosage form (e.g., sample derivatization, sample concentration, or sample clean up) are discussed in Chap. 9.

7.2 Step 1: Review Available API and Drug Product Information

7.2.1 API and Drug Product Information

Regardless of the phase of development for the dosage form, the sample preparation method development process should begin by collecting and reviewing available information on the API (e.g., structure, solubility in organic solvents and buffers, solution stability, pK_a, etc.) and dosage form (e.g., excipients, excipient compatibility information, excipient solubility information, formulation type, tablet disintegration time, etc.). API solubility and solution stability information is critical in helping determine potential extracting or dissolving solvents for the method. Solubility information for relevant impurities or degradants is also important as they can sometimes be different from that of the API. Excipient information and formulation type are important to help determine how to disperse the dosage form, thereby facilitating extraction and solubilization of the API. Additional information that would be useful, if available, are extraction and sample preparation methods for other formulations of the API (or similar APIs) as these might provide insight into conditions needed to dissolve or extract the drug. In addition, evaluation of extraction and sample preparation methods for similar formulation types (e.g., if the dosage form is a matrix controlled release tablet, look at other matrix controlled release methods) may provide insight into conditions needed to disintegrate or disperse this particular dosage form type.

7.2.2 Consider Potential for Drug–Excipient Interactions

On the basis of the structure of the API and the excipients present in the formulation, an assessment of potential drug–excipient interactions should be made at this time, as these interactions will impact the development of the method (e.g., selection of extraction solvent type or pH). Potential drug–excipient interactions fall into two categories: (1) chemical interactions that result in a covalent bond breaking or formation of the bond in the drug (e.g., covalent bond forms between the drug and excipient (or an impurity in the excipient) or excipient (or an impurity in the excipient) catalyzes a reaction of the drug); (2) physical interactions that result in low recovery of the drug during sample preparation. Chemical drug–excipient interactions are a concern and impact formulation stability and hence impact development and selection of the dosage form. These chemical interactions can also impact sample solution stability. Physical drug–excipient interactions are of concern and impact sample preparation method development. These interactions include adsorption or binding of drug onto nonsoluble excipients and the inclusion or trapping of drug by nonsoluble or gelling polymer excipients. These physical interactions can result in low recovery of the active during sample testing and/or delayed drug release during dissolution testing.

As mentioned earlier, one type of physical drug–excipient interaction is adsorption of drug onto an insoluble excipient in the dosage form and these interactions are typically due to weak van der Waals forces, hydrogen bonding, or electrostatic interactions. Literature references are available documenting adsorption of drug on microcrystalline cellulose (MCC, Avicel) (Franz and Peck 1982; Aboutaleb et al. 1986; Okada et al. 1987; Pramar and Gupta 1991; Senderoff et al. 1992; Al-Nimry et al. 1997; Qtaitat et al. 1998), carboxymethylcellulose sodium (Kennon and Higuchi 1956; Pramar and Gupta 1991) and croscarmellose sodium (Ac-di-sol) (Chien et al. 1981; Hollenbeck et al. 1983; Hollenbeck 1988; Pramar and Gupta 1991; Senderoff et al. 1992; Al-Nimry et al. 1997; Cory et al. 2004). MCC and croscarmellose are insoluble excipients, which are negatively charged at pH greater than 4 (McBurney 1954; Edelson and Hermans 1963; Mark et al. 1965) and 2 (Hollenbeck et al. 1983), respectively. If the drug is positively charged, there will be an electrostatic interaction between the oppositely charged insoluble excipient and drug. This interaction can be minimized or eliminated by adjusting the pH of the sample diluent such that the drug and the excipient are not oppositely charged or by increasing the ionic strength of the diluent so that the ions in the diluent compete with the drug to interact with the charged excipient. Thus, if a potential exists for drug–excipient interactions, this is important to know and will influence selection of the extraction solvent (e.g., solvent type, ionic strength, or pH). Literature references are also available reporting drug interactions with pregelatinized starch (Aboutaleb et al. 1986; Al-Nimry et al. 1997), calcium sulfate dehydrate (Aboutaleb et al. 1986), calcium phosphate dibasic dehydrate (Aboutaleb et al. 1986), kaolin (Qtaitat et al. 1998), colloidal silica (Czaja and Mielck 1982), sodium starch glycolate (Chien et al. 1981; Senderoff et al. 1992), and sodium steryl fumarate (Howard et al. 1994).

To test for potential drug–excipient interaction, one can "wet-spike" API solution onto excipient blends and prepare the sample using the diluent or diluents under consideration. If there is drug–excipient interaction, low drug recovery will be obtained. A case study demonstrating the effectiveness of the "wet-spike" approach to identify drug–excipient interactions is discussed in Sect. 7.3.2.

Another type of drug–excipient interaction is the entrapment of drug (e.g., undissolved drug or concentrated drops of drug solution) by swelling or gelling polymer excipients. This type of drug–excipient interaction can be minimized through diluent selection. If possible, select a diluent that does not swell the polymer but will solubilize the drug. In this case mechanical means, such as homogenization, may be needed to disperse the dosage form. Alternatively, a diluent that dissolves the polymer can be used. In this case, a step to remove the polymer (e.g., liquid–liquid extraction, solid phase extraction, centrifugation through molecular weight cut off filters) prior to analysis may be needed to avoid interferences or issues with the analysis (e.g., column fouling or interfering peaks in HPLC). Additional information on potential drug–excipient interactions and more case studies involving drug–excipient interactions are discussed in Chap. 6.

7.3 Step 2: Evaluate and Select Diluent to Extract and Dissolve API

Evaluation and selection of an appropriate dissolving/extraction diluent is perhaps the most important decision in the development of a reliable extraction and sample preparation method. If an appropriate diluent is not selected to provide adequate solubility and recovery of the active ingredient, then it will not be possible to achieve complete extraction and solubilization of the active ingredient regardless of any amount of mechanical dispersion or agitation used. In this section, strategies for the selection of the dissolving/extraction diluent are presented.

When selecting the dissolving/extraction diluents to evaluate, there are many options facing the method developer and many strategies that may be considered, for example:

- pH adjustment: Typically the following are considered: non-pH adjusted (i.e., one based on purified water), acidic media, basic media, and buffered media; there are, of course, many types of acids, bases, and salts that may be used.
- Organic solvents: Methanol, acetonitrile, ethanol, and propanol are perhaps the most commonly used organic solvents in the dissolving/extraction diluent (due to the ability to dissolve API and diluent compatibility with chromatographic analysis), but other solvents such as THF, propanone, butanone, DMSO, and DMF have also been used. The level of organic solvent used is, of course, a key consideration and can vary anywhere between 0 and 100%. Use of at least a minimal amount of organic solvent (e.g., 10%) is recommended even if the drug is water soluble to aid in solubilizing potential unknown impurities or degradation products (Wrezel et al. 2005).
- Ionic strength: This factor has been known to influence the effectiveness of the extraction procedure.

- Additives: Diluent additives such as surfactants, competitive binders, and excipient-solubility suppressors may also be considered.
- Multi-step strategies: It may sometimes be advantageous or necessary to design sample preparation procedures that use different solvents, diluents, and media for each step. For instance, the diluent most effective in disintegrating and dispersing the tablet matrix may not be the most effective diluent for dissolving and extracting the API, and this may not be the most appropriate diluent for the chromatographic method. It is therefore important to understand the relationship between the extraction and solubility of the API (Step 2) and the dispersion of the drug product (Step 3) in developing the sample preparation method.
- Concentration (i.e., choice of extraction volume): Although this is strictly not a diluent parameter, it is an important consideration when designing a reliable sample preparation. There should typically be a solubility margin (e.g., 2–3×) to ensure robust solubilization of the drug (e.g., in case of variation in laboratory temperature, etc.).
- Compatibility with analysis method: The final sample solution must be compatible with the analysis method (e.g., solvent strength differences between the sample diluent and mobile phase may lead to peak shape issues in HPLC analysis). Dilution of the sample solution with another diluent, sample clean up, or using a smaller injection volume may be needed to obtain a sample that is compatible with the analysis method.

From the above array of options, this section describes an approach to assess the performance of several dissolving/extraction diluents using a minimum number of key tests. The choice and number of dissolving/extraction diluents to be evaluated depends upon the review of available API and drug product information (Step 1) and on the phase of development (generally less time is invested in early-development drug candidates). For instance, it may be appropriate for an early-development drug candidate that has no concerns raised from the review of information, and that a single dissolving/extraction diluent is evaluated. However, the tests presented here for the evaluation of dissolving/extraction diluents are reasonably quick to perform, so screening a number of potential dissolving/extraction diluents should not be too onerous.

In general, the most important considerations for the dissolving/extraction diluent are (arguably) as follows: the level of organic solvent, diluent pH, and the type of organic solvent. Other factors, however, may be more important for certain drug products. The effects of these important factors may be evaluated using "screening experiments" before a more focused evaluation is carried out in which all factors are assessed.

7.3.1 Diluent Screening

A diluent screening experiment should be conducted to evaluate potential dissolving/extraction diluents. An example screening experiment applicable to many drug products would be to evaluate the following parameters using a "Design of

Table 7.1 Example of diluents to evaluate in a diluent DOE screening experiment

Number	Diluent
Diluent 1	Water/methanol, 50/50
Diluent 2	Water/acetonitrile, 20/80
Diluent 3	Water/methanol, 90/10
Diluent 4	0.05 M NaOH/acetonitrile, 20/80
Diluent 5	Water/acetonitrile, 90/10
Diluent 6	0.05 M HCl/acetonitrile, 90/10
Diluent 7	Water/methanol, 20/80
Diluent 8	0.05 M NaOH/methanol, 90/10
Diluent 9	0.05 M HCl/methanol, 20/80

Notes. The design is constructed using the method of collapsing levels. pH was constructed as a 4-level factor, orthogonal to the other main effects, and then collapsed into three levels (unadjusted pH ran more often). This has the advantage of keeping main effects uncorrelated (e.g., the effect of one factor will not bias the estimate of another) and gives more info about the unadjusted pH setting. There will be some aliasing between main effects and interactions (i.e., not all effects can be separately estimated). In some cases, scientific judgment may be needed to determine the true cause of an effect if there is more than one possible explanation

Experiment" (DOE) approach so that the effects and interactions of these main parameters can be determined:

- Three levels of organic solvent: "high," "medium," and "low" (e.g., 10, 50, and 80%, but the review of available API and drug product information should help decide which levels should be evaluated).
- Three levels of pH: acidic, neutral (or unadjusted), and basic (e.g., 0.05 M HCl, water (or 0.05 M phosphate buffer, pH 6.5), and 0.05 M NaOH, but review of available API solubility data may eliminate some of these conditions if the API is known to be insoluble at certain pHs).
- Acetonitrile or methanol used as the organic solvent (or other solvents as indicated during the review of available API information).

An example DOE using diluents consisting of mixtures of water, acidic buffer, basic buffer, methanol, and acetonitrile is listed in Table 7.1.

The diluents selected from the DOE (e.g., nine diluents in Table 7.1) are then used in three experiments to assess: API solubility, binding to excipients, and performance on drug product. These three assessments are discussed below.

Assessment 1: API solubility. The diluents are first assessed for their ability to dissolve API (in the absence of excipients). A reasonable test concentration is selected; for instance, the dose strength divided by a reasonable extraction volume (e.g., 25–500 mL, but sufficiently concentrated to provide an on-column load that allows a satisfactory limit-of-detection) is a reasonable rule-of-thumb. When selecting the extraction concentration used for this assessment, concentrations higher than the intended method concentration are typically used to ensure that a sufficient solubility margin is obtained. The required amount of API is added to each of the diluents and mildly agitated (e.g., stirred). Any diluents failing to provide sufficient API solubilization may be discounted. For those diluents that fully dissolve the API, the time taken to dissolve the API may be recorded (this may provide some discrimination when selecting a diluent for progression). Additional tests at this stage are to visually assess the solutions periodically (e.g., over several days) for any reprecipitation on standing and to assess the solution stability (i.e., drop in assay or change in impurity profile) over time.

Assessment 2: binding to excipients. The API solutions generated in Assessment 1 may be used for this experiment. An aliquot from each API solution is added onto a representative blend of excipients (at appropriate levels); the mixtures are agitated (e.g., for ~30 min) and clarified (e.g., by centrifugation to avoid potential binding issues with filters). The recovery from the excipient blend can be directly determined by assessment against the ingoing API solution. This test reveals if the API significantly binds to the excipients (and hence is not extracted) in the selected diluents. Our experience has showed that although satisfactory performance in this test does not guarantee that the diluent will perform well on the actual drug product, this test can quickly eliminate many unsuitable diluents. The solutions prepared in this assessment, prior to clarification, may be considered to represent the solutions that would be obtained if the sample solutions were thoroughly agitated and the drug product was perfectly dispersed. Therefore, poor performance on this test is independent of the agitation and therefore can only be rectified by changing the diluent; modification of the pH in combination with the organic solvent level has often been found to be key to improving the diluent's performance in this assessment. Solution stability of these solutions and ease of clarification should also be assessed.

Assessment 3: performance on drug product. An appropriate volume of each diluent is added to the drug product, and observed for its ability to disperse and disintegrate the drug product, using an appropriate type of agitation. The time taken to disintegrate and disperse the drug product and any other pertinent visual observations are recorded (this may provide some discrimination when selecting a diluent for progression). If possible, a batch of drug product of known potency and with a good content uniformity should be chosen. Multiple units should ideally be tested in each diluent to reduce the uncertainty due to unit-to-unit variability; a composite sample of many units in a single solution may be used to reduce the workload. The potency result should be determined for the drug product batch in each of the diluents studied. Solution stability of these solutions and ease of clarification should also be assessed. If possible, stressed or aged drug product should also be assessed here since this may alter the way in which the drug product behaves during the sample preparation (e.g., the disintegration time).

The diluent to be progressed needs to have performed well (or predicted to perform well) in each of these assessments. A statistical analysis of the results from this DOE may be used to predict the effects and interactions of factors such as pH, organic solvent type, and percentage, and therefore the performance of diluents not directly assessed in these experiments may be predicted. Table 7.2 provides an example template of the sort of information and criteria that may be used to help select diluents to be progressed; some criteria are "must-have" and some are "nice-to-have." For some

Table 7.2 Example template for results from diluent screening experiments

Example assessment criteria	Diluent #1	Diluent #2	Diluent "*N*"	Example limits
Solubility after "*n*" minutes mild agitation				Higher the better; "X" mg/mL minimum
Time taken to fully dissolve				The lower the better
Chromatographs of diluent blank, placebo blank and impurity profile: any interfering peaks? Generally acceptable? Rate 0 (poor) to 10 (good)				Higher the better
Peak shape problems (e.g., splitting)? Rate 0 (poor) to 10 (good)				Higher the better
Impurities methods: all expected impurities recovered from Assessment 1? Rate 0 (poor) to 10 (good)				Higher the better
%Recovery of API in Assessment 2				The closer to 100% the better Within: 98.0–102.0%
Impurities methods: all expected impurities recovered from Assessment 2? Rate 0 (poor) to 10 (good)				Higher the better
Ease of clarification. Rate 0 (poor) to 10 (good)				Higher the better
Disintegration time under mild agitation (Assessment 3)				The shorter the better
% Recovery from drug product (Assessment 3)				The closer to 100% the better Within: 97.0–103.0%

Solution stability of API (std) solution (% deg per day) e.g., at room temperature, at 5°C, in light and/or dark	Assay drop and imp growth: the lower the better Assay drop: <0.5% per day Imp growth: <0.06% per day
Solution stability of drug product solution (% deg per day) e.g., at room temperature, at 5°C, in light and/or dark	Assay drop and imp growth: the lower the better Assay drop: <0.5% per day Imp growth: <0.06% per day
Cost and Environmental impact (e.g., type and volume of solvent used per sample)	The lower the better
Total cycle time (e.g., per sample, per diluent preparation)	The lower the better

Note. It is generally good practice to have a sample diluent that is similar to or the same as the mobile phase at the point of injection. Having the mobile phase the same as the sample diluent may also have some advantages in terms of efficiency and convenience and may sometimes result in better peak shapes or smaller solvent fronts in the chromatography. However, some mobile phases do not perform well in the above assessments and some mobile phases are unnecessarily complex and time-consuming to prepare as a general-purpose dissolving/extraction diluent criteria, a rating (e.g., from 0 to 10) may be applied; this can facilitate the statistical analysis and may provide some differentiation between acceptable diluents. A case study illustrating this approached is provided in Sect. 7.3.2.

If good performance is obtained in Tests 1 and 2 but poor performance is obtained in Test 3, this warrants further investigation and is not necessarily an indication of an unsuitable dissolving/extraction diluent. In such cases, the visual observations should be scrutinized for potential causes (e.g., lack of disintegration, size of suspended particles, etc.). Also the integrity of the drug product samples analyzed should also be questioned (e.g., "Is this batch truly sub-potent?" and "What is the Content Uniformity of this batch?"). Generally, a difference in performance between Tests 2 and 3 is indicative of either insufficient agitation during extraction, or an indication that the selected diluent does not sufficiently dissolve or disperse certain key excipients (e.g., that may be coating the API particles). In such cases, further experiments on processed drug product intermediates (such as granulated blends) could be informative; dry-spike experiments may also have some value in shedding light on the problem. Generally, strategies for overcoming such problems typically involve increased energy of agitation (e.g., ultrasonication or mechanical homogenization) and/or selecting a diluent that can (partially) dissolve or disperse the excipient(s) that are impeding the access of the diluent to the API. Disintegration/dispersion of various types of solid oral dosage forms is discussed in Sect. 7.4.

If no diluents emerged from Assessments 1, 2, and 3 as a potentially viable basis for the sample preparation, then it is important to use the observations and results obtained in this screening experiment to redesign a second iteration of the screening experiment, based on such considerations as those listed in Table 7.3 where appropriate.

Once potentially viable diluent(s) have been found, these can now be further assessed and validated alongside other factors that could affect the robustness of the sample preparation procedure. For instance, the robustness around small perturbations in pH (e.g., ±0.5 pH units, where appropriate), ionic strength (e.g., ±0.01 M, where appropriate), and organic solvent level (e.g., ±5%) may be explored and validated. Chapters 10 and 11 provide additional information on this topic.

7.3.2 Case Study: Identification of an Appropriate Dissolving/Extraction Diluent for a Tablet Formulation of "Compound A"

Formulation development was underway to develop an immediate release (IR) tablet for a given "compound A" in early development. Several formulations were set up on stability to aid in formulation selection. Excipients used in these formulations included MCC, lactose, dibasic calcium phosphate anhydrous, mannitol, sodium starch glycolate, croscarmellose sodium, and magnesium stearate. The diluent used in the API method, pH 3.4 phosphate buffer (20 mM)/acetonitrile (80/20 v/v), was initially evaluated as an extraction diluent for the different formulations using 30 min ultrasonication and 30 min stirring to ensure adequate disintegration and agitation of the dosage forms. Poor recovery of the API was observed for all formulations

Table 7.3 Troubleshooting considerations in diluent selection

Issue observed	Next steps	Sections that may contain additional information
Poor API solubility	Change diluent, taking log *P*, log *D* and pK_a into consideration	Section 7.3
Poor wet spike recovery results	Consider – Changing the diluent (e.g., pH, organic solvent type and level). It is common to obtain low recoveries for API and impurities that are amines (particularly primary and secondary amines) in presence of certain excipients such as Avicel, Ac-Di-Sol and Explotab. pH in combination with organic solvent type and % level are important factors to adjust. Consider the pK_a of the API and impurities – Adding a competitive binder (e.g., an amine such as Et_2NH)	Chapter 6
Lack of disintegration of dosage form	Consider – Changing the diluent to improve the dispersion and solubility behavior of excipients – Cutting, scoring or grinding of the dosage form prior to extraction – Vigorous homogenization/agitation of solution – Multi-step diluent addition procedures (e.g., small amount of water to disintegrate the dosage form, then add different diluent to solubilize and extract the API)	Section 7.4 and Chap. 3
Poor recovery from dosage form but good wet-spike recovery (and good disintegration)	Consider – Vigorous homogenization/agitation of solution; extending agitation time – Changing diluent/solvent (e.g., to dissolve key excipients) – Integrity of dosage form – Investigative work on representative processed drug product intermediates	Section 7.5
Poor solution stability	– Consider changing diluent – Consider refrigeration of samples (e.g., use of refrigerated autosampler) – Investigate different types of storage glassware – For photostability issues, investigate protecting from light – For oxidation issues, investigate use of an antioxidant in the sample solution – Investigate alternative solution concentrations (e.g., higher sample concentrations, stock solutions) – In the case of an increase in potency check seals of containers to ensure no evaporation	

(continued)

Table 7.3 (continued)

Issue observed	Next steps	Sections that may contain additional information
Variability in % recovery (from drug product)	Consider – Content uniformity of the batch of drug product – Increasing the vigorousness of agitation (e.g., to ensure adequate mixing to achieve homogeneous solution) – Changing diluent/solvent (e.g., to dissolve key excipients)	
Variability in % recovery during validation known-weight "dry-spike" experiments	– Interrogate statistics for effects of experimental parameters – Investigate potential insufficient agitation – Investigate potential weighing problems (e.g., static issues, ensure quantities weighed are above the minimum weight of the balance)	
Analytical bias (towards super-potency)	Consider – Poor homogenization (shaking) of solutions – Poor container seals leading to evaporation of solution – The use of excessively volatile diluents leading to evaporation of solution – Possible delta volume issues (e.g., solvent expansion or contraction when mixing two diluent components, undissolved excipients affecting volume) – consider use of a suitable pipette or device to deliver an exact volume	
Difference in disintegration performance with stressed samples	– Ensure choice of agitation and diluent are compatible with all relevant sample types – Increase agitation – Change diluent (e.g., to (partially) dissolve key excipients) – Consider cutting, scoring or grinding the dosage form prior to extraction	Section 7.4
Difference in recovery with stressed samples (but disintegration performance is similar)	Consider – Integrity of stressed samples (may be true result) – Changing diluent/solvent (e.g., to (partially) dissolve key excipients)	

Peak shape problems (e.g., splitting)	Consider – Lowering injection volume – Lowering the diluent organic solvent level – Making a dilution of the sample solution to make it more compatible with the analysis method	
Premature column death	Consider – If noneluting excipients being solubilized by diluent – Alternative diluents that do not dissolve certain key excipients or, "crashing out" the excipients in a multi-step procedure – Liquid-liquid extraction (LLE) or solid-phase extraction (SPE) to remove problematic excipients	Chap. 4 and 9
Difficulty in clarification	Consider – Choosing diluents that dissolve or have a different dispersive effect upon certain key excipients – Centrifugation instead of filtration – Centrifugation or standing time followed by filtration – Molecular weight cut off centrifugation filters to remove polymer excipients – LLE or SPE to remove problematic excipients – Precipitating out problematic excipients	Section 7.6 and Chap. 9
Poor recovery of (low-level) analytes or poor recovery upon dilution	Consider – Evaluating higher ionic strength diluents – Evaluating alternative pH or organic solvents types and levels – Analyte–surface compatibility[a]	

[a]Some analytes in solution will interact with or adsorb on surfaces such as volumetric flasks, pipettes, autosampler vials, HPLC tubing, etc. Positively charged and amphiphilic compounds exhibit such behavior with glass surfaces due to the negative charges on the silanol groups in glass (Wrezel et al. 2005). If these interactions are not addressed with the appropriate sample diluent (e.g., appropriate pH to eliminate charge interactions; increased diluent ionic strength to compete for interaction with the surface sites; or increased organic content) analyte recovery will be reduced. Wrezel and coworkers recommend performing a transfer experiment with the low concentration solution of analyte to evaluate for potential analyte–surface interaction (Wrezel et al. 2005). The analyte solution is prepared in a flask using the selected diluent. An aliquot is immediately withdrawn and saved in an HPLC vial as the control sample. The remaining solution is then transferred to a second flask, then to a third flask, etc. After each transfer, an aliquot is transferred into an HPLC vial. The samples are then analyzed by HPLC. A decreasing trend in peak area vs. transfer indicates loss of analyte due to interactions with container surfaces

(90% recovery or less). Poor solubility was unlikely to be the cause of the lack of recovery since the saturated solubility of the API in the API diluent is many times higher than the working concentration of these drug product samples.

"Wet-Spike" experiments to investigate low recovery. A "wet-spike" experiment designed to determine the potential suitability of extraction diluents was performed as follows:

Step A: API is dissolved in the extraction diluent(s) and at the extraction concentration(s) to be evaluated. This is useful in ensuring that the API is sufficiently rapidly dissolving in the diluent studied.

Step B: An aliquot of the API solution from Step A is added to the dry excipient blend(s) of the drug product formulation(s) under development, and the solution is agitated for a sufficient time period (e.g., stirred for 20 min).

Step C: The "wet-spike" sample(s) containing excipients from Step B are clarified by filtration or centrifugation and the recovery is determined by quantification against the API solutions prepared in Step A above.

It should be noted that "wet-spike" experiments do not detect all causes of formulation-related recovery problems, such as lack of disintegration or API trapped inside insoluble polymeric excipients. However, the "wet-spike" approach is rapid and can be used to quickly screen out extraction solvents that are unsuitable because of solubility or excipient-binding issues. An additional advantage of the "wet-spike" approach is that many sources of analytical variability are eliminated by this approach; for instance, there are no sources of variability or error due to weighing or preparation of standards. As a result, the "wet-spike" approach has a greater sensitivity to detect subtle recovery issues. For instance, a recovery result between 97 and 99% would be concerning if it were obtained using a "wet-spike" approach, but could be justifiably argued to be due to experimental variability for other types of recovery experiments.

In the case of this particular API and the excipient blends, a "wet-spike" experiment with the API extraction diluent gave low recoveries consistent with those observed for the stability samples. This is an important observation. The API is fully dissolved in solution prior to the addition of excipients; when the excipients are added there is a significant decrease in the API in solution. This can be explained by either a suppression of the API's solubility by the excipients (unlikely in this case since the API concentration is many times lower than the saturated solubility), or more likely due to API binding with undissolved excipients that are removed at the point of clarification. As the dilution is increased, there is an improvement to the % recovery, but this never reaches 100%, perhaps indicating that the API is in equilibrium between the bound and the solvated states:

$$\text{API (bound to excipient)} \Leftrightarrow \text{API (solvated)}.$$

This experiment indicates that the composition of the extraction diluent at the point of clarification is the critical factor in the performance of the extraction procedure for this particular method and other factors such as the mode of agitation, its intensity, and duration are less important.

Diluent screening. Based on API solubility information, a range of diluents were selected for evaluation in a diluent screening study. As described in Sect. 7.3.1, a wet-spike approach (API dissolved in each diluent spiked onto dry excipient blend) was used to screen out unsuitable extraction diluents in a rapid series of experiments. The effect of diluent attributes such as pH, choice of organic modifier, and level of organic were evaluated. Table 7.4 shows the performance of the extraction solvents with the selected formulations. The conclusions from this part-systematic, part-iterative approach to finding a suitable diluent were as follows:

(a) The level of organic solvent (methanol or acetonitrile) is an important factor, with higher recoveries obtained with higher organic-content diluents and methanol performing generally better than acetonitrile.
(b) The diluent pH is also an important factor: acidic diluents (e.g., pH 3.4 or less) performed better than those at pH 6.8 or pH-nonadjusted diluents. Diluents at a very high pH (using 0.1 M NaOH) also produced excellent recoveries. pH is more important than ionic strength because even very low concentrations of HCl (e.g., 0.006 M) brought about marked improvements in the recovery results, and 20 mM phosphate at pH 3.4 performed better than 20 mM phosphate at pH 6.8.
(c) Neat strong solvents such as DMAC and DMSO gave excellent recoveries, but DMAC-water mixtures did not.

On the basis of these experiments, a number of potentially viable diluents emerged. Neat DMAC or neat DMSO was discounted because they produced large solvent fronts that could potentially interfere with early eluting impurities (in addition to the obvious operational difficulties working with such solvents). The high pH option (0.1 M NaOH) was discounted because of solution stability issues with the API. Acidified methanol/water (80/20 v/v) was the diluent selected on the basis of these results. Acidification using 0.05% v/v concentrated HCl (resulting in approximately 0.006 M HCl) was selected on the basis of cost and convenience. Experiments with individual excipients indicated that Ac-Di-Sol and Explotab were the excipients mostly responsible for the low recoveries.

The suitability of the chosen diluent was further confirmed by analyzing two tablets from a selection of the formulations: in each case good disintegration of the tablets was observed and the assay results indicated full recovery. The impurity profile of the drug product samples obtained using this diluent were identical to that of the ingoing API, suggesting that recovery of impurities should not be an issue. This extraction procedure was successfully validated for the formulation chosen for use in clinical studies.

7.4 Step 3: Evaluate and Select Appropriate Conditions to Disintegrate/Disperse the Dosage Form

Dosage form type (e.g., disintegrating/IR tablet, nondisintegrating controlled or sustained release tablet) and excipients present in the formulation impact disintegration/dispersion of the solid oral dosage form. Based on the formulation type, mechanisms and conditions should be evaluated and selected to disperse/

Table 7.4 Summary of results from diluent selection experiments carried out on "Compound A"

	Diluents containing pH adjustment with common organic solvent/aqueous mixtures				
Assessment criteria	20 mM phosphate buffer, pH 3.4/ACN (80/20 v/v)[a]	20 mM phosphate buffer, pH 6.8/MeOH (50/50 v/v)	$NaOH_{aq}$ (0.1 M)/MeOH (50/50 v/v)	20 mM phosphate buffer, pH 3.4/MeOH (50/50 v/v)	20 mM phosphate buffer, pH 3.4/MeOH (20/80 v/v)
Solubility of API after 15 min mild agitation	100%	100%	100%	100%	100%
Time taken to fully dissolve	Rapid	Rapid	Rapid	Rapid	Rapid
Ease of clarification	Easy	Easy	Easy	Easy	Easy
Chromatographs of diluent blank and placebo blank: any interfering peaks? Generally acceptable?	Good	Good	Good	Good	Good
Peak shape problems (e.g., splitting)?	None	None	None	None	None
%Recovery of API from "wet spike"	ca. 90%	ca. 90%	ca. 100%	ca. 94%	ca. 100%
Solution stability of API (std) solution	N/A	N/A	Poor	N/A	Good
Solution stability of drug product solution	N/A	N/A	Poor	N/A	Good
Disintegration time of tablet with stirring	Rapid	Rapid	Rapid	Rapid	Rapid
%Recovery from tablet	N/A	N/A	ca. 100%	N/A	ca. 100%

Shaded cells indicate an unsatisfactory performance in the assessment
N/A not assessed (usually because diluent did not perform satisfactorily in a previous assessment)
[a]Diluent used in the API method

tablets

		Diluents containing simple water/organic solvent mixtures			Diluents containing strong solvents				
0.05% HCl_{aq}/MeOH (80/20 v/v)	0.05% HCl_{aq}/MeOH (20/80 v/v)	Water/ACN (acetonitrile) (50/50 v/v)	Water/MeOH (methanol) (50/50 v/v)	Water/MeOH (20/80 v/v)	MeOH, neat	DMSO (dimethylsulphoxide), neat	DMAC (*N*, *N*-dimethy-lacetamide), neat	Water/DMAC (80/20 v/v)	MeOH/DMAC (80/20 v/v)
100%	100%	100%	100%	100%	100%	100%	100%	100%	100%
Rapid	Rapid	Rapid	Rapid	Rapid	Rapid	Rapid	Rapid	Rapid	Rapid
Easy	Easy	Easy	Easy	Easy	Easy	Easy	Easy	Easy	Easy
Good	Good	Good	Good	Good	Good	Very large solvent front	Very large solvent front	Large solvent front	Large solvent front
None	None	None	None	None	None	None	None	None	None
ca. 95%	ca. 100%	ca. 75%	ca. 90%	ca. 93%	ca. 100%	ca. 100%	ca. 100%	ca. 80%	ca. 100%
N/A	Good	N/A	N/A	N/A	N/A	N/A	N/A	N/A	N/A
N/A	Good	N/A	N/A	N/A	N/A	N/A	N/A	N/A	N/A
Rapid	Rapid	Rapid	Rapid	Rapid	None	Rapid	Rapid	Rapid	Rapid
N/A	ca. 100%	N/A	N/A	N/A	N/A	N/A	N/A	N/A	N/A

disintegrate the tablet or capsule as described later. It is also important to understand the relationship between the dispersion step (discussed in this section) and the extraction/solubilization step (discussed in Sect. 7.3) and how these will affect each other.

7.4.1 Tablets

Tablet dosage forms include IR tablets (uncoated and coated), extended release tablets, delayed release tablets (e.g., granules or particles covered with gastro-resistant coating, enteric coated tablets), orodispersible tablets (e.g., orally disintegrating tablets, buccal tablets, sublingual tablets), and chewable tablets. Tablet dosage formulations typically contain API and excipients such as binders, glidants, lubricants, disintegrants, sweeteners or flavoring agents, coloring agents, and coating components.

A common practice to ensure thorough dispersion of any of these types of tablets is to use mechanical means, such as manually grinding a tablet or a composite of tablets using a mortar and pestle, to obtain a fine powder prior to adding diluent. After grinding, the entire sample or a portion of it is transferred to a flask for further preparation. Numerous examples of this approach are available in the literature. In addition to manual grinding, other mechanical techniques such as milling (Kok and Debets 2001; Nickerson et al. 2008) or homogenization (Shamrock et al. 2000; Höller et al. 2003; Toro et al. 2004; Lee et al. 2007; Nickerson et al. 2008) are available and are discussed in Chap. 3. Although mechanical dispersion (e.g., grinding) can be simple to use, it can have limitations such as being labor intensive for large numbers of samples and requires cleaning of the equipment (e.g., mortar and pestle; homogenizer) between samples. It can also be challenging to quantitatively transfer the material for unit dose testing. Low potency results can also be obtained due to adsorption of drug to surfaces during the grinding (e.g., mortar and pestle surfaces) (Kirschbaum 1989). Variable results can be obtained due to segregation of particles during grinding.

An alternative to using mechanical techniques is to disintegrate or disperse the tablet in solution through use of an appropriate diluent. IR tablets contain disintegrants or superdisintegrants, and the addition of water causes the tablet to disintegrate. Commonly described mechanisms of disintegration for IR tablets (Lowenthal 1972; Kanig and Rudnic 1984; Melia and Davis 1989; Guyot-Herman 1992) are as follows: (1) disintegrant draws water up into the porous network of the tablet (via wicking or capillary action), breaking up intermolecular hydrogen bonding forces between particles/granules thereby leading to tablet disintegration; and (2) disintegrant swells after water uptake, causing build up in force and leading to tablet disintegration.

For sample preparation of IR tablets, disintegration is usually accomplished by placing the tablet in a flask and then adding water or an aqueous buffer and allowing the sample to shake or stir for a period of time (Joshi et al. 2010; USP Monograph for Cyclophosphamide Tablets 2010; USP Monograph for Theophylline Tablets

2010; USP Monograph for Trihexyphenidyl Hydrochloride Tablets 2010). If the drug is not water soluble, organic solvent can be added after the tablet has disintegrated. Alternatively, a water or aqueous buffer/organic solvent mixture can initially be added to disintegrate the tablet and extract the drug (USP Monograph for Flurbiprofen Tablets 2010; USP Monograph for Ranitidine Tablets 2010). If an aqueous/organic solvent mixture is desired, it should be evaluated to ensure that the organic type and content do not prevent or retard disintegration of the tablet. Even if the drug is soluble in water or an aqueous buffer, addition of organic or a subdilution with some organic in the diluent may be necessary to ensure compatibility with the analysis method (e.g., to ensure suitable chromatography with HPLC analysis).

Some methods use a combination of these two approaches, disintegrating an intact tablet in water or buffer for content uniformity testing and forming a fine powder of a composite of tablets then adding diluent for assay testing (USP Monograph for Leucovorin Calcium Tablets 2010; USP Monograph for Megestrol Acetate Tablets 2010; USP Monograph for Metoprolol Tartrate and Hydrochlothiazide Tablets 2010; USP Monograph for Tocainide Hydrochloride Tablets 2010; USP Monograph for Triprolidine Hydrochloride Tablets 2010).

Typically agitation, such as stirring, shaking (manually or on a mechanical shaker) (Owens et al. 2007; Joshi et al. 2010), vortexing (Mostafavi et al. 2009), or sonication (Basniwal et al. 2008; Shaikh et al. 2008; Kurade et al. 2009), is used to facilitate the disintegration process.

Other easily dispersible or disintegrating dosage forms, such as orally dispersive tablets (e.g., buccal tablets, sublingual tablets), are also designed to readily disperse or disintegrate in water. So a similar approach as used with IR tablets can be used to disperse these dosage forms using an aqueous or aqueous/organic diluent (USP Monograph for Ergotamine Tartrate Sublingual Tablets 2010; USP Monograph for Mirtazapine Orally Disintegrating Tablets 2010; USP Monograph for Nitroglycerin Sublingual Tablets 2010). In addition, chewable tablets are relatively soft tablets, and can usually be dispersed in aqueous, heated aqueous (USP Monograph for Thiabendazole Chewable Tablets 2010), aqueous/organic media or organic media (USP Monograph for Pseudoepherine Hydrochloride Tablets 2010) with agitation.

Delayed release tablets, such as enteric coated tablets (or enteric coated pellets), are designed to be stable and not release drug under the acidic pH conditions found in the stomach, but will release the drug at less acidic pH conditions found in the small intestine. These enteric coatings are therefore typically soluble at pH values above 6.5. An aqueous buffer at pH ≥6.5 with agitation can therefore be used to dissolve the enteric coating and disperse the dosage form. Using heated water as the sample diluent has also been reported (USP Monograph for Oxtriphylline Delayed-Release Tablets 2010). Alternatively use of organic solvents may be used to overcome the solubility dependence on pH. Examples of aqueous/organic diluents with agitation have been used to disperse delayed release tablets (Wardrop et al. 2000; USP Monograph for Diclofenac Sodium Delayed-Release Tablets 2010; USP Monograph for Divalproex Sodium Delayed-Release Tablets 2010; USP Monograph for Naproxen Delayed-Release Tablets 2010; USP Monograph for Pantoprazole Sodium Delayed-Release Tablets 2010). As previously mentioned, grinding or making a powder of delayed-release tablets is also an option (USP Monograph for Aspirin Delayed-Release Capsules 2010; USP Monograph for Mesalamine Delayed-Release Tablets 2010).

For extended release tablet dosage forms (e.g., prolonged release, sustained release, controlled release, modified release, etc.), which are designed to be nondisintegrating, an appropriate diluent may be able to disperse the tablet by swelling one of the excipients, such as a polymer, or dissolving the coating (USP Monograph for Acetaminophen Extended-Release Tablets 2010) to then allow the tablet to disintegrate. After the tablet has been dispersed, a second diluent can be added if needed to solubilize the API. Diluents used to disperse extended release tablet dosage forms include water (USP Monograph for Acetaminophen Extended-Release Tablets 2010; USP Monograph for Metoprolol Succinate Extended-Release Tablets 2010), alcohol (USP Monograph for Pseudoephredrine Hydrochloride Extended-Release Tablets 2010), acetonitrile (USP Monograph for Felodipine Extended-Release Tablets 2010), organic mixtures (USP Monograph for Oxycodone Hydrochloride Extended-Release Tablets 2010; USP Monograph for Procainamide Hydrochloride Extended-Release Tablets 2010; USP Monograph for Oxybutynin Chloride Extended-Release Tablets 2010), or aqueous/organic diluent (USP Monograph for Ferrous Fumarate and Docusate Sodium Extended Release Tablets 2010). Alternatively, mechanical means (e.g., grinding or milling) can be used to finely powder the dosage form (USP Monograph for Aspirin Delayed-Release Capsules 2010; USP Monograph for Carbamazepine Extended-Release Tablets 2010; USP Monograph for Etodolac Extended-Release Tablets 2010) or the dosage form can be crushed (Oliveira et al. 2009) or cut into pieces to expose the tablet core and allow disintegration in an appropriate solvent, or homogenization in the presence of diluent (USP Monograph for Bupropion Hydrochloride Extended-Release Tablets 2010; USP Monograph for Metformin Hydrochloride Extended-Release Tablets 2010; USP Monograph for Pseudoephredrine Hydrochloride Extended-Release Tablets 2010) can be used to disperse the tablet. Examples of dispersion methods used for extended release tablet dosage forms are discussed in Sect. 7.4.2.

7.4.2 Examples of Dispersion Strategies for Extended Release Tablet Formulations

Hydrophilic matrix tablet formulation #1: "Compound B" is formulated as a high dose controlled release matrix tablet formulation. A sample preparation method was developed, which involved placing one tablet into a 500 mL flask and adding ~250 mL acetonitrile. The solution is allowed to stand for 15 min and is then sonicated for 5 min to fully disperse the tablet and extract the active ingredient. At this point, the flask is diluted to volume with water and mixed well. A subdilution is then made by pipetting 7 mL into a 10 mL volumetric flask and diluting to volume with 50% water/50% acetonitrile. The sample solution is then filtered. In this case since this is a controlled release tablet rather than an IR tablet, acetonitrile instead of water is used to disperse the tablet. In this case one diluent, acetonitrile, is sufficient to disperse the tablet matrix and extract the active that is readily soluble in acetonitrile. Water is used to dilute the volume to make the sample solution more amenable to subsequent HPLC analysis.

Hydrophilic matrix tablet formulation #2: "Compound C" was formulated as a controlled release matrix tablet. Water was initially used to disperse the tablet and extract the water soluble drug. This procedure resulted in a viscous sample solution that was difficult to filter because a polymer exicipient in the formulation also dissolved, leading to the high viscosity. The sample diluent was then changed to methanol. A tablet was placed in a volumetric flask, methanol was added, and the tablet would disperse into fine particles with mixing. Later when this method was used on accelerated stability samples low recoveries were obtained and the degradation level observed did not account for the low recovery values.

When methanol was added to accelerated stability samples (e.g., stored at 70°C/75% RH for 1 week), the tablets did not disperse well. Instead, the tablets broke up into large particles or chunks and low recoveries were obtained (<75%). A suitable diluent to disperse and ensure complete extraction of the drug could not be found for the accelerated stability samples.

The sample preparation method was redeveloped and involved placing a tablet in a bottle and adding methanol. The tablet would disperse into chunks with mixing. A second diluent, 0.1 N HCl, was then added to increase the solubility of the drug in the solution, and the solution was homogenized for 1 min at 10,000 rpm using a homogenizer (Kinematica Polytron® PT 3100) to completely disperse the tablet and dissolve the drug. The vigorous agitation and shredding action of the homogenizer dispersed the tablet chunks and extracted the drug, and under these conditions complete extraction of the drug was achieved. A limitation of the method was the time needed to clean the homogenizer between samples, so this method is not ideal for high sample numbers. This case study does illustrate the value of using stressed stability samples to confirm method robustness during sample preparation method development.

Osmotic tablet formulation #1: An osmotic extended release tablet formulation was developed for "Compound D." This tablet formulation contains a nonwater soluble, polymeric coating; therefore, mechanical means are used during sample preparation to expose the tablet core to the dissolving/extraction solvent to allow tablet dispersion. A single tablet or a composite of tablets is crushed (e.g., with the blunt end of a mortaring pestle), transferred to a flask, and mixed with methanol to disperse the tablet(s) and facilitate subsequent extraction of the drug. Then 0.1 N HCl is added and the sample is shaken to fully extract the drug. The sample solution is then filtered and injected onto an HPLC for analysis.

Osmotic tablet formulation #2: "Compound E" was formulated as a 10 mg extended release osmotic tablet formulation. The tablet formulation has a polymer film coating consisting of cellulose acetate and polyethylene glycol (PEG). The tablet core consists of standard excipients. A sample preparation method was developed, which involved placing the coated tablet in a volumetric flask and adding an aliquot of 50% water/50% acetonitrile and shaking for 1 h. During this time, the tablet core swells and this causes the coating to split and expose the tablet core. The coating does not dissolve and can be seen in the flask. After shaking, 70% water/30% acetonitrile is added and the sample solution is sonicated for 5 min to completely dissolve the drug. The volumetric flask is then filled to volume, mixed well, and the solution is filtered and analyzed by HPLC.

Osmotic tablet formulation #3: For this osmotic tablet formulation of "Compound F," a diluent was identified, which could dissolve the polymer coating and facilitate tablet dispersion. Acetonitrile was added to the tablet in a volumetric flask and the solution was shaken for 2 h to dissolve the coating. Methanol was then added to the flask to disperse the table core and extract the drug using additional shaking for 3 h. A buffer was then added to the mark to make the sample solution more compatible with the mobile phase used for HPLC analysis.

7.4.3 Capsules

Many capsule shells used in solid oral dosage forms are made of gelatin. Gelatin-related peaks can be observed in HPLC analysis with UV detection, and these peaks can overlap with the signal from the API and/or its degradation products if not resolved during chromatographic analysis. Dissolution of the gelatin capsule during sample preparation can therefore lead to interfering peaks in chromatographic analysis. To avoid this issue, a common sample preparation practice for gelatin capsules is to remove the contents of the capsule for both IR capsules (Caviglioli et al. 1994; Srinivasu et al. 2000; Kartel 2001; Marin et al. 2002; Dias et al. 2005; Vijaykumar et al. 2006) and delayed or controlled release capsules (USP Monograph for Aspirin Delayed-Release Capsules 2010; USP Monograph for Pancrelipase Delayed-Release Capsules 2010; USP Monograph for Chlorpheniramine Maleate Extended-Release Capsules 2010; USP Monograph for Theophylline Extended-Release Capsules 2010). Hard gelatin capsules (HGCs) can be opened, while soft gelatin capsules (SGCs) can be cut to remove the capsule contents (USP Monograph for Theophylline Capsules 2010), which can then be quantitatively transferred into a flask. The contents of several capsules can be pooled and an aliquot of the sample is taken and used to prepare a sample solution. In the case of content uniformity, where only one capsule is used per sample, it can be challenging to remove the entire capsule blend if the contents adhere to the shell and care must be taken not to lose any of the contents when opening the shell. This can also be an issue for some formulations where sticking occurs for stressed samples or stability samples.

To ensure no loss of active due to contents adhering to the capsule shell, HGCs can be dissolved in an aqueous solution (or in an aqueous/organic solution with low organic content) with sufficient agitation, such as sonication (Tzanavaras et al. 2008). Shaking and stirring alone may not achieve dissolution of the capsule shell. In some cases, this approach is used for individual capsules, while emptying the contents of multiple capsules is performed for assay (Tzanavaras et al. 2008).

For purity analysis where gelatin may interfere with the chromatographic analysis, Zhao and coworkers report resolving potential interferences caused by gelatin in HPLC analysis of API in capsules by adding collagenase to the sample solution to break down the gelatin (Zhao et al. 2009). The procedure eliminates sample loss since the entire capsule is used in the sample preparation including any drug that may be adhered to the inner wall of the capsule shell. The procedure also extends column life by preventing fouling of the column by gelatin.

Soft gelatin or liquid/semi-solid filled capsules can also be dissolved during sample preparation. For soft gelatin or liquid/semi-solid filled capsules, aqueous media can be used to dissolve or break up the gelatin shell with the aid of shaking or sonication. Then if needed, an organic diluent can be added to solubilize the active ingredient.

Example sample preparation for SGC: A self-emulsifying SGC formulation was developed for "Compound G." An acidic aqueous diluent is used to rupture the SGC. One capsule is placed in acidic medium, 0.2% H_3PO_4 in water, and shaken for a period of time (e.g., 30 min) to break up the gelatin shell. The sample is visually checked to ensure that the gelatin shell has been perforated or broken apart to allow the fill to release prior to adding the second diluent, an organic solvent (acetonitrile), to solubilize the active ingredient. The solution generally turns cloudy when the capsule fill is released. In this case, a gelatin peak is present in the chromatogram but does not interfere with quantitation of the API or potential degradation products.

Alternatives to gelatin, such as hydroxypropyl methylcellulose (a plant-derived material) and pullulan (a water soluble polysaccharide), have been used to manufacture capsules for use in pharmaceutical dosage forms. HPMC and pullulan capsule shells are water soluble, thus these types of capsule formulations can be placed in water or an aqueous buffer to readily dissolve the capsule shell and disperse the capsule contents. This eliminates potential sample loss since the entire capsule is used. If the drug is not water soluble, organic solvent can be added either after the capsule shell has dissolved, or a water or aqueous buffer/organic solvent mixture can initially be added to dissolve the capsule shell and extract the drug. Typically agitation, such as shaking on a mechanical shaker, is used to facilitate capsule rupture or dissolution.

7.5 Step 4: Evaluate and Select Agitation Parameters

7.5.1 General Practices

Agitation is used to help efficiently disperse the dosage form and facilitate dissolution of the active and components of interest. Optimization of agitation conditions should be performed using representative drug product samples as well as stressed samples (e.g., samples on accelerated stability). For IR formulations, agitation need not be too

Table 7.5 Example agitation parameters that require optimization during development of a sample preparation method

Agitation techniques	Parameters to evaluate/optimize
Shaking – manual	Number of inversions Diluent volume in flask
Shaking – mechanical	Type of shaker (e.g., orbital, reciprocating, wrist action®) Time, speed Diluent volume Orientation of bottle or vessel if not using volumetric flask
Stirring	Size and type of stir bar Time, speed Dilute volume in flask
Sonication	Sonication type (e.g., horn, probe) Time Diluent volume in flask Water level inside bath Performance in different locations within a bath Bath-to-bath variability No degradation due to heating
Homogenization	Type of homogenizer (e.g., Polytron®) Speed, time, number of cycles Diluent volume Potential effects of heating (e.g., degradation) and solution evaporation (if no subsequent dilute to volume step) Cleaning procedure between samples to eliminate sample carry over

vigorous (shaking, stirring, or inverting should generally suffice). The general philosophy is to "let the extraction solvent do the work." The choice of the right extraction solvent should avoid agitation-critical extractions. Shaking and stirring are commonly used. Vortexing and sonication are also reported. Sometimes the choice of agitation technique may depend upon equipment availability. With each of these agitation techniques, evaluate and confirm the use of an appropriate rate and time.

Manual shaking or agitation can introduce analyst-to-analyst variability and impact the robustness of the method. For example, Kirschbaum reported on a method that stated to sonicate samples for "15 min with occasional shaking of the volumetric flasks." Analyst-to-analyst variability was observed: analysts who vigorously swirled samples obtained assay results that were 3% higher than analysts who swirled the flasks once (Kirschbaum 1989).

Since extended release dosage forms are designed to be nondisintegrating, agitation can play a critical role in dispersing the dosing form and facilitating solubilization of the API. Extended release dosage forms may require more vigorous agitation (e.g., homogenization) or agitation for longer periods of time compared with IR dosage forms. Some agitation parameters to consider evaluating and optimizing during method development are listed in Table 7.5.

Sonication is a common technique reported in the literature to prepare sample solutions. Heating of the sample solution may occur during sonication and this can be an issue for heat labile compounds. In addition, there can be variability in performance between sonication baths and between locations within a given sonication bath. Performance can vary depending on the number of flasks within the sonication bath or depending on the level of water in the sonication bath. This variability can be an issue with method ruggedness and transferability. This may be more of an issue with extended release dosage forms where agitation may be critical for dispersion of the dosage form and extraction of the active. Reliance on just sonication for agitation in the sample preparation step, especially for controlled and sustained release formulations (e.g., nondisintegrating formulations) is therefore not recommended.

7.5.2 Potential Issues During Agitation

After stirring or shaking a sample solution in a volumetric flask, many methods say something like, "Dilute to volume and mix well." At this point, it is important to ensure adequate mixing to obtain a homogeneous solution. This can be problematic in two cases. One case is for viscous solutions that are difficult to mix, as is the case for some controlled release formulations that contain polymers. The polymers can swell leading to viscous solutions. If a homogeneous solution is not obtained, low or variable potency results can be obtained. The second case is when preparing a large number of samples, intensity of manual mixing may not be consistent. This is especially true for large sample solution volumes that may be required for composite samples of high dose formulations. It is important to evaluate if mixing will be an issue. To address this potential issue, methods may explicitly state "mix by inversion" or specify the number of inversions needed to ensure adequate mixing. Case studies 1 and 3 in Chap. 13 illustrate this issue.

Several other issues may be encountered during sample agitation. One potential issue is when sample solution containing undissolved material may splash into the neck of the volumetric flask while on a shaker. API in the undissolved material may not be extracted and be lost if it is not rinsed down the neck when diluting to volume or inverting the flask to mix. This may pose a more significant issue with low dose products or compounds that need longer times to dissolve.

Another potential issue is having insoluble material or undissolved excipients at the top of the solution in the neck of the flask. The top layer of solution may not be homogeneous with the rest of the solution and one may need to decant the top layer.

Yet another potential issue is having sample solutions foam while on the shaker or being stirred. This may make it difficult to dilute to the mark accurately. A possible means to resolve this issue would be to increase the organic content of the diluent or consider alternative, milder agitation conditions. Another approach would be to dispense a known quantity of diluent into a bottle (e.g., using a pipet or bottle top dispenser), adding the dosage form and shaking, instead of using a volumetric flask and diluting to volume.

7.6 Step 5: Evaluate and Select Appropriate Means to Remove Insoluble Components

7.6.1 Removal of Insoluble Components

After the dosage form is dispersed and the components of interest are extracted and dissolved, there are typically insoluble excipient components in the sample solution. These insoluble components may cause interferences in the sample analysis method and are typically removed by filtration or centrifugation. Filtration, if feasible, is preferred in many laboratories due to ease of use. When using filtration, a study should be conducted to identify an appropriate filter (e.g., drug is not retained by the filter) and to ensure the use of an appropriate filter pass through volume prior to collection of sample solution for analysis. Examples of drug absorbing to filter membranes and leading to low assay values have been published in the literature (Nordling et al. 1973; Yahya et al. 1988; Guilfoyle et al. 1990; Carlson and Thompson 2000; Kiehm and Dressman 2008). Considerations for filter selection include cost, compatibility with the sample solution in terms of recovery and lack of interfering leachables.

Centrifugation of sample solutions in tubes is used if sample volumes are limited or if the solution is extremely viscous. Centrifugation of sample solutions in molecular weight cut off filters is also an option for extremely viscous sample solutions containing high molecular weight polymeric excipients. Appropriate centrifugation conditions (e.g., relative centrifugal force, time) should be established. When using centrifugation, specifying relative centrifugal force (rcf) instead of revolutions per minute (rpm) will reduce the likelihood of differences in performance if different centrifuges are used. Other options include the following: using solid phase extraction to clarify viscous sample solutions; precipitating out the polymer excipient(s) after the API have been extracted and dissolved to obtain a less viscous solution for filtration; or using automation to filter the sample solutions. Additional information on clarification of samples solutions is provided in Chap. 9.

7.7 Step 6: Confirm Sample Preparation Method Works

The sample preparation method in its entirety should be evaluated or validated to show that it works as expected and completely extracts the drug from the dosage form. Ideally, this verification is performed using the final dosage form and samples with a known amount of drug. If possible, use a variety of samples to evaluate the robustness of the sample preparation method. These samples can include aged or stressed samples, IR tablets with higher hardness values, CR tablets with thicker functional coating, etc. These types of samples may disperse/extract differently and can help to evaluate the robustness of the method. It is important that these stressed or "aberrant" samples are not too extreme.

For dosage forms in later stages of development where there is high sample throughput or where the product will be tested by multiple laboratories, method robustness and ruggedness is important. Manual portions of the sample preparation and extraction method such as manual shaking and manual grinding may pose potential ruggedness issues under high throughput conditions. It is therefore recommended that robustness and/or ruggedness studies for the sample preparation portion of the method be performed prior to transfer to manufacturing or quality control laboratories. Investment of time and effort at this stage of the process will save significant time, effort, money, and headache during scale up and subsequent activities. This topic is discussed in detail in Chap. 11.

7.8 Additional Considerations

There are some additional considerations to take into account to help make the sample preparation process more efficient. For example, sample preparation often involves the use of volumetric flasks. The volumetric flask is often partially filled with diluent and then the dosage form is added. After agitation, the volumetric flask is diluted to volume and mixed. It can be time consuming and tedious to perform this procedure with a large number of samples. An alternative is to use a different type of vessel and means of dispensing diluent. For example, diluent can be dispensed into a bottle using a bottle top dispenser on the diluent bottle. The bottle can be capped and then placed on a shaker for agitation, or alternatively a stir bar can be added and the bottle can be placed on a magnetic stirrer. In this case, the full volume of diluent can be dispensed at one time, and there is no need to "dilute to volume" after the agitation step. This process reduces the total number of sample preparation steps, and dispensing the diluent can occur very rapidly.

Another way to minimize sample preparation time is to avoid the need to make subdilutions of the sample solutions. This can be done by selecting solution volumes, sample concentrations, and detection parameters (e.g., detection wavelength for UV detection) to avoid a sample dilution step if possible to save time and effort in manually preparing samples. A dilution step may not be avoidable for composite samples of high dose products. In these cases consider the use of solvent dispensers or pipettors to prepare sample solutions in nonvolumetric glassware or containers (e.g., bottles) to minimize reagent volume and make the dilution step rapid to perform. Consider the potential for solvent contraction/expansion if doing a two-step diluent addition process when dispensing reagents in this manner.

Consider using additional sample types during sample preparation method development to evaluate method ruggedness. These samples may include aged (e.g., accelerated stability samples) or "aberrant" samples (e.g., IR tablets with higher hardness values, CR tablets with thicker coatings). These types of samples may disperse or extract differently, so they are a good test of the sample preparation method. Although it is important to challenge the method, care should be taken to not use samples that are too extreme or unrealistic and may cause additional method development that is unwarranted.

An additional consideration is to ensure that the method is clearly written with sufficient detail to avoid issues during testing. Missing information (e.g., details of actions required), unclear wording, or wording subject to interpretation may lead to method issues.

7.9 Conclusions

An approach to developing extraction and sample preparation methods for solid oral dosage forms is presented in this chapter. Using a systematic approach to develop sample preparation methods will result in the development of robust methods and minimize the likelihood of low recovery issues during use of the method. If low recovery issues are encountered, some troubleshooting guidance is provided in Chap. 13. An additional factor for consideration, especially for compounds that have high sample throughput, is automation of the sample preparation method and this is discussed in Chap. 12.

Acknowledgments The authors acknowledge the following individuals who were involved in developing the sample preparation methods described in the examples of this chapter: William Aripko, Eddie Ebrahimi, Alanya Engtrakul, Justin Ringling, Jennifer Vogt, Hannah Jupe, Craig Donnelly, Pierre Barratt, and John Warzeka.

References

Aboutaleb AE, Rahman AA, Saleh SI, Ahmed MO (1986) Factors affecting the release of diazepam from directly compressed tablets: I. Interaction of diazepam with certain directly compressible vehicles. STP Pharma 2:23–27

Al-Nimry SS, Assaf SM, Jalal IM, Najib NM (1997) Adsorption of ketotifen onto some pharmaceutical excipients. Int J Pharm 149:115–121

Basniwal PK, Shrivastava PK, Jain D (2008) Hydrolytic degradation profile and RP-HPLC estimation of cilostazol in tablet dosage form. Indian J Pharm Sci 70(2):222–224

Carlson M, Thompson RD (2000) Analyte loss due to membrane filter adsorption as determined by high-performance liquid chromatography. J Chromatogr Sci 38(2):77–83

Caviglioli G, Parodi B, Cafaggi S, Bignardi G, Romussi G (1994) Stability indicating HPLC assay for retinoic acid in hard gelatin capsules containing lactose and as bulk drug substances. Drug Dev Ind Pharm 20(15):2395–2408

Chien YW, van Nostrand P, Hurwitz AR, Shami EG (1981) Drug-disintegrant interactions: binding of oxymorphone derivatives. J Pharm Sci 70(6):709–710

Cory W, Field K, Wu-Linhares D (2004) Is it the method or the process – separating the causes of low recovery. Drug Dev Ind Pharm 30(8):891–899

Czaja J, Mielck JB (1982) Solid-state degradation kinetics of nitrazepam in the presence of colloidal silica. Pharm Acta Helv 57(5–6):144–153

Dias CL, Rossi RC, Donato EM, Bergold AM, Froehlich PE (2005) LC determination of ritonavir, a HIV protease inhibitor, in soft gelatin capsules. Chromatographia 62:589–593

Edelson MR, Hermans J (1963) Flow of gels of cellulose microcrystals. II. Effect of added electrolyte. J Polym Sci C 2:145–152

Franz AM, Peck GE (1982) In-vitro adsorption-desorption of fluphenazine dihydrochloride and promethazine hydrochloride by microcrystalline cellulose. J Pharm Sci 71:193–1199
Guilfoyle DE, Roos R, Carito SL (1990) An evaluation of preservative adsorption onto nylon membrane filters. J Parenter Sci Technol 44(6):314–419
Guyot-Herman AM (1992) Tablet disintegration and disintegrating agents. STP Pharma Sci 2(6):445–462
Hollenbeck RG (1988) Bioavailability of phenylpropanolamine HCl from tablet dosage forms containing croscarmellose sodium. Int J Pharm 47:89–93
Hollenbeck RG, Mitrevej KT, Fan AC (1983) Estimation of the extent of drug-excipient interactions involving croscarmellose sodium. J Pharm Sci 72(3):325–327
Höller U, Brodhad C, Hofman P, Spitzer V (2003) Automated determination of selected water-soluble vitamins in tablets using a bench-top robotic system coupled to reversed phase (RP-18) HPLC with UV detection. J Pharm Biomed Anal 31(1):151–158
Howard AL, Shah MC, Ip DP, Brooks MA, Strode JT III, Taylor LT (1994) Use of supercritical fluid extraction for sample preparation of sustained-release felodipine tablets. J Pharm Sci 83(11):1537–1542
Joshi SJ, Karbhar PA, Bhoir SI, Bindri KS, Das C (2010) RP-HPLC method simultaneous estimation of bisoprolol fumarate and hydrochlorothiazide in tablet formulation. J Pharm Biomed Anal 52(3):362–371
Kanig JL, Rudnic EM (1984) The mechanisms of disintegrant action. Pharm Technol 8:50–63
Kartel M (2001) LC method for the analysis of paracetamol, caffeine and codeine phosphate in pharmaceutical preparations. J Pharm Biomed Anal 26(5–6):857–864
Kennon L, Higuchi T (1956) Interaction studies of cationic drugs with anionic polyelectrolytes I. Sodium carboxymethylcellulose. J Am Pharm Assoc Sci 45:157–160
Kiehm K, Dressman JB (2008) Evaluation of drug absorption to membrane filters under biowaver test conditions. Diss Technol 15(4):13–17
Kirschbaum JJ (1989) Inter-laboratory transfer of HPLC methods: problems and solutions. J Pharm Biomed Anal 7(7):813–833
Kok SJ, Debets AJ (2001) Fast sample preparation for analysis of tablets and capsules: the ball-mill extraction method. J Pharm Biomed Anal 26(4):599–604
Kurade VP, Pai MG, Gude R (2009) RP-HPLC estimation of ramipril and telmisartan in tablets. Indian J Pharm Sci 71(2):148–151
Lee C, Gallo J, Arikpo W, Bobin V (2007) Comparison of extraction techniques for spray dried dispersion tablet formulations. J Pharm Biomed Anal 45(4):565–571
Lowenthal W (1972) Disintegration of tablets. J Pharm Sci 61(11):1695–1711
Marin A, García E, García A, Barbas C (2002) Validation of a HPLC quantification of acetaminophen, phenylephrine and chlorpheniramine in pharmaceutical formulations: capsules and sachets. J Pharm Biomed Anal 29(4):701–714
Mark HF, Gaylord NG, Bikales NM (1965) The encyclopedia of polymer science and technology, vol 3. Wiley, New York, pp 170–178
McBurney LF (1954) High polymers, vol. 5, part 2: cellulose and cellulose-derivatives. Wiley, New York
Melia CD, Davis SS (1989) Review article: mechanisms of drug release from tablets and capsules. I: Disintegration. Aliment Pharmacol Ther 3:223–232
Mostafavi A, Abedi G, Jamshidi A, Afzali D, Talebi M (2009) Development and validation of a HPLC method for the determination of buprenorphine hydrochloride, naloxone hydrochloride and noroxymorphone in a tablet formulation. Talanta 77(4):1415–1419
Nickerson B, Arikpo WB, Berry MR, Bobin VJ, Houck TL, Mansour HL, Warzeka J (2008) Leveraging elevated temperature and particle size reduction to extract API from various tablet formulations. J Pharm Biomed Anal 47(2):268–278
Nordling S, Miettinen H, Saxén L (1973) Adsorption of dactinomycin to nitrocellulose filters. Exp Cell Res 82(2):469–471
Okada S, Nakahara H, Isaka H (1987) Adsorption of drugs on microcrystalline cellulose suspended in aqueous solutions. Chem Pharm Bull 35:761–768

Oliveira PR, Bernardi LS, Mendes C, Cardoso SG, Sangoi MS, Silva MAS (2009) Liquid chromatographic determination of norfloxacin in extended-release tablets. J Chromatogr Sci 47:739–744

Owens PK, Wikström H, Någård S, Karlsson L (2007) Development and validation of a capillary electrophoresis method for ximelagatran assay and related substance determination in drug substance and tablets. J Pharm Biomed Anal 27(3–4):587–598

Pramar Y, Gupta VD (1991) Quantitation of scopolamine hydrobromide when adsorbed onto microcrystalline cellulose and sodium carboxymethyl cellulose. Drug Dev Ind Pharm 17:401–2407

Qtaitat MA, Zughul MB, Badwan AA (1998) Bromhexine hydrochloride adsorption by some solid excipients used in the formulation of tablets. Drug Dev Ind Pharm 14:415–429

Senderoff RI, Mahjour M, Radebaugh GW (1992) Characterization of adsorption behavior by solid dosage form excipients in formulation development. Int J Pharm 83:65–72

Shaikh K, Patil S, Devkhile A (2008) Development and validation of a reversed-phase HPLC method for simultaneous estimation of ambroxol hydrochloride and azithromycin in tablet dosage form. J Pharm Biomed Anal 48(5):1481–1484

Shamrock WF, Reilly K, Lloyd DK (2000) Automated sample preparation of Roxifiban tablets: transfer of a manual method to an automated workstation. J Pharm Biomed Anal 21(6):1225–1232

Srinivasu MK, Lakshmi Varayana C, Sreenivas RD, Om RG (2000) A validated LC method for the quantitative determination of celecoxib in pharmaceutical dosage forms and purity evaluation in bulk drugs. J Pharm Biomed Anal 22(6):949–956

Toro I, Dulsat JF, Fabregas JL, Claramunt J (2004) Development and validation of a fully automated method for the chromatographic determination of content uniformity of drug tablets. J Pharm Biomed Anal 36(1):57–63

Tzanavaras PD, Zacharis CK, Rigas P (2008) Novel automated assay for the quality control of mexiletine hydrochloride formulations using sequential injection and on-line dilution. J Pharm Biomed Anal 48(4):1254–1260

USP Monograph for Acetaminophen Extended-Release Tablets (2010). In: United States Pharmacopeia 2010, vol 33 reissue. USP-NF Online. http://www.uspnf.com

USP Monograph for Aspirin Delayed-Release Capsules (2010). In: United States Pharmacopeia 2010, vol 33 reissue. USP-NF Online. http://www.uspnf.com

USP Monograph for Bupropion Hydrochloride Extended-Release Tablets (2010). In: United States Pharmacopeia 2010, vol 33 reissue. USP-NF Online. http://www.uspnf.com

USP Monograph for Carbamazepine Extended-Release Tablets (2010). In: United States Pharmacopeia 2010, vol 33 reissue. USP-NF Online. http://www.uspnf.com

USP Monograph for Chlorpheniramine Maleate Extended-Release Capsules (2010). In: United States Pharmacopeia 2010, vol 33 reissue. USP-NF Online. http://www.uspnf.com

USP Monograph for Cyclophosphamide Tablets (2010). In: United States Pharmacopeia 2010, vol 33 reissue. USP-NF Online. http://www.uspnf.com

USP Monograph for Diclofenac Sodium Delayed-Release Tablets (2010). In: United States Pharmacopeia 2010, vol 33 reissue. USP-NF Online. http://www.uspnf.com

USP Monograph for Divalproex Sodium Delayed-Release Tablets (2010). In: United States Pharmacopeia 2010, vol 33 reissue. USP-NF Online. http://www.uspnf.com

USP Monograph for Ergotamine Tartrate Sublingual Tablets (2010). In: United States Pharmacopeia 2010, vol 33 reissue. USP-NF Online. http://www.uspnf.com

USP Monograph for Etodolac Extended-Release Tablets (2010). In: United States Pharmacopeia 2010, vol 33 reissue. USP-NF Online. http://www.uspnf.com

USP Monograph for Felodipine Extended-Release Tablets (2010). In: United States Pharmacopeia 2010, vol 33 reissue. USP-NF Online. http://www.uspnf.com

USP Monograph for Ferrous Fumarate and Docusate Sodium Extended Release Tablets (2010). In: United States Pharmacopeia 2010, vol 33 reissue. USP-NF Online. http://www.uspnf.com

USP Monograph for Flurbiprofen Tablets (2010). In: United States Pharmacopeia 2010, vol 33 reissue. USP-NF Online. http://www.uspnf.com

USP Monograph for Leucovorin Calcium Tablets (2010). In: United States Pharmacopeia 2010, vol 33 reissue. USP-NF Online. http://www.uspnf.com

USP Monograph for Megestrol Acetate Tablets (2010). In: United States Pharmacopeia 2010, vol 33 reissue. USP-NF Online. http://www.uspnf.com
USP Monograph for Mesalamine Delayed-Release Tablets (2010). In: United States Pharmacopeia 2010, vol 33 reissue. USP-NF Online. http://www.uspnf.com
USP Monograph for Metformin Hydrochloride Extended-Release Tablets (2010). In: United States Pharmacopeia 2010, vol 33 reissue. USP-NF Online. http://www.uspnf.com
USP Monograph for Metoprolol Succinate Extended-Release Tablets (2010). In: United States Pharmacopeia 2010, vol 33 reissue. USP-NF Online. http://www.uspnf.com
USP Monograph for Metoprolol Tartrate and Hydrochlothiazide Tablets (2010). In: United States Pharmacopeia 2010, vol 33 reissue. USP-NF Online. http://www.uspnf.com
USP Monograph for Mirtazapine Orally Disintegrating Tablets (2010). In: United States Pharmacopeia 2010, vol 33 reissue. USP-NF Online. http://www.uspnf.com
USP Monograph for Naproxen Delayed-Release Tablets (2010). In: United States Pharmacopeia 2010, vol 33 reissue. USP-NF Online. http://www.uspnf.com
USP Monograph for Nitroglycerin Sublingual Tablets (2010). In: United States Pharmacopeia 2010, vol 33 reissue. USP-NF Online. http://www.uspnf.com
USP Monograph for Oxtriphylline Delayed-Release Tablets (2010). In: United States Pharmacopeia 2010, vol 33 reissue. USP-NF Online. http://www.uspnf.com
USP Monograph for Oxybutynin Chloride Extended-Release Tablets (2010). In: United States Pharmacopeia 2010, vol 33 reissue. USP-NF Online. http://www.uspnf.com
USP Monograph for Oxycodone Hydrochloride Extended-Release Tablets (2010). In: United States Pharmacopeia 2010, vol 33 reissue. USP-NF Online. http://www.uspnf.com
USP Monograph for Pancrelipase Delayed-Release Capsules (2010). In: United States Pharmacopeia 2010, vol 33 reissue. USP-NF Online. http://www.uspnf.com
USP Monograph for Pantoprazole Sodium Delayed-Relcase Tablets (2010). In: United States Pharmacopeia 2010, vol 33 reissue. USP-NF Online. http://www.uspnf.com
USP Monograph for Procainamide Hydrochloride Extended-Release Tablets. In: United States Pharmacopeia 2010, vol 33 reissue. USP-NF Online. http://www.uspnf.com
USP Monograph for Pseudoepherine Hydrochloride Tablets (2010). In: United States Pharmacopeia 2010, vol 33 reissue. USP-NF Online. http://www.uspnf.com
USP Monograph for Pseudoephredrine Hydrochloride Extended-Release Tablets (2010). In: United States Pharmacopeia 2010, vol 33 reissue. USP-NF Online. http://www.uspnf.com
USP Monograph for Ranitidine Tablets (2010). In: United States Pharmacopeia 2010, vol 33 reissue. USP-NF Online. http://www.uspnf.com
USP Monograph for Theophylline Capsules (2010). In: United States Pharmacopeia 2010, vol 33 reissue. USP-NF Online. http://www.uspnf.com
USP Monograph for Theophylline Extended-Release Capsules (2010). In: United States Pharmacopeia 2010, vol 33 reissue. USP-NF Online. http://www.uspnf.com
USP Monograph for Theophylline Tablets (2010). In: United States Pharmacopeia 2010, vol 33 reissue. USP-NF Online. http://www.uspnf.com
USP Monograph for Thiabendazole Chewable Tablets (2010). In: United States Pharmacopeia 2010, vol 33 reissue. USP-NF Online. http://www.uspnf.com
USP Monograph for Tocainide Hydrochloride Tablets (2010). In: United States Pharmacopeia 2010, vol 33 reissue. USP-NF Online. http://www.uspnf.com
USP Monograph for Trihexyphenidyl Hydrochloride Tablets (2010). In: United States Pharmacopeia 2010, vol 33 reissue. USP-NF Online. http://www.uspnf.com
USP Monograph for Triprolidine Hydrochloride Tablets (2010). In: United States Pharmacopeia 2010, vol 33 reissue. USP-NF Online. http://www.uspnf.com
Vijaykumar K, Murthy PB, Kannababu S, Syamasundar B, Subbaraju GV (2006) Quantitative determination of corosolic acid in *lagerstroemia speciosa* leaves, extracts and dosage forms. Int J Appl Sci Eng 4(2):103–114
Wardrop J, Ficker D, Franklin S, Groski RJ (2000) Determination of erythromycin and related substances in enteric-coated tablet formulations by reversed-phase liquid chromatography. J Pharm Sci 89(9):1097–1105

Wrezel PW, Chion I, Pakula R (2005) Key factors in sample diluent selection for HPLC assays of active pharmaceutical ingredients. LCGC North America 29:107–111

Yahya AM, McElany JC, D'Arcy PF (1988) Drug sorption to glass and plastics. Drug Metabol Drug Interact 6(1):1–45

Zhao C, Furukawa S, Ohki Y (2009) A novel collagenase-assisted extraction of active pharmaceutical ingredients from gelatin products for quantitative analysis by high performance liquid chromatography. J Chromatogr A 1216:4524–4528

Chapter 8
Sample Preparation for Select Nonsolid Dosage Forms

Xin Bu, Sachin Chandran, John Spirig, and Qinggang Wang

Abstract Solid oral doses intended for the patient to swallow are the most common form of drug delivery, but formulations for other mechanisms of delivery exist to support patient needs. The state of the patient (e.g., inability to swallow), abilities of the patient to reliably self-dose (e.g., pediatric population, geriatric population), or even the physical properties of the drug (e.g., biologics) often require a nonsolid oral formulation. This chapter reviews sample preparation methods of dosage forms classified as oral liquids, semi-solids, nonoral solids, and parenterals. Products that fall into these classes include oral suspensions, syrups, oral solutions, creams, ointments, gels, topical powders, suppositories, transdermal systems, and injectables.

8.1 Introduction

Solid oral doses intended for the patient to swallow are the most common form of drug delivery (Hilfiker 2006), but formulations for other mechanisms of delivery exist to support patient needs. The state of the patient (e.g., inability to swallow), abilities of the patient to reliably self-dose (e.g., pediatric population, geriatric population), or even the physical properties of the drug (e.g., biologics) often require a nonsolid oral formulation. The following section reviews the special needs for sample preparation methods of dosage forms classified as oral liquids, semi-solids, nonoral solids, and parenterals. Products that fall into these classes include oral suspensions, syrups, oral solutions, creams, ointments, gels, topical powders, suppositories, transdermal systems, and injectables. To distinguish between these varied formulations, products are

X. Bu (✉)
Analytical Research and Development, Bristol-Myers Squibb Company,
1 Squibb Drive, New Brunswick, NJ 08903, USA
e-mail: xin.bu@bms.com

B. Nickerson (ed.), *Sample Preparation of Pharmaceutical Dosage Forms: Challenges and Strategies for Sample Preparation and Extraction*, DOI 10.1007/978-1-4419-9631-2_8,

grouped by both USP classification (USP General Chapter on Pharmaceutical Dosage Forms <1151> 2009) and suitable API extraction method.

Sample preparation is a critical part of dose analysis and directly impacts the quality of results. The purpose of sample preparation is to obtain a representative portion of the sample in a medium suitable for subsequent analysis. The extent of sample preparation required is highly dependent on the analytical technique; the more specific the analytical technique, the less sample clean up is required. Analytical techniques used to determine the quantity of the API (i.e., assay test) include chromatographic methods such as liquid chromatography (LC), gas chromatography (GC), and capillary electrophoresis (CE); spectroscopic methods, such as mass spectrometry (MS), ultraviolet (UV), and fluorescence spectroscopy, infrared (IR), and Raman spectroscopy; atomic absorption (AA), titration as well as microbial assays. Among these techniques, chromatographic methods, especially LC, are especially popular due to readily achievable specificity, accuracy, precision and robustness, as well as a reduced requirement on sample clean up. In addition to analytical technique, the following information are generally needed when developing sample preparation procedures: (1) the properties of the API, such as solubility in water and solvents, pH/solubility relationship (if ionizable), stability at ambient and elevated temperatures; and (2) the properties of the excipients, such as the composition and solubility of each formulation component.

8.2 Oral Liquids

Oral suspensions, syrups, and solutions are generally for pediatric patients or adults who experience difficulty swallowing. Commonly used excipients in the formulation of oral liquids include emulsifiers, thickening agents, suspending agents, stabilizers, buffers, preservatives, and flavorants. The solubility and physical properties of excipients may be different from that of the API, impairing effective extraction from the matrix. Sample preparation techniques for oral liquid formulations range from direct dissolution (Deicke and Süverkrüp 2000; Koundourellis et al. 2000; Aghazadeh and Kazemifard 2001; Heinänen and Barbas 2001; Suntornsuk 2001; Hood and Cheung 2003; Galli and Barbas 2004; Culzoni et al. 2005; Mishal and Sober 2005; Ali et al. 2006; Samanidou et al. 2006; El-Gindy et al. 2006, 2007; El-Sherbiny et al. 2007; Grosa et al. 2006; Johnson et al. 2007; Malesuik et al. 2008; Tagliari et al. 2008; Louhaichi et al. 2009; USP Monograph for Clindamycin Hydrochloride Oral Solution 2009; USP Monograph for Neomycin Sulfate Oral Solution 2009; USP Monograph for Ranitidine Oral Solution 2009; USP Monograph for Propoxyphene Napsylate Oral Suspension 2009; USP Monograph for Acetaminophen Oral Suspension 2009; USP Monograph for Cefpodoxime Proxetil for Oral Suspension 2009; USP Monograph for Famotidine for Oral Suspension 2009; USP Monograph for Ergocalciferol Oral Solution 2009) to liquid–liquid extraction (LLE) (Basavaiah and SriLatha 1999; Pires et al. 2004; Abdallah 2006; USP Monograph for Amantadine Hydrochloride Oral Solution 2009; USP Monograph for Chlorpromazine Hydrochloride Syrup 2009; USP Monograph for

Dihydrotachysterol Oral Solution 2009; USP Monograph for Erythromycin Estolate and Sulfisoxazole Acetyl Oral Suspension 2009; USP Monograph for Griseofulvin Oral Suspension 2009; USP Monograph for Haloperidol Injection 2009; USP Monograph for Simethicone Oral Suspension 2009; USP Monograph for Thioridazine Hydrochloride Oral Solution 2009) and solid phase extraction (SPE) (Uysal and Tuncel 2006; USP Monograph for Levocarnitine Oral Solution 2009). LLE and SPE are effective when direct spectroscopic detection of API is performed without first chromatographing the sample to separate API from the matrix. In general, multiple steps are involved in LLE and SPE, consuming time and resources and creating more chances for the analyst to make an error.

8.2.1 Suspension

Doses formulated as suspensions can be divided into oral suspensions, powder for oral suspensions, or suspensions prepared from powdered tablets or capsule contents. Inactive components in suspension formulations often include suspending/thickening agents, sweeteners, flavorants, colorants, buffers, and preservatives.

8.2.1.1 Sampling Procedure

Unlike traditional oral solid doses, such as tablets and capsules, where each unit dose is discrete, suspensions are often dispensed to patients in multi-dose containers. Patients measure each dose using a sample-measuring device. The oral suspension may settle to the bottom and powder for oral suspension may segregate in multi-dose containers after manufacture. Therefore, it becomes crucial to obtain representative samples precisely and reproducibly for accurate determination of potency of the product.

For oral suspensions or suspensions prepared from tablets/capsules, obtaining uniformity prior to sampling is necessary. Hand shaking, mechanical shaking, vortexing, or sonication is often necessary to ensure uniformity. Samples may then be taken volumetrically or gravimetrically (Mishal and Sober 2005; Ali et al. 2006; Samanidou et al. 2006; Johnson et al. 2007; Tagliari et al. 2008; USP Monograph for Acetaminophen Oral Suspension 2009; USP Monograph for Cefpodoxime Proxetil for Oral Suspension 2009; USP Monograph for Cefprozile for oral Suspension 2009; USP Monograph for Famotidine for Oral Suspension 2009; USP Monograph for Ganciclovir Oral Suspension 2009; USP Monograph for Ibuprofen Oral Suspension 2009; USP Monograph for Tetracycline Oral Suspension 2009). Volumetric sampling is comparatively less complex as a pipette can reliably extract an accurate volume. A positive displacement pipette may be used when a suspension is viscous, but is limited to small volume transfer (e.g., 1 mL or less). If the suspension is too viscous, however, pipetting may be less accurate than samples taken gravimetrically, especially if air bubbles are introduced by shaking or sonication. The advantage of gravimetric sampling is that potencies will not be biased due to suspension viscosity and entrapment of air bubbles. The weight of a unit dose sample is calculated based on formulation composition or an accurate measurement of suspension density.

To prepare samples from powders intended for oral suspension, weighing the equivalent to one unit dose of powder is the simplest sampling method (Deicke and Süverkrüp 2000; Aghazadeh and Kazemifard 2001; Malesuik et al. 2008; USP Monograph for Cefpodoxime Proxetil for Oral Suspension 2009). If the powder tends to segregate in its multi-dose container, however, it may be necessary to first constitute and form a suspension, then mix to obtain uniformity. Samples may then be taken from the resulting suspension using the volumetric or gravimetric methods noted (USP Monograph for Bacampicillin Hydrochloride for Oral Suspension 2009; USP Monograph for Famotidine for Oral Suspension 2009).

8.2.1.2 Extraction Methods

The API in suspension formulations is often intentionally dispersed as solid particles, rather than dissolved in aqueous medium, to mask the unpleasant taste of the API or afford better chemical stability. Addition of organic solvent (e.g., methanol, acetonitrile, etc.) in the diluent is, therefore, often necessary to dissolve suspended API. Solubility of API in the selected organic solvent should be sufficient to ensure complete dissolution at the target working concentration. pH modifiers (e.g., hydrochloric acid or trifluoroacetic acid) may also be used to increase API solubility.

Many excipients in suspensions are water soluble, such as sweeteners (e.g., sucrose), preservatives (e.g., parabens), and buffers (e.g., inorganic salts). Examples of insoluble or partially soluble excipients include stearic acid, iron oxide, titanium dioxide, and magnesium stearate. The diluent, often a mixture of aqueous and organic solvents that provides optimum extraction of API, also dissolves most of the excipients as well. If the dissolved excipients do not interfere with the subsequent analysis, no further modification to the diluent is necessary. However, if high quantities of dissolved excipients are incompatible with the subsequent analysis, it is necessary to adjust the solvent mixture or sample preparation procedure to extract API with a minimum amount of excipients in solution. Sometimes, several extraction and dilution steps are necessary to achieve quantitative extraction and minimize excipient interferences. An example of such an effect would be an API complexed with stearic acid in a powder for oral suspension. Stearic acid, a fatty acid, is insoluble in water but highly soluble in alcohol, such as methanol (The Merck Index 2006). Addition of pure methanol will dissolve the stearic acid and release the API from the complex. The API/methanol solution can then be subsequently diluted with aqueous diluent to induce precipitation of stearic acid. Filtration would then render such a solution suitable for LC analysis. The dilution with aqueous diluent to remove stearic acid is necessary to prevent it from precipitating on column and to match the solvent strength to that of mobile phase for the HPLC analysis.

Moderate force may be required to assist the dispersion and dissolution process. Mechanical shakers, sonicators, and vortexers are all suitable apparati. Special caution should be taken when dissolving unconstituted powder in diluent directly.

Powders for oral suspensions are usually very fine in order to obtain good uniformity and physical stability upon constitution (Ansel et al. 1999). Samples should be agitated immediately upon addition of diluent to prevent powder from caking, resulting in an incomplete extraction of API. Undissolved excipients that remain following agitation of the sample solution could be eliminated by filtration (for larger particles) (USP Monograph for Acetaminophen Oral Suspension 2009; USP Monograph for Tetracycline Oral Suspension 2009), centrifugation (for fine particles) (USP Monograph for Bacampicillin Hydrochloride for Oral Suspension 2009), or centrifugation followed by filtration (for bimodal particle distributions).

When a suitable diluent to extract API directly from the sample matrix cannot be identified, LLE can be performed (Basavaiah et al. 1999; Abdallah 2006; USP Monograph for Chlorothiazide Oral Suspension 2009; USP Monograph for Erythromycin Estolate and Sulfisoxazole Acetyl Oral Suspension 2009; USP Monograph for Mebendazole Oral Suspension 2009; USP Monograph for Simethicone Oral Suspension 2009; USP Monograph for Thioridazine Oral Suspension 2009). As noted previously, LLE procedures are often employed when subsequent analysis is a spectroscopic method, such as UV or IR. However, LLE followed by HPLC has also been reported (USP Monograph for Erythromycin Estolate and Sulfisoxazole Acetyl Oral Suspension 2009). The LLE process usually involves extraction from suspension with an organic solvent, phase separation, evaporation of organic solvent (if active is extracted into the organic layer), and reconstitution with suitable diluent for analysis by HPLC or UV spectroscopy.

8.2.2 Syrup

Medicated syrups are viscous solutions containing API and a concentrated sugar or sugar-substitute that may or may not contain flavorants. Since the active is already dissolved and distributed homogeneously in the formulation, sample preparation is generally straightforward and a quantitative dilution with appropriate diluent is often sufficient.

8.2.2.1 Sampling Procedure

Unlike suspensions, uniformity is not an issue in syrup formulations because the active ingredient is already in solution. Therefore, mixing prior to sampling is not necessary and not called for in the USP. Volumetric sampling by pipetting an aliquot of syrup is the most often used technique (USP Monograph for Chlorothiazide Hydrochloride Syrup 2009; USP Monograph for Docusate Sodium Syrup 2009; USP Monograph for Promazine Hydrochloride Syrup 2009). However, if the sample is viscous, gravimetric sampling may be more accurate and repeatable. If sampling gravimetrically, the density of the syrup should be determined to sample the equivalent of one dose (USP Monograph for Methenamine Mandelate for Oral Solution 2009).

8.2.2.2 Extraction Method

Similar to suspensions, diluents should be compatible with the sample and the subsequent analysis method. Syrups are aqueous formulations and direct dilutions with water or mobile phase may be used (Galli and Barbas 2004; Heinänen and Barbas 2001; Koundourellis et al. 2000). Acid, base, or organic modifiers can also be added to improve sample solution stability and to prevent precipitation (Hood and Cheung 2003; Culzoni et al. 2005).

Adequate forces are necessary to provide sufficient mixing of syrups with diluent to obtain a uniform solution. Frequently used mixing methods include hand shaking, vortexing, and sonication.

Similar to suspensions, when spectroscopic methods are used without chromatographic separation, LLE can be used to eliminate interference from the sample matrix (USP Monograph for Chlorpromazine Hydrochloride Syrup 2009; USP Monograph for Docusate Sodium Syrup 2009; USP Monograph for Promazine Hydrochloride Syrup 2009). A speedier alternative to LLE, SPE, can also be used to clean up the sample matrix if the presence of excipients is known to interfere with the detection method (Uysal and Tuncel 2006).

8.2.3 Oral Solution

Oral solutions are dispensed either as solution or as powder to be constituted with commercially available vehicles. Most oral solutions contain flavorants and colorants in addition to the active ingredient in an aqueous medium (Ansel et al. 1999). They may also contain preservatives and stabilizers.

8.2.3.1 Sampling Procedure

When reversed-phase HPLC is used for analysis, typical sample preparations in USP monographs for constituted solutions involve a quantitative dilution with appropriate diluent, such as water or mobile phase. Since the API is homogeneously distributed in solution, no mixing prior to sampling is required.

Oral solutions dispensed as powder may be prepared by either directly weighing an adequate amount of powder (USP Monograph for Methenamine Mandelate for Oral Solution 2009), or first constituting the powder in vehicle and then sampling from the resultant solution (USP Monograph for Clindamycin Hydrochloride Oral Solution 2009).

8.2.3.2 Extraction Method

The choice of extraction method depends on the analytical techniques employed for API detection. When HPLC or titration is the analytical method of choice, a simple dilution of the oral solution is often sufficient, as reflected in numerous USP monographs (USP Monograph for Caffeine Citrate Oral Solution 2009; USP Monograph for Clindamycin Hydrochloride Oral Solution 2009; USP Monograph for Cyclosporine Oral Solution 2009; USP Monograph for Fluphenazine Hydrochloride Oral Solution 2009; USP Monograph for Loperamide Hydrochloride Oral Solution 2009; USP Monograph for Phenobarbital Oral Solution 2009; USP Monograph for Prednisone Oral Solution 2009; USP Monograph for Ranitidine Oral Solution 2009). LLE is often performed to clean up sample matrices when spectroscopic methods are used for analysis (USP Monograph for Doxylamine Succinate Oral Solution 2009; USP Monograph for Haloperidol Injection 2009; USP Monograph for Levocarnitine Oral Solution 2009; USP Monograph for Mesoridazine Besylate Oral Solution 2009; USP Monograph for Mibolerone Oral Solution 2009; USP Monograph for Thiorisazine Hydrochloride Oral Solution 2009; USP Monograph for Valproic Acid Oral Solution 2009). Sample preparations using LLE are also reported in several USP monographs when chromatographic techniques such as GC or normal phase HPLC were used for analysis (USP Monograph for Dihydrotachysterol Oral Solution 2009; USP Monograph for Methadone Hydrochloride Oral Solution 2009; USP Monograph for Mibolerone Oral Solution 2009; USP Monograph for Valproic Acid Oral Solution 2009). Examples of extraction solvents include chloroform (USP Monograph for Mesoridazine Besylate Oral Solution 2009; USP Monograph for Phenobarbital Oral Solution 2009), methylene chloride (USP Monograph for Amantadine Hydrochloride Oral Solution 2009), ether (USP Monograph for Haloperidol Oral Solution 2009; USP Monograph for Methadone Hydrochloride Oral Solution 2009), and heptane (USP Monograph for Valproic Acid Oral Solution 2009).

8.3 Solid Nonoral Dosage Forms

Sample preparation of dosage forms that are solid at room temperature, but not administered orally, are discussed in the following section. Matrices for such dosage forms vary in complexity, thus sample preparations vary from direct dilutions to multi-step extractions requiring careful sample treatment. Once the API has been extracted, however, the resulting solutions are generally suitable for chromatography, titration, mass spectroscopy, or spectroscopic detection. Several minor exceptions are noted below.

8.3.1 Topical Powders

Topical powders are typically formulated with excipients such as zinc oxide, talc, starch, and kaolin (USP General Chapter on Pharmaceutical Dosage Forms <1151> 2009). These bulk ingredients serve to enhance topical coverage, absorb moisture, or act as palliatives when applied to the skin. Sample preparation procedures for topical powders are comparatively simple to other formulations as centrifugation or filtration tends to cleanly separate dissolved API from undissolved excipients.

8.3.1.1 Sampling Procedures

Topical powders are blended and packaged in containers for self-application by the patient. Blending powder to ensure uniformity of a marketed product is typically not required and not called for in any USP assay for topical powders. Early in product development, however, it is often necessary to test the content uniformity of a powder blend to ensure there has been adequate mixing. When testing uniformity, a powder riffler affords accurate sampling of the bulk powder (Venables and Wells 2002).

8.3.1.2 Extraction Methods

Excipients in topical powders are typically insoluble in most common diluents of pharmaceutical molecules. In some instances, diluent/excipient combinations will induce gelling of the sample matrix, e.g., water and starch (Newman 1996). Gelling will not necessarily impede analysis of the API, but will often require more mechanical force to separate dissolved API from gelled solutions.

Significant force is often required to ensure adequate separation of API from the powder matrix. Stirring (Cavrini et al. 1989; Costi et al. 2006), shaking (Cavrini et al. 1982; Mason and Crozier 1988; Omar and Abdelmageed 2006), sonication (Benjamin et al. 1983), and rotation (USP Monograph for Tolnaftate Topical Powder 2009) are all suitable methods to promote API dissolution into the diluent. Undissolved excipients that remain in solution must be segregated from API, as they are likely to harm chromatographic instrumentation and cause unpredictable scatter in spectrophotometric instrumentation (Blanco et al. 1999). For powders blended into a vehicle of uniformly large particle size excipients, gravimetric filtration over a sintered glass filter will adequately segregate the sample (USP Monograph for Neomycin Sulfate, Isoflupredone Acetate, and Tetracaine Hydrochloride Topical Powder 2009). Typically, however, additional force is required to obtain a clean sample solution. The majority of published preparations use centrifugation (USP Monograph for Miconazole Nitrate Cream 2009; USP Monograph for Miconazole Nitrate Topical Powder 2009), filtration (Cavrini et al. 1982, 1989; Costi et al. 2006; Mason and Crozier 1988; Omar and Abdelmageed 2006), or a combination of both (Benjamin et al. 1983) to obtain a solution of API free of insoluble excipients.

8.3.2 Suppositories

Suppositories contain API dispersed in either a hydrophilic or lipophilic matrix and may be used to deliver system-wide or local doses through dissolution into the anogenital or urethral mucosa. Sample preparation depends on the suppository base; suppositories formulated from polyethylene glycol (PEG), glycerin, or surfactant bases typically can be analyzed with minimal sample treatments while suppositories formulated from cocoa butter or cocoa butter substitutes require treatment to separate the API from the nonpolar matrix (USP Monograph for Progesterone Vaginal Suppositories 2009).

8.3.2.1 Sampling Procedures

Sampling techniques for suppositories are similar to that of oral tablets, with some additional steps taken for formulations prepared from fatty acid bases. The most straightforward sampling procedure is to place a single suppository into an extraction vessel (USP Monograph for Morphine Sulfate Suppositories 2009). If the formulation has a melting point reasonably above room temperature, multiple doses can be ground or mashed in a mortar and pestle and the resulting mass portioned for analysis (McEvoy et al. 2007; Li et al. 2008; USP Monograph for Ergotamine Tartrate and Caffeine Suppositories 2009). Should the suppository be particularly greasy or difficult for the analyst to create a uniform grind/mash, the suppositories can be submitted to a procedure where a number of doses are melted together, homogenized by stirring, and then cooled. Upon cooling, a portion of the melt is separated from the bulk and weighed for analysis (McEvoy et al. 2008). To facilitate handling, the melt can be placed in a refrigerator or freezer for a short period of time to create a brittle mass that will be easier for the analyst to break off sections to sample (Mohammadi et al. 2004; Zhang et al. 2007).

Finally, some PEG-based suppository preparations are intended for patient use only after first wetting with water. Early in development, it is advisable to assay both pre-wet and unadulterated suppositories to ensure that the wetting procedure does not adulterate the dose.

8.3.2.2 Extraction Methods

Direct extraction from suppositories prepared from both lipophilic and hydrophilic bases is possible. When the drug is extractable in a solvent in which the suppository base is soluble, it is possible to generate a clean sample solution by direct dilution (Haney and Dash 1997). Even if the suppository appears to dissolve completely in the selected solvent, it is advised that the preparation be agitated by sonication or shaking to break up API/base agglomerates that may have formed during storage, especially for lipophilic preparations in early development. If the drug is robust against heat degradation and extracted in a solvent in which the base is insoluble, heating is recommended to disperse as much of the base as possible and expose API to the diluent (Guneri et al. 2004). Should heating in the extraction solvent produce a large quantity of insoluble material, the remaining lumps can be removed from the initial extraction vessel and washed with solvent to ensure maximum recovery (Pedraza et al. 2006). To reduce sample preparation times and ensure uniform sample treatment, a focused microwave assisted extraction (FMAE) unit can be used to heat the sample and rapidly disperse the suppository base (Labbozzetta et al. 2005). Filtration is recommended for any method of direct extraction to avoid interference with measurement methods or damage to analytical instrumentation (Zhang et al. 2007).

LLE, wherein two immiscible solvents are introduced to solvate and separate the hydrophilic API from the lipophilic base, are particularly useful for suppositories created from fatty acids. The classical LLE, executed with a single dosage-equivalent in a separatory funnel, remains the recommended procedure for numerous products in the USP (USP Monograph for Indomethacin Suppositories 2009). Of interest to the analyst with pressures to handle a large volume of samples is an ever-expanding (Silvestre et al. 2009) list of LLE-based automation methods (Priego-Capote and Luque de Castro 2003). A variant of the classic LLE procedure is to first dissolve the suppository base and disperse the API in a nonpolar diluent. After agitating the preparation, the nonpolar solvent is removed by pipette and the remainder is evaporated by aspiration in air (USP Monograph for Miconazole Nitrate Vaginal Suppositories 2009) or introduction of vacuum (Zhang et al. 2007). The resulting mass contains dispersed API and can be extracted in an appropriate solvent and filtered for analysis.

A refinement of traditional supercritical fluid extraction (SFE), inverse SFE, is also well-suited for the analysis of lipophilic-based suppositories as the supercritical solvent is used to remove the nonpolar base, leaving behind polar API that can be collected for analysis (Almodovar et al. 1998). Although technically more complex, the authors note that inverse SFE is faster than LLE methods where an emulsion is known to form in the separator, such as the USP method for acetaminophen (USP Monograph for Acetaminophen Oral Suspension).

Finally, an emerging set of surfactant-stabilized microemulsion-based methods offer the promise of rapid extraction with minimal sample treatment. Microemulsions can solubilize both polar and nonpolar compounds as they express the solvent properties of both oils and water. Suppositories with lipophilic bases can be dissolved directly without having to separate partitioned layers or undissolved bulk. Microemulsion liquid chromatography (MELC) (Marsh et al. 2005; McEvoy et al. 2007) and microemulsion electrokinetic chromatography (MEEKC) (McEvoy et al. 2008; Ryan et al. 2009) allow for the direct analysis of suppositories dissolved in the aforementioned manner, offering the simplicity of direct extraction as well as a reduction in total analysis time.

8.3.3 Transdermal Systems

API introduced via a transdermal system is intended for system-wide dosing and the rate of introduction is controlled by the drug migration from the system through the skin and into the bloodstream. Tapes and plasters are solid systems with a dose that is delivered topically, rather than system-wide, but share common sample preparation procedures to transdermal systems.

Sample preparation of transdermal systems tends to differ by system formulation; drug-in-adhesive type, matrix type, and reservoir-type system contain API suspended in different types of matrices. As the name suggests, drug-in-adhesive transdermal systems contain a dose of API suspended in the skin adhesive. In matrix-type systems, the API is suspended in a polymer matrix. Reservoir type systems contain a discrete compartment wherein the API is exposed to a polymeric rate-controlling membrane. Note that research into transdermal drug delivery is expanding and there are numerous other types of specialized designs (Aggarwal and Dhawan 2009). If analyzing a dose from a transdermal system that defies classification amongst the aforementioned types, pick a method of analysis based on the matrix in which the API is suspended, be it polymeric, adhesive, or simply bound by the container.

8.3.3.1 Sampling Procedures

Although there are a wide variety of transdermal systems, sampling procedures are relatively plain. Preparing samples from systems where the API is extracted directly or the API and the suspension medium arc extracted simultaneously, the sampling procedure is to simply remove the protective layer and place the dose in the extraction vessel (Carlisle et al. 1992; Walters et al. 1995; Takashina et al. 2009).

In matrix-type systems, removing the protective layer and cutting the system into small pieces prior to extraction improves the rate at which solvent can penetrate the polymer matrix and solubilize the API (Van Nimmen and Veulemans 2007). Cutting the patch can also be useful for drug-in-adhesive as it inhibits folding when the dose is exposed to extraction solvent (USP Monograph for Estradiol Transdermal System 2009; Edwardson and Gardner 1990). Cleaning-verification type analyses (Liu and Pack 2007), wherein cutting implements are swabbed with extraction solvent, should be performed to determine whether the cutting process transfers API to the cutting medium.

Finally, special care must be taken early in development, especially in analysis of stressed stability samples, to test every packaging component in which the dose has been in contact for the presence of active. The true potency of the dose is a sum of the drug extracted from the backing laminate, release liner, membrane/scrim, or other packaging. Typically, if a component (most typically for marketed product, the liner) is known to retain API, methods will be written wherein extraction takes place on both the liner and the system simultaneously (Klaffenbach et al. 1998).

8.3.3.2 Extraction Methods

Extraction methods vary widely, but can be roughly grouped together by both type of extraction (direct, LLE, etc.) and type of patch (drug-in-adhesive, matrix, reservoir).

The most straightforward of the direct extractions is that of API from reservoir-type systems. Some patch formulations feature a drug reservoir that can be probed with a syringe, allowing the drug to be extracted as-received, or after first infusing the reservoir chamber with a compatible solvent (Takashina et al. 2009). When using either method, it is recommended to flush the reservoir chamber numerous times with solvent infusions to ensure maximum API recovery. For reservoirs that cannot be probed by syringe, the reservoir and rate-controlling membrane can be digested in an appropriate solvent or combination of solvents until only the backing laminate remains (USP Monograph for Clonidine Transdermal System 2009). The resulting solution will contain a complex mixture of drug and excipients and should be cleaned by filtration/centrifugation to obtain a solution suitable for detection.

Direct extraction from drug-in-adhesive types can be performed with or without dissolving the adhesive along with API. Adhesives tend to dissolve in nonpolar solvents and would require a LLE to cleanup if LC is the intended detection mechanism. By employing GC, the adhesive and API can be extracted simultaneously in a mixture of polar/nonpolar organic diluents and analyzed without further sample preparation (Klaffenbach et al. 1998). Should the drug be only marginally soluble in the adhesive, it is possible to perform an extraction by selecting a solvent in which the API is soluble, but in which the adhesive in insoluble (Li 2002). In extraction where the adhesive is not solubilized, it is often necessary to employ long sonication or shake times to ensure maximal API recovery. Note that as solvent removes API or adhesive from drug-in-adhesive type transdermal systems, there is a propensity for the dose to fold and lose contact with the extraction solution. If excessive folding or warping of the patch is observed, a different solvent system should be selected, or the extraction should be performed in a container where the solvent completely engulfs the folded patch.

Matrix-type transdermal systems can also be extracted directly, but the manner in which that extraction proceeds depends on the sampling procedure. For patches that have been left intact, the matrices are swelled with an API-solubilizing solvent and sonicated/shaken for long periods to ensure maximal API recovery (Li 2002; Li et al. 2008). If the patches are first sectioned by cutting into small pieces, the extraction can be performed in a similar manner, but likely in a shorter period as there is less of a diffusion barrier for the solvent to be exposed to API suspended in the polymer matrix (Mittal et al. 2009; Van Nimmen and Veulemans 2007). Although the solution will likely contain less undissolved mass than patches that have been digested thoroughly, filtration is still recommended to avoid interference with detection. Note that if the API is robust against heat degradation, gentle heating can be used during the extraction to hasten release from matrix-type patches (Gao et al. 2009).

LLEs can be used for matrix (Walters et al. 1995), drug-in-adhesive (Edwardson and Gardner 1990; Carlisle et al. 1992), tapes (Murakami et al. 2008), or plasters (Liu et al. 2008). In each case, the layer containing the suspended API, either adhesive or polymer, is extracted first in a compatible nonpolar solvent. Then, API is extracted using a compatible polar solvent. As with other LLE methods noted earlier, filtration/centrifugation is recommended to remove undissolved adhesive/matrix prior to detection.

A more recent procedure for reservoir-type patches employs SPE to separate API from a complex matrix (Van Nimmen and Veulemans 2007). In this work, patches were first sectioned by cutting with scissors. To ensure that drug was not transferred to the scissors during sectioning, the scissors were wiped with solvent-soaked wipes that were added to the SPE cartridge along with the system. By using SPE to extract the API from the all-in-one matrix, the authors avoided having to create multiple samples from each lab implement in which the patch came into contact during sample preparation.

8.4 Semi-Solid Dosage Forms

Semi-solid dosage forms include ointments, creams, gels, and pastes. They are mainly for topical application to the skin or mucous membranes, and can be used to treat either dermatological ailments or provide systemic therapy (Shah et al. 1992). The APIs in semi-solid formulations are mainly small molecule compounds, with a few exceptions such as inorganic materials (e.g., zinc oxide) or proteins (e.g., becaplermin). The classification of semi-solid dosage forms is based on the base used in formulation. Ointment bases are classified into four general classes in USP: the hydrocarbon bases, the absorption bases, the water-removable bases, and the water-soluble bases (USP General Chapter on Pharmaceutical Dosage Forms 2009). While formulations under the first two classes are still referred to as ointments, the water-removable bases are more correctly referred to as creams and the water-soluble bases are more correctly referred to as gels.

8.4.1 Ointments

Ointments are semisolid formulations with hydrocarbon or absorption bases. Hydrocarbon bases are made of oleaginous materials. White Petrolatum USP and White Ointment USP are two typical examples of hydrocarbon bases. Absorption bases may be further divided into two subgroups: the first group includes those that permit the incorporation of aqueous solutions to form a water-in-oil emulsion (e.g., Hydrophilic Petrolatum USP); and the second group includes those that are water-in-oil emulsions, and permit the incorporation of additional quantities of aqueous solutions (e.g., Lanolin USP). Such water-in-oil emulsions can also be referred to as cold creams.

8.4.1.1 Sampling Procedure

Owing to their high viscosity, ointment samples containing a representative amount of API typically are measured gravimetrically into a suitable container. Sample homogeneity is generally not an issue, and neither special treatment of the ointment sample is needed before weighing nor called for in any USP procedure.

8.4.1.2 Extraction Methods

LLE is the most commonly used technique to extract API from ointment samples (Izumoto et al. 1997; Wang et al. 2008; Huidobro et al. 2009; Zheng et al. 2009; USP Monograph for Alclometasone Dipropionate Ointment 2009; USP Monograph for Flurandrenolide Ointment 2009; USP Monograph for Neomycin Sulfate Ointment 2009). Nonpolar lipophilic materials, such as petrolatum and lanolin, are the major components in these ointment bases, and are practically insoluble in water and polar organic solvents. Certain nonpolar solvents, such as hexane, can be added to dissolve the oily material. Alternatively, as the melting point is between 38 and 60°C for petrolatum, and 38–44°C for lanolin, heating the ointment sample in a water bath slightly above its melting point can also be used to disperse the base (Nina et al. 1988; Tjornelund and Hansen 1997; Dallet et al. 2009; Zheng et al. 2009; USP Monograph for Dibucaine Ointment 2009; USP Monograph for Nitroglycerin Ointment 2009; USP Monograph for Fluticasone Propionate Ointment 2009).

Extraction solvents are selected predominantly on their ability to solubilize API. Polar organic solvents, such as methanol, acetonitrile, or their mixtures with water, are used commonly for extraction. If the API is an acid or a base, aqueous solubility varies significantly with pH. Therefore, the pH value of the extraction solvent can be adjusted to enhance extraction efficiency. For example, Dibucaine, a basic compound with a pK_a of 8.7, is extracted from the ointment sample in 0.1 N hydrochloric acid, as described in the USP monograph (USP Monograph for Dibucaine Ointment 2009). Besides pH adjustment, ion-pairing reagents have also been investigated to improve the extraction efficiency for ionic compounds (Hoogewijs and Massart 1983).

After addition of extraction solvent, mechanical forces such as vortexing or vigorous shaking may be used to prompt extraction before phase separation. If heating is used, the mixture typically is cooled down to allow the oily material to congeal. Centrifugation could be used to assist phase separation, as well as to break down any emulsion that may form. Filtration with a suitable membrane filter may be necessary to further clean up the solution before chromatographic analysis. If the extraction solvent is not compatible with the subsequent analysis, it may be evaporated and the residuals dissolved in a compatible solvent.

It may be necessary to repeat the extraction steps several times to achieve satisfactory recovery. If chromatographic methods are used for subsequent analysis, an internal standard can be used to compensate for an incomplete extraction. Two factors should be considered when selecting an internal standard. First, the internal standard should have a similar distribution coefficient with the API to mimic the loss of the API during the extraction process. A structurally similar analog is commonly used. Second, the internal standard should not introduce any interference in the subsequent chromatographic analysis.

A typical LLE procedure is used to prepare Clobetasol Propionate ointment for a reversed-phase LC-UV method, as described in the USP monograph (USP Monograph for Clobetasol Propionate Ointment 2009). The ointment is dispersed in hexane and extracted by shaking with methanol containing Beclomethasone Dipropionate as an internal standard. After collecting the methanol layer, the remaining hexane is subjected to two additional extractions with mobile phase.

All three extracts are combined, quantitatively diluted with mobile phase, and filtered before analysis.

When less discriminative techniques, such as spectroscopic methods or titration are used, a multi-step LLE procedure might be needed to remove matrix interference. For example, a solution of Tetracaine ointment is assayed with UV spectroscopy at 310 nm after a three-step LLE (USP Monograph for Tetracaine ointment 2009). For APIs without chromophores, derivatization is needed for direct UV spectroscopic analysis or LC-UV. Such a procedure was described for mechlorethamine in three types of ointment formulations (Reepmeyer 2005). Mechlorethamine hydrochloride was derivatized with benzenethiol to form the disubstitution product (a tertiary amine). After derivatization, the product was extracted into an acidified aqueous phase to remove oil-soluble excipients, then back-extracted into heptane by alkalifying the aqueous phase to remove water-soluble excipients before analysis by normal phase LC.

The main disadvantage of LLE is that it is labor-intensive. Besides LLE, other extraction methods have also been used or evaluated. Direct solvent extraction is desired, if feasible. For example, Diflorasone Diacetate ointment can be directly dissolved in chloroform, centrifuged, and then analyzed by normal phase LC (USP Monograph for Diflorasone Diacetate ointment 2009).

Solid-phase extractions (SPE) using disposable cartridges filled with different types of sorbents also have been studied for ointment sample preparations (Nguyen et al. 1986; Cavrini et al. 1989; Bonazzi et al. 1995; Di Pietra et al. 1992, 1996; Cardoso et al. 2000). Because of the presence of a large amount of lipophilic material in ointments, normal-phase type sorbents (e.g., silica, diol, and aminipropyl) are typically used. Mixtures of hexane and methylene chloride can be used as loading and washing solvents as they are weak solvents for normal-phase sorbents and provide good solubility for lipophilic materials. Polar solvents, such as methanol, can be used as eluting solvents.

SFEs with carbon dioxide or carbon dioxide/alcohol mixtures can reduce the consumption of hazardous solvents, but have found very limited application in the analysis of semi-solid formulations (Karlsson et al. 1997). Most pharmaceutical compounds are polar and therefore exhibit low recovery when extracted with nonpolar carbon dioxide. To improve recovery, an inverse SFE procedure was developed where the matrix is removed from the polar API with supercritical carbon dioxide (Messer and Taylor 1994; Moore and Taylor 1994). In a similar approach, diatomaceous earth powder was added to retain polar analytes, Retinol Plamitate and Tocopherol Acetate, while the matrix was removed with SFE using pure carbon dioxide. Analytes were then eluted from the trap using ethanol-modified carbon dioxide (Masuda et al. 1993).

8.4.2 Creams

Creams are traditionally referred to as semi-solid dosage forms that possess a relatively fluid consistency and are formulated as either water-in-oil emulsions or oil-in-water emulsions. More recently, however, creams have been restricted to products consisting of oil-in-water emulsions that are characterized by an aqueous external phase and an oily internal phase (USP General Chapter on Pharmaceutical Dosage Forms 2009). In addition to water and oily material, common inactive ingredients include emulsifying agents, stiffening agents, preservatives, and antioxidants. An example of such a cream base is Hydrophilic Ointment USP. It contains methylparaben and propylparaben as preservatives, sodium lauryl sulfate as the emulsifying agent, stearyl alcohol as the stiffening agent, white petrolatum to form the oily internal phase, and water for the external phase.

8.4.2.1 Sampling Procedure

Similar to ointment samples, gravimetric sampling is commonly used and neither a pretreatment is needed before weighing, nor called for in any USP cream procedure.

8.4.2.2 Extraction Methods

Creams are oil-in-water emulsions and are much more hydrophilic than ointments. Direct extractions with polar solvents or their aqueous mixtures are widely used to extract polar APIs from cream samples (Garcia et al. 2005; Gupta et al. 2005; Hamoudova and Pospisilova 2006; Kuehl et al. 2006; Novakova et al. 2006; USP Monograph for Hydrocortisone Acetate Cream 2009; USP Monograph on Lidocaine and Prilocaine Cream 2009; USP Monograph for Meclocycline Sulfosalicylate Cream 2009; USP Monograph for Miconazole Nitrate Cream 2009; USP Monograph for Tretinoin Cream 2009). Factors that need to be considered in selection of extraction solvent include API solubility, solubility of the oily internal phase, and compatibility with subsequent analysis. Commonly used polar organic solvents for the extraction of API from creams include methanol, acetonitrile, tetrahydrofuran, or their mixtures with water. The pH of extraction solvents can be adjusted to improve the solubility of acidic or basic APIs. If the solubility of the oily internal phase in the extraction solvent is low, lumps may form that can often lead to lower recovery. A comparison of the solubility of 11 commercial placebo bases in tetrahydrofuran, methanol, and acetonitrile, indicated that tetrahydrofuran was superior to the other two (Haikala et al. 1991). Gently heating the sample after addition of extraction solvents can also help to avoid lump formation (USP Monograph for Clotrimazole Cream 2009; USP Monograph for Fluocinonide Cream 2009; USP Monograph for Naftifine Hydrochloride Cream 2009). Mechanical forces, such as vortexing, shaking, or sonication, are often necessary to disperse cream matrices and assist extraction. Alternatively, a FMAE unit can be employed to more rapidly extract API from the sample matrix (Labbozzetta et al. 2008). Centrifugation or filtration may be used to clean up the sample after extraction. If necessary, a second dilution with mobile phase or water can be used to decrease the diluent solvent strength when reversed-phase LC is used for analysis (USP Monograph for Miconazole Nitrate Cream 2009).

A typical direct solvent extraction procedure is used to prepare Tretinoin cream sample for reversed-phase LC analysis, as described in the USP monograph (USP Monograph for Tretinoin Cream 2009). The cream sample is first dispersed in tetrahydrofuran by shaking, then subjected to filtration to remove any undissolved materials. After filtration, a portion of the sample solution is diluted further with a mixture of tetrahydrofuran and 1% phosphoric acid to improve compatibility with the starting mobile phase conditions during chromatographic analysis.

LLE procedures are also used for cream samples (USP Monograph for Mafenide Acetate Cream 2009; USP Monograph for Piroxicam Cream 2009). An example of the utility of LLE in analyzing multiple cream formulations is a screening procedure for the detection of 49 corticosteroids in topical pharmaceutical products, such as creams and gels by reversed-phase LC-UV (Reepmeyer 2001). Two sequential extraction procedures were developed to isolate corticosteroids from both water soluble and lipid soluble matrix components. The first procedure was performed with ethyl acetate and 0.1 M citric buffer saturated with sodium chloride to remove water soluble excipients and leave the corticosteroids in ethyl acetate. The second procedure was performed with acetonitrile/water (9/1) and heptane to remove lipid soluble excipients and leave the corticosteroids in acetonitrile/water.

SPE have also been studied for preparation of cream samples (Bonazzi et al. 1995; Di Pietra et al. 1996; Cardoso et al. 2000). Because of the presence of a large amount of water in the ointment samples, reversed-phase type sorbents, such as C18, and ion-exchange type sorbents are particularly useful. Mixtures of water and small amounts of methanol are used as loading and washing solvents, while mixtures with increased amounts of methanol are often used as eluting solvents. The pH of both loading and eluting solvents can be adjusted accordingly to facilitate the retention of API during loading/washing cycle or the elution of the API during the eluting cycle.

8.4.3 Gels

Gels are semi-solid systems consisting of a suspension of small distinct particles or large organic molecules interpenetrated by a liquid. The jelly-like feature of gels is due to the presence of a gelling agent. Common gelling agents include synthetic macromolecules, such as carbomer and cellulose derivatives, or natural gums, such as tragacanth (USP General Chapter on Pharmaceutical Dosage Forms 2009). Besides water and gelling agent, other inactive ingredients include solvents, preservatives, antioxidants, and stabilizers.

8.4.3.1 Sampling Procedure

Similar to ointment and cream samples, gravimetric sampling is commonly used and neither a pretreatment is needed before weighing, nor called for in any USP gel procedure.

8.4.3.2 Extraction Methods

Most excipients used in gel formulations are water soluble. Direct solvent extraction with polar solvents or mixtures with water are the most commonly used technique to prepare gel samples for assay testing (Garcia et al. 2005; Hamoudova and Pospisilova 2006; USP Monograph on Dyclonine Hydrochloride Gel 2009; USP Monograph for Erythromycin Topical Gel 2009; USP Monograph for Metronidazole Gel 2009; USP Monograph on Tolnaftate Gel 2009; DiNunzio and Gadde 1992). Extraction solvents are selected predominantly on their ability to solubilize API. Polar organic solvents, such as methanol, acetonitrile, or their mixtures with water are commonly used. The pH value of the extraction solvents can be adjusted to improve the solubility of acidic or basic APIs. In many cases when an LC analysis method is used, mobile phase can be used for extraction. Mechanical forces, such as vortexing, vigorous shaking, or sonicating, can be used to disperse the gel matrix and to prompt extraction. Centrifugation or filtration can be used to clean up the sample after extraction.

A typical direct solvent extraction procedure is used to prepare Clindamycin Phosphate Gel samples for reversed-phase LC analysis (USP Monograph for Clindamycin Phosphate Gel 2009). The gel sample is diluted with mobile phase. After shaking by mechanical means, a portion is centrifuged and filtered before analysis.

LLE is occasionally used for gel samples, especially when less discriminative techniques, such as spectroscopic methods are used (USP Monograph for Tolnaftate Topical Powder 2009).

8.5 Parenteral Dosage Forms

Parenteral dosage forms are preparations intended for injection through the skin or another external boundary tissue so that the active substances contained are administered directly into a blood vessel, organ, tissue, or lesion (USP General Chapter on Injections <1> 2009). These systems offer unique options for the delivery of lipophilic and poorly bioavailable drugs that cannot be administered by other routes. The most common parenteral routes are subcutaneous, intravenous, and intramuscular, the choice of which depends on the drug and required mode of delivery. Parenteral systems can consist of water-soluble liquid injections, solids that can be constituted such that they conform to requirements of an injection, injectables emulsion, injectable suspension and lastly, drug solids that upon addition of vehicles conform to requirements for an injectable suspension. Typical sample analysis and quantitation methods for parenterals are chromatography (reverse-phase/affinity), optical techniques (UV absorbance/fluorescence/circular dichroism), or spectroscopic techniques such as FTIR or Raman spectroscopy.

8.5.1 Analysis of Water-Soluble Parenterals

Water-soluble parenterals form a broad class of molecules ranging from inorganic salts such as $MgSO_4$ and NaCl, small molecules such as bupivacaine and histamine phosphate, vitamins such as folic acid, peptides such as oxytocin, glycosaminoglycans such as heparin, and proteins such as antithrombin III human. Despite the different classes of molecules involved in the analysis, all of these classes of molecules exhibit good solubility in primarily aqueous buffers.

8.5.1.1 Sampling Procedures

Aqueous-soluble parenterals are sterile solutions in water or buffer, or lyophilized powders that have to be reconstituted at the time of use. If the injection is in the form of a sterile solution, a relevant volume can be pipetted into the extraction vessel. If the sample is in the form of a lyophilate powder, then a suitable amount can be weighed out and constituted in aqueous buffer.

8.5.1.2 Extraction Procedures

Water-soluble injectables have appreciable solubility in aqueous buffers and therefore, extraction procedures are typically not elaborate. The primary concern of the analyst is the choice of buffer, the role of properties such as pH, osmolarity, and ionic strength of the extraction solvent.

In the case of injectable inorganic salts, colorimetric assays following complexation with ionochromic dyes are available for determination of metallic ions (Durham and Walton 1983; Hattori and Yoshida 1986, 1987). Owing to the simplicity of the matrix, sample preparation for colorimetric assays is limited to preparative work involved for a titration. For example, the assay for Ca^{2+} in a calcium chloride injection (USP Monograph for Calcium Chloride Injection 2009) involves the accurate transfer of a known amount of calcium chloride, followed by addition of acid and water with a colorimetric titration using sodium edetate.

For small molecules such as bupivacaine that can be constituted in aqueous buffers, stirring or sonication may be needed to dissolve the API (USP Monograph on Bupivacaine Hydrochloride in Dextrose Injection 2009). Occasionally, for molecules such as histamine phosphate where the buffer interferes with the API/dye complexation, sample preparation requires the additional step of removal of the original buffer on a steam bath and reconstitution of the dry residue into an analysis buffer (USP Monograph on Histamine Phosphate in Dextrose Injection 2009). For water-soluble vitamins, HPLC methods have been described that involve dissolution of the vitamins in water followed by direct injection (van der Horst et al. 1989). HPLC methods are a convenient approach for determination with a considerable number of examples in literature (USP Monograph on Bupivacaine Hydrochloride in Dextrose Injection 2009; USP Monograph for Haloperidol Injection 2009). For proteins such as Antithrombin III Human that are injected parenterally, sample analysis can be carried out by UV absorbance (USP Monograph for Antithrombin III Human 2009). Sample preparation would involve simply dissolving the protein sample in water or suitable buffer. Care should be taken to control pH since the extinction coefficient can significantly vary with pH. In the case that UV is not a suitable approach, LC may also prove to be a useful technique.

8.5.2 Analysis of Parenterals That Are Suspensions and Emulsions

In certain cases, the injectable formulation is dosed either as a suspension or an emulsion. In rare cases, the injectable formulation is a nonaqueous solution in which case it has to be suitably extracted before analysis. Like the aqueous-soluble parenterals, injectables that fall into this class come in a wide range from silicone-coated iron oxide to small molecules dissolved in oils. However, unlike the water-soluble systems, sample preparation and analysis require additional steps to extract API.

8.5.2.1 Sampling Procedures

Sampling is typically carried out by weighing the requisite amount for the assay if the drug product is present in solid form. If the sample is present as a suspension or emulsion, the lower viscosities allow for samples to be withdrawn volumetrically for further extraction.

8.5.2.2 Extraction Procedures

Often a small molecule has very poor solubility in water and is formulated in an organic vehicle to achieve solubility (Nema et al. 1997; Strickley 2004). Some of these vehicles such as PEG 300, ethanol, Tween 80, and Cremophor EL are miscible in water and do not pose a large challenge for sample analysis. In these cases, the choice of a proper organic solvent to water ratio is required such that the entire sample can be dissolved.

Extraction procedures for suspensions and emulsions fall under a few different classes. In certain cases, no significant extraction is required. For example, ferumoxsil oral (USP Monograph for Ferumoxsil Oral Suspension 2009), a silicone-coated superparamagnetic iron oxide requires dissolution in a mixture of oral suspension and water, followed by gentle inversion mixing. UV absorbance can be determined directly from the dissolved API. The preparation for Isophane Insulin Human suspension requires centrifugation of about 10 mL of the suspension at

1,500 × *g* for 10 min to remove the zinc that is present in the form of zinc-insulin crystals. Agitation is followed by the evaluation of the supernatant by UV-absorbance or HPLC to determine the amount of insulin (USP Monograph for Isophane Insulin Human Suspension 2009). In rare cases, when the injection is formulated as a suspension or emulsion that can be completely dissolved into an extraction solvent (e.g., phytonadione injectable emulsion), the sample is dissolved in dehydrated alcohol for further analysis by UV-absorbance or HPLC (USP Monograph for Phytonadione Injectable Emulsion 2009).

Some extraction procedures involve the removal of the liquid content in the suspension by filtration so that the solid can be analyzed. For example, in the case of dexamethasone acetate injectable suspension (USP Monograph for Dexamethasone Acetate Injectable Suspension 2009), the contents of the container are transferred to a fine-porosity, sintered-glass vacuum filter, and filtered by washing several times with 10 mL portions of water to remove the water-soluble excipients, after which the powder is air-dried. The chemical stability of the API should be considered before heat is used for drying.

For systems that are emulsions or solubilized in oil, the extraction procedure will be required for separating the polar API from nonpolar matrix for analysis. Solvents used to solubilize the nonpolar matrix for LLE of injectables include methylene chloride, chloroform, and ether (USP Monograph for Haloperidol Injection 2009; USP Monograph for Hydrocortisone Injectable Suspension 2009; USP Monograph for Valproate Sodium Injection 2009). An example is that of hydrocortisone injectable suspension, which is extracted using chloroform, after which the chloroform layer is separated and evaporated, and the residue dissolved in alcohol for analysis (USP Monograph Hydrocortisone Injectable Suspension 2009). Valproate sodium injection, a sterile solution extracted by LLE, employs strong shaking for extraction of the drug (USP Monograph for Valproate Sodium Injection 2009). A similar approach is useful for cephapirin benzathine, which is dissolved in oil (USP Monograph for Cephapirin Benzathine Intramammary Infusion 2009) or haloperidol (USP Monograph for Haloperidol Injection 2009). Care should be taken such that an emulsion does not form in these cases, and that they should be broken up either by stirring with a rod, or by ultra-centrifugation before analysis. In some cases, precipitates may form during the extraction process and they should be filtered. While filtering, enough sample should be discarded to account for adsorption onto the filter. The filtrate can then be subsequently analyzed by a suitable analysis.

8.5.3 Sample Preparation and Analysis for Novel Polymeric Injectables

In recent history, drug delivery has seen the advent of novel polymeric devices for the injection of hydrophobic drugs into the human body (Duncan 2003). The advantage of these systems is that they provide a viable technique for sustained release of the drug, in addition to mechanisms based on targeting ligands that can uniquely bind to various targets within the body. However, such polymeric systems require slightly different sample preparation when compared with the systems described in the earlier sections as the API may be either chemically bound to a large polymer, or entrapped within a polymeric structure, and may therefore not be directly available in a free form.

8.5.3.1 Sampling Procedures

Typically, polymeric injectables are either soluble or form suspensions in aqueous buffer. In certain cases, the dosage form exists as a suspension, in which case, a representative sample can be withdrawn by rapid agitation followed by pipetting from the aqueous layer.

8.5.3.2 Extraction Procedures

PEG has been the polymer of choice for the attachment of small molecules and proteins with the primary goal of increasing solubility and circulatory half-life in vivo (Caliceti and Veronese 2003; Pasut and Veronese 2009). PEG is soluble in water, methanol, benzene, acetonitrile, chloroform, and dichloromethane. Aqueous solubilization of hydrophobic small molecules or proteins bound to PEG is achieved by the PEG arm of the conjugate driving the rest of the molecule into solution (Caliceti and Veronese 2003; Duncan 2003; Greenwald et al. 2000). Sample preparation for these and other water soluble polymer-bound systems begins by dissolution in water or aqueous buffer aided by stirring or sonication. Direct detection by UV absorption is possible for solutions prepared by direct extraction. In certain cases, PEG-based systems may be bound to the API via ester or carbamate linkages that can degrade in an aqueous milieu and care should be taken to minimize the effects of degradation by reducing any time delay between sample preparation and UV analysis (Greenwald et al. 2000).

In multivalent polymer conjugates, such as systems that employ polymers such as 2-hydroxypropyl methacrylate (HPMA) or polyglutamic acid (PGA), API loading is higher than in monovalent polymers. Polymer conjugates can be dissolved in deuterated organic solvents and quantitated by 1H-NMR or UV spectroscopy (Chandran et al. 2007). Dissolution in deuterated solvents can also be used for PEGylated systems if low molecular weight PEGs are used for conjugation.

In the case of conjugates involving amino acids as linkers, amino acid analysis can be used for its determination. The presence of the amino acids provides a technique relatively independent of the amount of API. In such systems, the amino acid linker is first bound to the API followed by attachment onto a polymer, and the ratio of amino acids to API is therefore fixed. For amino acid analysis, samples are usually digested in highly acidic or basic conditions, which cleaves the amine linkages, followed by GC/LC to determine the amount of amino acids present (Barkholt and

Jensen 1989; Yamada et al. 1991). The amino acid analysis is useful for as the estimated potency can be made independently of the structure of API.

Nanoparticle/microparticle-based systems offer the advantage of delivering a sizeable payload of a drug in bolus form, but require additional care during sample preparation. The FDA-approved molecule polylactic-*co*-glycolic acid (PLGA) has traditionally been used to encapsulate hydrophobic API molecules in its core. PLGA is an extremely hydrophobic molecule that is soluble in organic solvents such as acetone and acetonitrile (Avgoustakis 2004; Mundargi et al. 2008). For characterization work, samples are first weighed, and then immersed in acetonitrile to dissolve both API and PLGA. API concentration can then be determined chromatographically. Since the exact weight of the mixture is known, quantitation by HPLC allows for a calculation of weight/weight percent loading. Theoretically, PLGA will make a very poor encapsulator for hydrophilic molecules, so the analyst should not be too wary of confronting such a system.

For systems such as Abraxane® (Gardner et al. 2008), a commercially-available albumin-based nanoparticle, or liposomal systems, acetonitrile could be used to extract paclitaxel from the protein core. This could be followed by an ultracentrifugation or filtration step to remove undissolved protein. The supernatant can be evaluated by HPLC or UV for API content. Liposomes are another choice of delivery for insoluble compounds, and these are thermodynamically stable structures of lipids such as phosphotidylcholine (Schiffelers et al. 2003). For sample preparation, they bear similarities to above-mentioned polymer-based systems in the sense that their vesicular structure comprising the liposomal bilayer can be disintegrated by the use of methanol (Chen et al. 2000; Schiffelers et al. 2003; Drummond et al. 2010). Consequently, the sequestered drug could dissolve in the organic matrix and further analysis could be carried out by UV-absorbance or HPLC.

References

Abdallah RM (2006) HPLC chromatographic methods for simultaneous determination of pholcodine and ephedrine HCl with other active ingredients in antitussive-antihistamine oral liquid formulations. Nat Prod Sci 12(1):55–61

Aggarwal GD, Dhawan S (2009) Development, fabrication and evaluation of transdermal drug delivery system – a review. Pharm Rev 7(5)

Aghazadeh A, Kazemifard G (2001) Simultaneous determination of amoxycillin and clavulanic acid in pharmaceutical dosage forms by LC with amperometric detection. J Pharm Biomed Anal 25(2):325–329

Ali MS, Ghori M, Khatri AR (2006) Stability indicating simultaneous determination of domperidone (DP), methylparaben (MP) and propylparaben by high performance liquid chromatography (HPLC). J Pharm Biomed Anal 41(2):358–365

Almodovar RA, Rodriguez RA, Rosario O (1998) Inverse supercritical extraction of acetaminophen from suppositories. J Pharm Biomed Anal 17(1):89–93

Ansel HC, Allen LV Jr, Popovich NG (1999) Pharmaceutical dosage forms and drug delivery systems, 7th edn. Lippincott Williams & Wilkins, Philadelphia

Avgoustakis K (2004) Pegylated poly(lactide) and poly(lactide-co-glycolide) nanoparticles: preparation, properties and possible applications in drug delivery. Curr Drug Deliv 1:321–333

Barkholt V, Jensen AL (1989) Amino acid analysis: determination of cysteine plus half-cystine in proteins after hydrochloric acid hydrolysis with a disulfide compound as additive. Anal Biochem 177:318–322

Basavaiah K, SriLatha J, Swamy JM (1999) Spectrophotometric determination of ceterizine hydrochloride with alizarin red S. Talanta 50(4):887–892

Benjamin EJL, Lee M, Tom J, Lin L, Henesian M, Wu D (1983) Stabilization of sulconazole nitrate in a topical powder formulation. Int J Pharm 14:209–211

Blanco MC, Cocollo J, Iturriaga H, Maspoch S, Villegas N (1999) Kinetic spectrophotometric determination of hydrocortisone acetate in a pharmaceutical preparation by use of partial least-squares regression. Analyst 124(6):911–915

Bonazzi D, Andrisano V, Gatti R, Cavrini V (1995) Analysis of pharmaceutical creams: a useful approach based on solid-phase extraction (SPE) and UV spectrophotometry. J Pharm Biomed Anal 13(11):1321–1329

Caliceti P, Veronese FM (2003) Pharmacokinetic and biodistribution properties of poly(ethylene glycol)-protein conjugates. Adv Drug Deliv Rev 55:1261–1277

Cardoso CAL, Vilegas W, Honda NK (2000) Rapid determination of furanocoumarins in creams and pomades using SPE and GC. J Pharm Biomed Anal 22(2):203–214

Carlisle MRC, Chicoine ML, Wygant MB (1992) A stability-indicating high-performance liquid chromatographic assay for nicotine in transdermal patches. Int J Pharm 80:227–242

Cavrini V, Di Pietra AM, Raggi MA (1982) High-pressure liquid chromatographic (HPLC) analysis of imidazole antifungals in commercial dosage forms. Int J Pharm 10(2):119–124

Cavrini V, Di Pietra AM, Gatti R (1989) Analysis of miconazole and econazole in pharmaceutical formulations by derivative UV spectroscopy and liquid chromatography (HPLC). J Pharm Biomed Anal 7(12):1535–1543

Chandran SS, Nan A, Rosen DM, Ghandehari H, Denmeade SR (2007) A prostate-specific antigen activated N-(2-hydroxypropyl) methacrylamide copolymer prodrug as dual-targeted therapy for prostate cancer. Mol Cancer Ther 6:2928–2937

Chen D, Cole DL, Srivatsa GS (2000) Determination of free and encapsulated oligonucleotides in liposome formulated drug product. J Pharm Biomed Anal 22:791–801

Costi EM, Sicilia MD, Rubio S, Perez-Bendito D (2006) Determination of cationic surfactants in pharmaceuticals based on competitive aggregation in ternary amphiphile mixtures. Anal Chim Acta 577(2):257–263

Culzoni MJ, De Zan MM, Robles JC, Mantovani VE, Goicoechea HC (2005) Chemometrics-assisted uv-spectroscopic strategies for the determination of theophylline in syrups. J Pharm Biomed Anal 39(5):1068–1074

Dallet P, Kauss T, Langlois MH, Fawaz F, Mossalayi DM, Dubost JP (2009) A validated RP-LC method for the determination of isosorbide 5-mononitrate in a leishmanicidal ointment. Chromatographia 69(9–10):1129–1131

Deicke A, Süverkrüp R (2000) Dose uniformity and redispersibility of pharmaceutical suspensions 2: assessment of three commercial erythromycin ethyl succinate oral liquids. Eur J Pharm Biopharm 49(1):73–78

Di Pietra AM, Cavrini V, Andrisano V, Gatti R (1992) HPLC analysis of imidazole antimycotic drugs in pharmaceutical formulations. J Pharm Biomed Anal 10(10–12):873–879

Di Pietra AM, Andrisano V, Gotti R, Cavrini V (1996) Online post-column photochemical derivatization in liquid chromatographic diode-array detection analysis of binary drug mixtures. J Pharm Biomed Anal 14(8–10):1191–1199

DiNunzio JE, Gadde RR (1992) Development of stability indicating assay methods for the determination of hydroxamic acids in topical formulations. J Chromatogr 594(1–2):209–216

Drummond DC, Noble CO, Guo Z, Hayes ME, Connolly-Ingram C, Gabriel BS, Hann B, Liu B, Park JW, Hong K, Benz CC, Marks JD, Kirpotin DB (2010) Development of a highly stable and targetable nanoliposomal formulation of topotecan. J Control Release 141:13–21

Duncan R (2003) The dawning era of polymer therapeutics. Nat Rev Drug Discov 2:347–360

Durham AC, Walton JM (1983) A survey of the available colorimetric indicators for Ca^{2+} and Mg^{2+} ions in biological experiments. Cell Calcium 4:47–55

Edwardson PA, Gardner RS (1990) Problems associated with the extraction and analysis of triamcinolone acetonide in dermatological patches. J Pharm Biomed Anal 8(8–12):935–938

El-Gindy A, Emara S, Mostafa A (2006) Application and validation of chemometrics-assisted spectrophotometry and liquid chromatography for the simultaneous determination of six-component pharmaceuticals. J Pharm Biomed Anal 41(2):421–430

El-Gindy A, Emara S, Shaaban H (2007) Development and validation of chemometrics-assisted spectrophotometric and liquid chromatographic methods for the simultaneous determination of two multicomponent mixtures containing bronchodilator drugs. J Pharm Biomed Anal 43(3):973–982

El-Sherbiny DT, El-Enany N, Belal FF, Hansen SH (2007) Simultaneous determination of loratadine and desloratadine in pharmaceutical preparations using liquid chromatography with a microemulsion as eluent. J Pharm Biomed Anal 43(4):1236–1242

Galli V, Barbas C (2004) High-performance liquid chromatographic analysis of dextromethorphan, guaifenesin and benzoate in a cough syrup for stability testing. J Chromatogr A 1048(2):207–211

Gao Y, Liang J, Liu J, Xiao Y (2009) Double-layer weekly sustained release transdermal patch containing gestodene and ethinylestradiol I. Int J Pharm 377(1–2):128–134

Garcia PL, Santoro MIRM, Kedor-Hackman ERM, Singh AK (2005) Development and validation of a HPLC and a UV derivative spectrophotometric methods for determination of hydroquinone in gel and cream preparations. J Pharm Biomed Anal 39(3–4):764–768

Gardner ER, Dahut W, Figg WD (2008) Quantitative determination of total and unbound paclitaxel in human plasma following abraxane treatment. J Chromatogr B Analyt Technol Biomed Life Sci 862:213–218

Greenwald RB, Conover CD, Choe YH (2000) Poly(ethylene glycol) conjugated drugs and prodrugs: a comprehensive review. Crit Rcv Ther Drug Carrier Syst 17:101–161

Grosa G, Grosso ED, Russo R, Allegrone G (2006) Simultaneous, stability indicating, HPLC-dad determination of guaifenesin and methyl and propyl-parabens in cough syrup. J Pharm Biomed Anal 41(3):798–803

Guneri TA, Arici M, Ertan G (2004) Preparation and diffusional evaluation of sustained-release suppositories containing ibuprofen microspheres. FABAD J 29:177–184

Gupta A, Stratton SP, Myrdal PB (2005) An HPLC method for quantitation of perillyl alcohol in a topical pharmaceutical cream formulation. J Pharm Biomed Anal 37(3):447–452

Haikala VE, Heimonen IK, Vuorela HJ (1991) Determination of ibuprofen in ointments by reversed-phase liquid chromatography. J Pharm Sci 80(5):456–458

Hamoudova R, Pospisilova M (2006) Determination of ibuprofen and flurbiprofen in pharmaceuticals by capillary zone electrophoresis. J Pharm Biomed Anal 41(4):1463–1467

Haney PW, Dash AK (1997) Simple liquid chromatographic method for the analysis of 5-aminosalicylic acid and its degradation product. J Chromatogr A 765:233–239

Hattori T, Yoshida H (1986) Thermometric titration of the dodecyl sulphate ion with metal-phenanthroline complexes. Talanta 33:167–169

Hattori T, Yoshida H (1987) Thermometric titration of cadmium with sodium diethyldithiocarbamate, with oxidation by hydrogen peroxide as indicator reaction. Talanta 34:733–735

Heinänen M, Barbas C (2001) Validation of an hplc method for the quantification of ambroxol hydrochloride and benzoic acid in a syrup as pharmaceutical form stress test for stability evaluation. J Pharm Biomed Anal 24(5–6):1005–1010

Hilfiker R (2006) Polymorphism in the pharmaceutical industry. Wiley-VCH Verlag, Weinheim

Hood DJ, Cheung HY (2003) A chromatographic method for rapid and simultaneous analysis of codeine phosphate, ephedrine HCl and chlorpheniramine maleate in cough-cold syrup formulation. J Pharm Biomed Anal 30(5):1595–1601

Hoogewijs G, Massart DL (1983) Development of a standardized analysis strategy for basic drugs using ion-pair extraction and high-performance liquid chromatography – I. Philosophy and selection of extraction technique. J Pharm Biomed Anal 1(3):321–329

Huidobro AL, Garcia A, Barbas C (2009) Rapid analytical procedure for neomycin determination in ointments by CE with direct UV detection. J Pharm Biomed Anal 49(5):1303–1307

Izumoto SI, Machida Y, Nishi H, Nakamura K, Nakai H, Sato T (1997) Chromatography of crotamiton and its application to the determination of active ingredients in ointments. J Pharm Biomed Anal 15(9–10):1457–1466

Johnson CE, Cober MP, Ludwig JL (2007) Stability of partial doses of omeprazole sodium bicarbonate oral suspension. Ann Pharmacother 41(12):1954–1961

Karlsson L, Torstensson A, Taylor LT (1997) The use of supercritical fluid extraction for sample preparation of pharmaceutical formulations. J Pharm Biomed Anal 15(5):601–611

Klaffenbach PK, Kronenfeld D, Peterson TA (1998) Stability indicating method for the analysis of nicotine in transdermal drug delivery systems by means of gas chromatography. J High Resol Chrom 21(12):649–652

Koundourellis JE, Malliou ET, Broussali TA (2000) High performance liquid chromatographic determination of ambroxol in the presence of different preservatives in pharmaceutical formulations. J Pharm Biomed Anal 23(2–3):469–475

Kuehl PJ, Angersbach BS, Stratton SP, Myrdal PB (2006) Development of an HPLC method for the analysis of apomine in a topical cream formulation. J Pharm Biomed Anal 40(4):975–980

Labbozzetta S, Valvo L, Bertocchi P, Manna L (2005) Focused microwave-assisted extraction and LC determination of the active ingredient in naproxen-based suppositories. J Pharm Biomed Anal 39(3–4):463–468

Labbozzetta S, Valvo L, Bertocchi P, Alimonti S, Gaudiano MC, Manna L (2008) Focused microwave-assisted extraction and LC determination of ketoprofen in the presence of preservatives in a pharmaceutical cream formulation. Chromatographia 69(3/4):365–368

Li J (2002) Elimination of polymer interference in chromatographic analysis of estradiol degradation products in a transdermal drug delivery formulation by proper selection of extraction solvents. J Pharm Sci 91(8):1873–1879

Li J, Jiang Y, Sun T, Ren S (2008) Fast and simple method for assay of ciclopriox olamine by micellar electrokinetic capillary chromatography. J Pharm Biomed Anal 47:929–933

Liu L, Pack BW (2007) Cleaning verification assays for highly potent compounds by high performance liquid chromatography mass spectrometry: strategy, validation, and long-term performance. J Pharm Biomed Anal 43(4):1206–1212

Liu XX, Wang L, Chen XQ, Deng XT, Cao Y, Wang Q (2008) Simultaneous quantification of both triterpenoid and steroidal saponins in various yunnan bai-yao preparations using HPLC-UV and HPLC-MS. J Sep Sci 31(22):3834–3846

Louhaichi MR, Jebali S, Loueslati MH, Adhoum N, Monser L (2009) Simultaneous determination of pseudoephdrine, pheniramine, guaifenisin, pyrilamine, chlorpheniramine and dextromethorphan in cough and cold medicines by high performance liquid chromatography. Talanta 78(3):991–997

Malesuik M, Cardoso S, Steppe M (2008) Development of a validated stability-indicating LC method for nitazoxanide in pharmaceutical formulations. Chromatographia 67(1): 131–136

Marsh A, Clark BJ, Altria KD (2005) A review of the background, operating parameters and applications of microemulsion liquid chromatography (MELC). J Sep Sci 28:2023–2032

Mason DJ, Crozier B (1988) An analytical study of formulations containing malathion. Pestic Sci 22:317–332

Masuda M, Koike S, Handa M, Sagara K, Mizutani T (1993) Application of supercritical fluid extraction and chromatography to assay fat-soluble vitamins in hydrophobic ointment. Anal Sci 9(1):29–32

McEvoy E, Donegan S, Power J, Altria K (2007) Optimisation and validation of a rapid and efficient microemulsion liquid chromatographic (MELC) method for the determination of paracetamol (acetaminophen) content in a suppository formulation. J Pharm Biomed Anal 44: 137–143

McEvoy E, Donegan S, Power J, Altria K (2008) Application of MELC and MEEKC for the analysis of paracetamol and related impurities in suppositories. Chromatographia 68:49–56

Messer DC, Taylor LT (1994) Inverse analytical supercritical fluid extraction of zovirax ointment 5%. Anal Chem 66(9):1591–1592

Mishal A, Sober D (2005) Stability indicating reversed-phase liquid chromatographic determination of metronidazole benzoate and diloxanide furoate as bulk drug and in suspension dosage form. J Pharm Biomed Anal 39(3–4):819–823

Mittal A, Sara US, Ali A, Mohammed A (2009) Design, development, physico-chemical, in vitro and in vivo evaluation of monolithic matrix type transdermal patches containing nitrendipine. Pharm Dev Tech 14(4):422–434

Mohammadi A, Kanfer I, Walker RB (2004) A capillary zone electrophoresis (CZE) method for the determination of cyclizine hydrochloride in tablets and sup-positories. J Pharm Biomed Anal 35(1):233–239

Moore WN, Taylor LT (1994) Analytical inverse supercritical fluid extraction of polar pharmaceutical compounds from cream and ointment matrices. J Pharm Biomed Anal 12(10):1227–1232

Mundargi RC, Babu VR, Rangaswamy V, Patel P, Aminabhavi TM (2008) Nano/micro technologies for delivering macromolecular therapeutics using poly(D, L-lactide-co-glycolide) and its derivatives. J Control Release 125:193–209

Murakami T, Kawasaki T, Takemura A, Fukutsu N, Kishi N, Kusu F (2008) Identification of degradation products in loxoprofen sodium adhesive tapes by liquid chromatography-mass spectrometry and dynamic pressurized liquid extraction-solid-phase extraction coupled to liquid chromatography-nuclear magnetic resonance spectroscopy. J Chromatogr A 1208(1–2): 164–174

Nema S, Washkuhn RJ, Brendel RJ (1997) Excipients and their use in injectable products. PDA J Pharm Sci Technol 51:166–171

Newman AW, Vitez IM, Kiesnowski C, Mueller RL (1996). Starches and starch derivatives. In: Encyclopedia of pharmaceutical technology, Swarbrick J, Boylan JC (eds.), Marcel Dekker, Inc., New York, pp. 223–248

Nguyen TT, Kringstad R, Rasmussen KE (1986) Use of extraction columns for the isolation of desonide and parabens from creams and ointments for high-performance liquid chromatographic analysis. J Chromatogr 366:445–450

Nina M, Edgardo AC, Amelia AR (1988) Determination of betamethasone 17-benzoate in lipophylic vehicles by reversed-phase high-performance liquid chromatography. J Pharm Sci 77(7):638–639

Novakova L, Matysova L, Solich P (2006) Advantages of application of UPLC in pharmaceutical analysis. Talanta 68(3):908–918

Omar Mahmoud A, Abuo-Rahma Gamal El-Din AA, Abdelmageed Osama H (2006) Colorimetric determination of certain antifungals in pure forms and in their pharmaceutical formulations. Bull Pharm Sci (Assiut Univ-Egypt) 29(2):201–519

Pasut G, Veronese FM (2009) PEG conjugates in clinical development or use as anticancer agents: an overview. Adv Drug Deliv Rev 61:1177–1188

Pedraza A, Sicilia MD, Rubio S, Perez-Bendito D (2006) Pharmaceutical quality control of avid and neutral drugs based on competitive self-assembly in amphiphilic systems. Analyst 131:81–89

Pires AE, Honda NK, Cardoso CAL (2004) A method for fast determination of psoralens in oral solutions of phytomedicines using liquid chromatography. J Pharm Biomed Anal 36(2):415–420

Priego-Capote F, Luque de Castro MD (2003) Ultrasound-assisted continuous liquid-liquid extraction without phase separation and hydrolysis of paracetamol in suppositories. Anal Chim Acta 489(2):223–232

Reepmeyer JC (2001) Screening for corticosteroids in topical pharmaceuticals by HPLC with a scanning ultraviolet detector. J Liq Chromatogr Relat Technol 24:693–709

Reepmeyer JC (2005) Analysis of the nitrogen mustard mechlorethamine in topical pharmaceutical preparations by high-performance liquid chromatography. J Chromatogr A 1085(2):262–269

Ryan R, Donegan S, Power J, McEvoy E, Altria K (2009) Recent advances in the methodology, optimisation and application of MEEKC. Electrophoresis 30(1):65–82

Samanidou VF, Evaggelopoulou EN, Papadoyannis IN (2006) Development of a validated hplc method for the determination of four 1,4-benzodiazepines in human biological fluids. J Sep Sci 29(11):1550–1560

Schiffelers RM, Koning GA, ten Hagen TL, Fens MH, Schraa AJ, Janssen AP, Kok RJ, Molema G, Storm G (2003) Anti-tumor efficacy of tumor vasculature-targeted liposomal doxorubicin. J Control Release 91:115–122

Shah VP, Behl CR, Flynn GL, Higuchi WI, Schaefer H, Barry BW, Behl CR, Conners DP, Evans CC, Flynn GL, Franz TJ, Gans EH, Higuchi WI, Kail N, Krueger GG, Leyden J, Maibach HI, Malick AW, Nacht S, Ng S, Peck CC, Pershing LK, Potts RO, Poulsen BJ, Schaefer H, Scott RC, Seqeira JA, Shah VP, Sharma D, Skelly JP, Wu MS (1992) Principles and criteria in the development and optimization of topical therapeutic products: sponsored by the American Association of Pharmaceutical Scientists (AAPS) and U.S. Food and Drug Administration (FDA). Int J Pharm 82(1–2):21–28

Silvestre CI, Santos JL, Lima JL, Zagatto EA (2009) Liquid-liquid extraction in flow analysis: a critical review. Anal Chim Acta 652(1–2):54–65

Strickley RG (2004) Solubilizing excipients in oral and injectable formulations. Pharm Res 21:201–230

Suntornsuk L (2001) Direct determination of s-carboxymethyl-l-cysteine in syrups by reversed-phase high-performance liquid chromatography. J Pharm Biomed Anal 25(1):165–170

Tagliari M, Stulzer H, Murakami F, Kuminek G, Valente B, Oliveira P, Segatto Silva M (2008) Development and validation of a stability-indicating lc method to quantify hydrochlorothiazide in oral suspension for pediatric use. Chromatographia 67(7):647–652

Takashina Y, Naito T, Mino Y, Kagawa Y, Kawakami J (2009) Validated LC coupled to ESI-ms/ms analysis for fentanyl in human plasma and UV analysis in applied reservoir transdermal patches using a simple and rapid procedure. J Clin Pharm Ther 34(5):523–529

The Merck Index (2006). 14th ed. Merck Research Laboratories.

Tjornelund J, Hansen SH (1997) Validation of a simple method for the determination of oxytetracycline in ointment by non-aqueous capillary electrophoresis. J Pharm Biomed Anal 15(8):1077–1082

USP General Chapter on Injections <1> (2009). In: United States Pharmacopeia 2009, vol 32, United States Pharmacopeial Convention, p 3

USP General Chapter on Pharmaceutical Dosage Forms <1151> (2009). In: United States Pharmacopeia 2009, vol 32, United States Pharmacopeial Convention, p 663

USP Monograph on Acetaminophen Oral Suspension (2009). In: United States Pharmacopeia 2009, vol 32, United States Pharmacopeial Convention, p 1390

USP Monograph on Alclometasone Dipropionate Ointment (2009). In: United States Pharmacopeia 2009, vol 32, United States Pharmacopeial Convention, p 1440

USP Monograph on Amantadine Hydrochloride Oral Solution (2009). In: United States Pharmacopeia 2009, vol 32, United States Pharmacopeial Convention, p 1505

USP Monograph on Antithrombin III Human (2009). In: United States Pharmacopeia 2009, vol 32, United States Pharmacopeial Convention, p 1570

USP Monograph on Bacampicillin Hydrochloride for Oral Suspension (2009). In: United States Pharmacopeia 2009, vol 32, United States Pharmacopeial Convention, p 1619

USP Monograph on Bupivacaine Hydrochloride in Dextrose Injection (2009). In: United States Pharmacopeia 2009, vol 32, United States Pharmacopeial Convention, p 1719

USP Monograph on Caffeine Citrate Oral Solution (2009). In: United States Pharmacopeia 2009, vol 32, United States Pharmacopeial Convention, p 1743

USP Monograph on Calcium Chloride Injection (2009). In: United States Pharmacopeia 2009, vol 32, United States Pharmacopeial Convention, p 1761

USP Monograph on Cefpodoxime Proxetil for Oral Suspension (2009). In: United States Pharmacopeia 2009, vol 32, United States Pharmacopeial Convention, p 1854

USP Monograph on Cefprozil for Oral Suspension (2009). In: United States Pharmacopeia 2009, vol 32, United States Pharmacopeial Convention, p 1856

USP Monograph on Cephapirin Benzathine Intramammary Infusion (2009). In: United States Pharmacopeia, 2009, vol 32, United States Pharmacopeial Convention, p 1876

USP Monograph on Chlorothiazide Oral Suspension (2009). In: United States Pharmacopeia 2009, vol 32, United States Pharmacopeial Convention, p 1905

USP Monograph on Chlorpromazine Hydrochloride Syrup (2009). In: United States Pharmacopeia 2009, vol 32, United States Pharmacopeial Convention, p 1915
USP Monograph on Clindamycin Hydrochloride Oral Solution (2009). In: United States Pharmacopeia 2009, vol 32, United States Pharmacopeial Convention, p 1968
USP Monograph on Clindamycin Phosphate Gel (2009). In: United States Pharmacopeia, 2009, vol 32, United States Pharmacopeial Convention, p 1970
USP Monograph on Clobetasol Propionate Ointment (2009). In: United States Pharmacopeia, 2009, vol 32, United States Pharmacopeial Convention, p 1977
USP Monograph on Clonidine Transdermal System (2009). In: United States Pharmacopeia 2009, vol 32, United States Pharmacopeial Convention, p 1991
USP Monograph on Clotrimazole Cream (2009). In: United States Pharmacopeia, 2009, vol 32, United States Pharmacopeial Convention, p 1998
USP Monograph on Cyclosporine Oral Solution (2009). In: United States Pharmacopeia 2009, vol 32, United States Pharmacopeial Convention, p 2047
USP Monograph on Dexamethasone Acetate Injectable Suspension (2009). In: United States Pharmacopeia, 2009, vol 32, United States Pharmacopeial Convention, p 2088
USP Monograph on Dibucaine Ointment (2009). In: United States Pharmacopeia, 2009, vol 32, United States Pharmacopeial Convention, p 2119
USP Monograph on Diflorasone Diacetate Ointment (2009). In: United States Pharmacopeia, 2009, vol 32, p 2142
USP Monograph on Dihydrotachysterol Oral Solution (2009). In: United States Pharmacopeia 2009, vol 32, United States Pharmacopeial Convention, p 2155
USP Monograph on Docusate Sodium Syrup (2009). In: United States Pharmacopeia 2009, vol 32, United States Pharmacopeial Convention, p 2196
USP Monograph on Doxylamine Succinate Oral Solution (2009). In: United States Pharmacopeia 2009, vol 32, United States Pharmacopeial Convention, p 2216
USP Monograph on Dyclonine Hydrochloride Gel (2009). In: United States Pharmacopeia, 2009, vol 32, United States Pharmacopeial Convention, p 2222
USP Monograph on Ergocalciferol Oral Solution (2009). In: United States Pharmacopeia 2009, vol 32, United States Pharmacopeial Convention, p 2270
USP Monograph on Ergotamine Tartrate and Caffeine Suppositories (2009). In: United States Pharmacopeia 2009, vol 32, United States Pharmacopeial Convention, p 2281
USP Monograph on Erythromycin Estolate and Sulfisoxazole Acetyl Oral Suspension (2009). In: United States Pharmacopeia 2009, vol 32, United States Pharmacopeial Convention, p 2290
USP Monograph on Erythromycin Topical Gel (2009). In: United States Pharmacopeia, 2009, vol 32, United States Pharmacopeial Convention, p 2284
USP Monograph on Estradiol Transdermal System (2009). In: States Pharmacopeia 2009, vol 32, United States Pharmacopeial Convention, p 2303
USP Monograph on Famotidine for Oral Suspension (2009). In: United States Pharmacopeia 2009, vol 32, United States Pharmacopeial Convention, p 2344
USP Monograph on Ferumoxsil Oral Suspension (2009). In: United States Pharmacopeia, 2009, vol 32, United States Pharmacopeial Convention, p 2368
USP Monograph on Fluocinonide Cream (2009). In: United States Pharmacopeia, 2009, vol 32, United States Pharmacopeial Convention, p 2401
USP Monograph on Fluphenazine Hydrochloride Oral Solution (2009). In: United States Pharmacopeia 2009, vol 32, United States Pharmacopeial Convention, p 2424
USP Monograph on Flurandrenolide Ointment (2009). In: United States Pharmacopeia, 2009, vol 32, United States Pharmacopeial Convention, p 2427
USP Monograph on Fluticasone Propionate Ointment (2009). In: United States Pharmacopeia, 2009, vol 32, United States Pharmacopeial Convention, p 4058
USP Monograph on Ganciclovir Oral Suspension (2009). In: United States Pharmacopeia 2009, vol 32, United States Pharmacopeial Convention, p 2482
USP Monograph on Griseofulvin Oral Suspension (2009). In: United States Pharmacopeia 2009, vol 32, United States Pharmacopeial Convention, p 2532

USP Monograph on Haloperidol Injection (2009). In: United States Pharmacopeia, 2009, vol 32, United States Pharmacopeial Convention, p 2548
USP Monograph on Haloperidol Oral Solution (2009). In: United States Pharmacopeia 2009, vol 32, United States Pharmacopeial Convention, p 2548
USP Monograph on Histamine Phosphate in Dextrose Injection (2009). In: United States Pharmacopeia 2009, vol 32, United States Pharmacopeial Convention, p 2558
USP Monograph on Hydrocortisone Acetate Cream (2009). In: United States Pharmacopeia, 2009, vol 32, United States Pharmacopeial Convention, p 2578
USP Monograph on Hydrocortisone Injectable Suspension (2009). In: United States Pharmacopeia, 2009, vol 32, United States Pharmacopeial Convention, p 2575
USP Monograph on Ibuprofen Oral Suspension (2009). In: United States Pharmacopeia 2009, vol 32, United States Pharmacopeial Convention, p 2608
USP Monograph on Indomethacin suppositories (2009). In: United States Pharmacopeial Convention, vol 32, p 2635
USP Monograph on Isophane Insulin Human Suspension (2009). In: United States Pharmacopeia, 2009, vol 32, United States Pharmacopeial Convention, p 2646
USP Monograph on Levocarnitine Oral Solution (2009). In: United States Pharmacopeia 2009, vol 32, United States Pharmacopeial Convention, p 2770
USP Monograph on Lidocaine and Prilocaine Cream (2009). In: United States Pharmacopeia, 2009, vol 32, United States Pharmacopeial Convention, p 2785
USP Monograph on Loperamide Hydrochloride Oral Solution (2009). In: United States Pharmacopeia 2009, vol 32, United States Pharmacopeial Convention, p 2802
USP Monograph on Mafenide Acetate Cream (2009). In: United States Pharmacopeia, 2009, vol 32, United States Pharmacopeial Convention, p 2821
USP Monograph on Mebendazole Oral Suspension (2009). In: United States Pharmacopeia 2009, vol 32, United States Pharmacopeial Convention, p 2853
USP Monograph on Meclocycline Sulfosalicylate Cream (2009). In: United States Pharmacopeia, 2009, vol 32, United States Pharmacopeial Convention, p 2860
USP Monograph on Mesoridazine Besylate Oral Solution (2009). In: United States Pharmacopeia 2009, vol 32, United States Pharmacopeial Convention, p 2899
USP Monograph on Methadone Hydrochloride Oral Solution (2009). In: United States Pharmacopeia 2009, vol 32, United States Pharmacopeial Convention, p 2915
USP Monograph on Methenamine Mandelate for Oral Solution (2009). In: United States Pharmacopeia 2009, vol 32, United States Pharmacopeial Convention, p 2923
USP Monograph on Metronidazole Gel (2009). In: United States Pharmacopeia, 2009, vol 32, United States Pharmacopeial Convention, p 2972
USP Monograph on Morphine Sulfate Suppositories (see section: assay for suppositories compounded in polyethylene glycol base) (2009). United States Pharmacopeial Convention, vol 32, p 3010
USP Monograph on Mibolerone Oral Solution (2009). In: United States Pharmacopeia 2009, vol 32, United States Pharmacopeial Convention, p 2980
USP Monograph on Miconazole Nitrate Cream (2009). In: United States Pharmacopeia, 2009, vol 32, United States Pharmacopeial Convention, p 2982
USP Monograph on Miconazole Nitrate Topical Powder (2009). In: United States Pharmacopeia, 2009, vol 32 United States Pharmacopeial Convention, p 2982
USP Monograph on Naftifine Hydrochloride Cream (2009). In: United States Pharmacopeia, 2009, vol 32, United States Pharmacopeial Convention, p 3023
USP Monograph on Neomycin Sulfate, Isoflupredone Acetate, and Tetracaine Hydro-Chloride Topical Powder (2009). In: United States Pharmacopeial Convention, vol 32, p 3055
USP Monograph on Neomycin Sulfate Ointment (2009). In: United States Pharmacopeia, 2009, vol 32, United States Pharmacopeial Convention, p 3046
USP Monograph on Neomycin Sulfate Oral Solution (2009). In: United States Pharmacopeia 2009, vol 32, United States Pharmacopeial Convention, p 3047
USP Monograph on Nitroglycerin Ointment (2009). In: United States Pharmacopeia, 2009, vol 32, United States Pharmacopeial Convention, p 3097

USP Monograph on Phenobarbital Oral Solution (2009). In: United States Pharmacopeia 2009, vol 32, United States Pharmacopeial Convention, p 3271
USP Monograph on Phytonadione Injectable Emulsion (2009). In: United States Pharmacopoeia, 2009, vol 32, United States Pharmacopeial Convention, p 3303
USP Monograph on Piroxicam Cream (2009). In: United States Pharmacopeia, 2009, vol 32, United States Pharmacopeial Convention, p 3318
USP Monograph on Prednisone Oral Solution (2009). In: United States Pharmacopeia 2009, vol 32, United States Pharmacopeial Convention, p 3377
USP Monograph on Progesterone Vaginal Suppositories (see section: assay for sup-positories compounded in fatty acid base) (2009). In: United States Pharmacopeial Convention, vol 32, p 3401
USP Monograph on Promazine Hydrochloride Syrup (2009). In: United States Pharmacopeia 2009, vol 32, United States Pharmacopeial Convention, p 3405
USP Monograph on Propoxyphene Napsylate Oral Suspension (2009). In: United States Pharmacopeia 2009, vol 32, United States Pharmacopeial Convention, p 3422
USP Monograph on Ranitidine Oral Solution (2009). In: United States Pharmacopeia 2009, vol 32, United States Pharmacopeial Convention, p 3477
USP Monograph on Simethicone Oral Suspension (2009). In: United States Pharmacopeia 2009, vol 32, United States Pharmacopeial Convention, p 3556
USP Monograph on Tetracaine Ointment (2009). In: United States Pharmacopeia, 2009, vol 32, United States Pharmacopeial Convention, p 3689
USP Monograph on Tetracycline Oral Suspension (2009). In: United States Pharmacopeia 2009, vol 32, United States Pharmacopeial Convention, p 3694
USP Monograph on Thioridazine Hydrochloride Oral Solution (2009). In: United States Pharmacopeia 2009, vol 32, United States Pharmacopeial Convention, p 3729
USP Monograph on Thioridazine Oral Suspension (2009). In: United States Pharmacopeia 2009, vol 32, United States Pharmacopeial Convention, p 3728
USP Monograph on Tolnaftate Gel (2009). In: United States Pharmacopeia, 2009, vol 32, United States Pharmacopeial Convention, p 3772
USP Monograph on Tolnaftate Topical Powder (2009). In: United States Pharmacopeia, 2009, vol 32, United States Pharmacopeial Convention, p 3772
USP Monograph on Tretinoin Cream (2009). In: United States Pharmacopeia, 2009, vol 32, United States Pharmacopeial Convention, p 3779
USP Monograph on Valproate Sodium Injection (2009). In: United States Pharmacopeia, 2009, vol 32, United States Pharmacopeial Convention, p 3837
USP Monograph on Valproic Acid Oral Solution (2009). In: United States Pharmacopeia 2009, vol 32, United States Pharmacopeial Convention, p 3839
Uysal UD, Tuncel M (2006) Validated capillary electrophoresis study for the determination of cetirizine in pharmaceutical forms. J Liq Chromatogr Relat Technol 29(12):1781–1792
van der Horst A, Martens HJ, de Goede PN (1989) Analysis of water-soluble vitamins in total parenteral nutrition solution by high pressure liquid chromatography. Pharm Weekbl Sci 11:169–174
Van Nimmen NF, Veulemans HA (2007) Validated GC-MS analysis for the determination of residual fentanyl in applied durogesic reservoir and durogesic d-trans matrix transdermal fentanyl patches. J Chromatogr B 846(1–2):264–272
Venables HJ, Wells JI (2002) Powder sampling. Drug development and industrial pharmacy. Drug Dev Ind Pharm 28(2):107–117
Walters DL, Strong CR, Green SV, Curtis MA (1995) Development and validation of a gradient reversed-phase high-performance liquid chromatographic as-say for s(-)-2-(n-propyl-n-2-thienylethylamino)-5-hydroxytetralin (n-0923) from a transdermal delivery system. J Chromatogr B 670(2):299–307
Wang LF, Peng JD, Liu LM (2008) A reversed-phase high performance liquid chromatography coupled with resonance Rayleigh scattering detection for the determination of four tetracycline antibiotics. Anal Chim Acta 630(1):101–106

Yamada H, Moriya H, Tsugita A (1991) Development of an acid hydrolysis method with high recoveries of tryptophan and cysteine for microquantities of protein. Anal Biochem 198:1–5

Zhang Q, Li J, Wang C, Sun W, Zhang Z, Cheng W (2007) A gradient HPLC method for the quality control of chlorogenic acid, linarin and luteolin in flos chrysanthemi indici suppository. J Pharm Biomed Anal 43(2):753–757

Zheng J, Patel D, Tang Q, Markovich RJ, Rustum AM (2009) Comparison study of porous, fused-core, and monolithic silica-based C18 HPLC columns for Celestoderm-V Ointment analysis. J Pharm Biomed Anal 50(5):815–822

Chapter 9
Postextraction Considerations

Ivelisse Colón

Abstract After extracting APIs and components of interest from dosage form samples, the nature of the resulting sample extracts often requires additional pretreatment steps prior to quantitative analysis. This complexity is a direct result of the presence of excipients with different physical properties in the dosage forms. This chapter summarizes the factors that should be taken into consideration prior to the quantitative analysis of a pharmaceutical dosage form extract. Current strategies to overcome, troubleshoot, and minimize potential problems are also presented through the discussion of relevant case studies.

9.1 Introduction

Ideally, the successful extraction of any target analyte from a given formulation will yield a stable extract that can be analyzed without further processing to obtain an accurate quantitative result. However, many factors can complicate the sample preparation of pharmaceutical dosage forms. Some of these factors are presented in the "mind map" shown in Fig. 9.1 and will be discussed further in this chapter.

The active extraction of pharmaceutical dosage forms is complicated predominantly by the presence of multiple excipients, each one with very different properties. Therefore, in most cases, these extracts need to undergo further treatment for accurate detection and quantitation. Furthermore, the extraction procedure in itself may require large amounts of solvent mixtures that present challenges for analyte detection in terms of sensitivity and quantitation limit. This is particularly important for formulations containing lower drug loading and for the detection of low level

I. Colón (✉)
Quality: CMC & QbD, Vertex Pharmaceuticals Inc.,
130 Waverly Street, Cambridge, MA 02139, USA
e-mail: ivelisse_colon-rivera@vrtx.com

B. Nickerson (ed.), *Sample Preparation of Pharmaceutical Dosage Forms: Challenges and Strategies for Sample Preparation and Extraction*, DOI 10.1007/978-1-4419-9631-2_9,

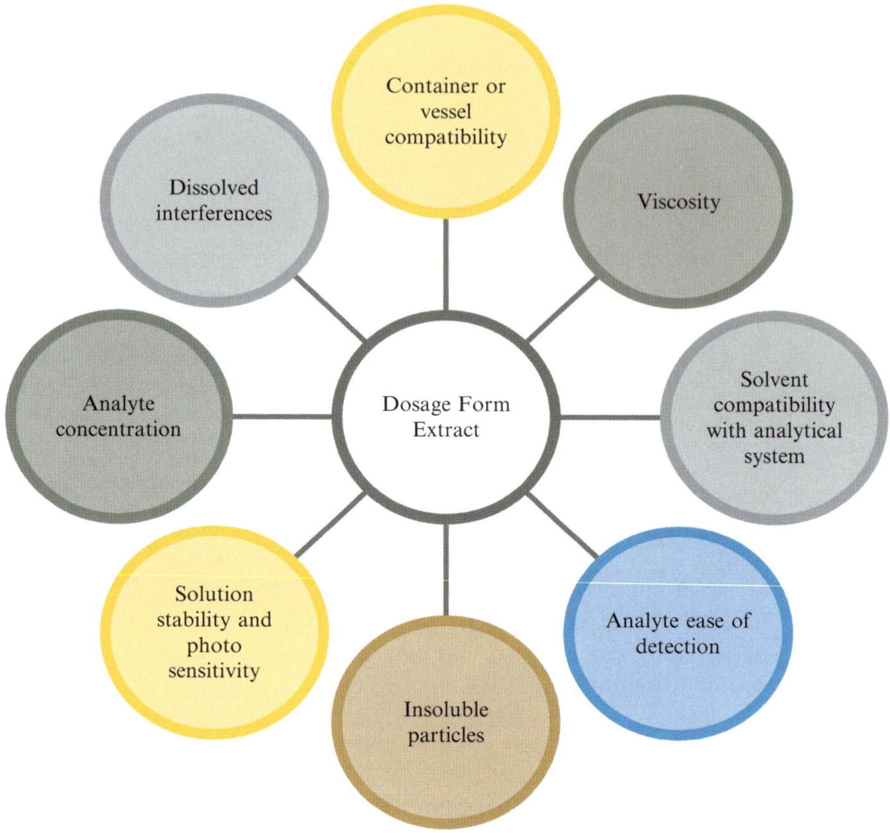

Fig. 9.1 Some factors to consider prior to analysis of dosage forms extracts

impurities, where a preconcentration step might be required. In addition, extractions sometimes require solvents that might not be compatible with the chromatography system downstream. In these cases, dilutions and solvent exchanges might be required. As an example, topical formulations might require nonpolar solvents for extractions (e.g., methylene chloride, chloroform) that would not be compatible with a reversed-phase chromatographic system.

Postextraction treatments do present some challenges and this is why it is preferred to move straight from extraction to quantitative analysis. The concerns associated with sample manipulations are related to factors such as additional validation exercises to demonstrate the suitability of the steps taken, stability or contamination concerns resulting from additional sample handling, additional training for analysts conducting the assays and concerns for potential problems with method transfers and robustness issues. Also, the cost of consumables and hands-on analyst time can be a factor. In summary, postextraction steps should be minimized. However, with current advances in alternatives available, these treatments prior to analysis can be of routine use without any major hurdles and the benefits outweigh the effort to demonstrate suitability so that they do not interfere with the accurate and precise determination of the desired analytes. For example, with appropriate sample extraction and clean-up, chromatographic analysis might not be necessary and determinations using simple UV technology could be feasible. Each of the factors shown in Fig. 9.1 is discussed in detail in the sections below.

9.2 Eliminating Insoluble Particles

In a majority of extractions of pharmaceutical dosage forms, there are insoluble components remaining in the extract that would make them unsuitable for immediate quantification by an analytical system. Having undissolved material is not necessarily a bad situation. In fact, this might be the desired state of a final sample extract. Ideally, the extraction of the active and related substances should be complete but the matrix components (i.e., excipients) should stay in the solid state, as it is easier to remove these undissolved components downstream. Eliminating undissolved material from the extracts will extend the life of consumables (i.e., columns, frits, internal filters, etc.), eliminate unnecessary baseline noise, and prevent pressure build-up in the analytical system in the case of chromatographic determinations.

As the first step, centrifugation should be considered due to its simplicity and low cost. Sample extracts are transferred into a conical shaped tube and centrifuged at a predetermined speed according to the size of the particles present. Centrifugation is usually achieved in less than 10 min. Lately, analytical laboratories have been upgrading their centrifuges to accommodate trays that hold HPLC vials directly, so that the transfer to a centrifuge tube is eliminated. In this case, the depth of the LC needle is manipulated so that it does not pick-up any of the undissolved material in the bottom of the vial.

If the particles are too fine, centrifugation might not offer the desired results and filtration might be necessary. Filters are sometimes seen as merely a screen or a sieve, when in reality they have a three-dimensional structure and act more like a labyrinth. Several considerations are important when selecting a filter:

(a) Nature of the analyte
(b) Extract solvent
(c) Extractables and leachables
(d) Pore size – based on size/amount of undissolved material and the particle size of the analytical column

Table 9.1 summarizes filter recommendations in terms of analyte nature and the extract solvent.

Each filter membrane is available in different formats for manual or automated use. The syringe filters are the most popular for analysis of pharmaceutical preparations.

Table 9.1 Filter membrane selection based on analyte and solvent properties

Filter	Membrane	Analyte properties	Aqueous solutions?	Incompatible organic solvents
GHP	GH polypropylene (hydrophilic)	Acids or bases	Yes	None
PVDF	Polyvinylidene fluoride (hydrophilic)	Acids	Yes	Acetone, DMF, DMSO, MEK, THF
Nylon	Nylon	Bases	Yes	None
PTFE	Polytetrafluoroethylene (hydrophobic)	Acids or bases	No	Chloroform[a]
PES	Polyethersulfone (hydrophilic)	Acids or bases	Yes	Most organic solvents
Glass	Borosilicate glass fiber	Acids or bases	Yes	THF[a]
CA	Cellulose acetate	Acids or bases	Yes	None
RC	Regenerated cellulose	Acids or bases	Yes	None

[a]Limited compatibility

Filters with pore size distribution around 0.45 μm are suitable for most applications, although high efficiency LC columns with particles of <3 μm usually require filtration of particles of ≤0.2 μm. This is especially important for the growing application of UPLC technology. In many cases, the amount or size of undissolved particles necessitates filters with bigger pore size (i.e., 1 μm) to avoid filter clogging as a first step, followed by a second filtration with filters of decreasing pore size.

It is important that the selected filter is validated during the validation exercise to ensure that the active or related impurities are not lost to the filter and that no significant interferences are generated by the selected filter. As part of this process, one needs to study the effect of wetting the filter prior to sample collection. Usually the first 5 mL of the eluate are discarded to obtain acceptable and reproducible recoveries. This validation is usually incorporated in the comparison of linearity samples with and without the presence of excipients.

The filtration step can be automated when using a tablet processing workstation (TPW) as described in Chap. 12 and by Opio et al. for the automation of the sample preparation of powders for oral suspension (Opio et al. 2011). In addition, the use of precolumn filters might help when double filtration is necessary. Zacharis et al. described on-line filtration for the analysis of acyclovir formulations by sequential injection chromatography (Zacharis et al. 2009). In this application, the authors used a flow injection analysis (FIA) system equipped with a manifold to automatically filter through a 0.45 μm filter as the samples (composite of tablets ground to a fine powder and dispersed) were aspirated into the FIA system. A monolithic column was used to separate acyclovir and its major impurity guanine prior to UV detection. With this arrangement, the authors were also able to perform automated dilutions of the extract. As in this example, the filtration step could be automated in the laboratory using readily available syringe pumps.

9.3 Reducing Extract Viscosity

Some excipients do not dissolve in the solvents commonly used in the extraction procedures (e.g., acetonitrile, methanol, tetrahydrofuran, etc.). In the ideal case, these excipients are completely insoluble and can be filtered out of the extract as described in the previous section. Unfortunately, in many cases, excipients either partially dissolve in the selected solvent or swell considerably. This is the case for many excipients used in modified release formulations. If swelling occurs, the viscosity of the resulting extracts might be too high to analyze as is. Molecular weight filtration should be considered in these cases where high molecular weight excipients are present in the extracts. Commercially available centrifugal devices equipped with cellulose membranes of different molecular weight cut-offs (MWCO) can be used (see schematic in Fig. 9.2). In our laboratories, this type of filtration has been successful in reducing the presence of carbomers in extracts of topical gels and of coagulant polyethylene oxide from extracts of osmotic tablets. For the topical gel, the formulation was dissolved in a 2% $MgCl_2$ solution and vortexed to break up the gel. This extract was placed into the centrifugal device with a 10 kDa MWCO membrane and spun for 10 min at a relative centrifugal force (RCF) of ~2,000g. These filters can dramatically reduce extract viscosity and subsequent problems with the LC system. In the case of osmotic tablet extracts (with coagulant polyethylene oxide present), the viscosity was dramatically reduced from 5.5 to 1.8 cP by using 10 kDa MWCO filters. It is important to note that the use of these filters might not be trivial. RCF instead of rpm should be the measurement used and it can be challenging to obtain the same RCF from different centrifuges, presenting problems during method transfers. Also, it can take more than 30 min to centrifuge enough sample for a LC analysis. If wetting of the filter is necessary for acceptable recovery, this type of filters might increase the analysis time as the centrifuge has to be stopped, the contents of the device discarded to waste, and new sample placed into the same device before restarting. These MWCO filters also increase the overall costs of the sample preparation procedure, as each device is approximately $5 USD. Solid-phase extraction might be a sensible alternative for reducing viscosity as discussed in Sect. 9.4.2.

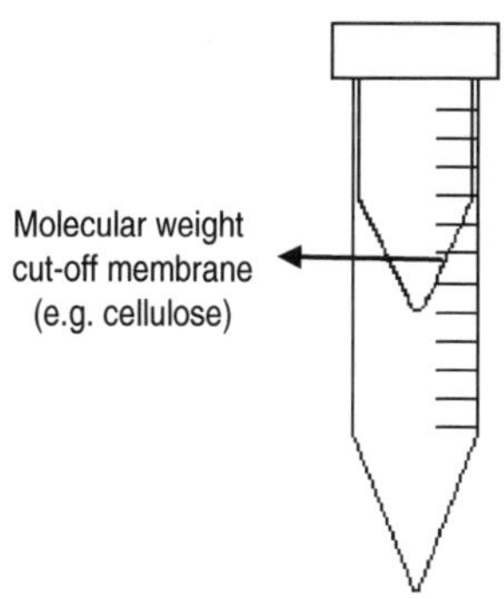

Fig. 9.2 Schematic diagram of a centrifugal device for molecular weight filtration

9.4 Increasing Signal to Noise by Enhancing Analyte Concentration and Reducing Matrix Effects

Achieving the desired limits of quantitation and detection (LOQ and LOD, respectively) for a method can be challenging, especially in the case of low dose formulations or for those formulations where a large diluent volume had to be used to ensure a complete extraction. To decrease the LOQ and LOD of a particular method, two main approaches must be followed as these are directly dependent on both the signal due to the analyte concentration in the extract and a reduction in the "noise." The first approach is to increase the analyte concentration in the sample extracts prior to analysis, and the second approach is to minimize the matrix "noise." In the case of sample preparation, this "noise" mostly refers to interferences from the formulation matrix, including all excipients and API related impurities. Both of these approaches are interrelated in pharmaceutical sample preparation and will be discussed jointly, since many approaches accomplish both goals at the same time.

9.4.1 "Classic" Approaches

Evaporation of organic solvents in an extract has traditionally been employed to increase the concentration of an analyte prior to analysis. This method can be performed by either exposing a sample to a flow of a highly pure inert gas (e.g., nitrogen) or by using a vacuum system. This method is not commonly used for pharmaceutical preparations due to the high content of aqueous solvent in the extracts, although several USP monographs follow this approach for assay methods. The USP monograph for butabarbital assay in oral solution is an example where pure nitrogen is used to reduce the total volume of the sample for better detection. After extraction with chloroform, a 2 mL aliquot of the extract was mixed with 2 mL of an internal solution and its total volume reduced from 4 to 1 mL to improve detection (USP Monograph for Butabarbital Oral Solution 2009). Although time consuming, this approach has also been successful when a complete solvent change is necessary for compatibility with the downstream analytical technique. This is the case for the erythromycin identification test described in the USP Monograph for Erythromycin Estolate Oral Suspension. After extraction with chloroform, the extract was evaporated to dryness and the residue dissolved in methanol prior to spotting on a thin-layer chromatography (TLC) plate (USP Monograph for Erythromycin Estolate Oral Suspension 2009). The approach is also common when isolating impurities by preparative HPLC for further identification by MS and NMR (Prabhu et al. 1992; Pan et al. 2006). It is important to consider that the drying process may require multiple container transfers that can increase the risks of sample contamination.

Another method based on physical properties is selective precipitation based on the solubility and pK_a properties of the API, excipients, and impurities. This method is rarely used, but can be very efficient in isolating the component of interest from an extract. Alsante et al. discussed a case study where a photodimer degradant was isolated from a sample by selective precipitation with acetone (Alsante et al. 2011). This procedure allowed the isolation of this degradant for complete characterization by NMR. Precipitation has been a commonly used methodology for minimizing matrix interferences in biological sample preparation, particularly blood analysis. Prior to extraction, samples are treated with an organic solvent (usually acetonitrile) to denature proteins and precipitate them out of the matrix (Polson et al. 2003). This same concept has been applied to the removal of polymeric excipients from formulation extracts. This approach has been employed for the analysis of niacin and niacinamide in vitamin preparations, where the excipients are precipitated with ethanol prior to analysis by TLC (Sherma and Ervin 1986; Sarangi et al. 1985). In our experience, complete extraction of an active can be achieved with a 100% organic system after dispersing the formulation with a minimal amount of water. This extract is subsequently diluted with an aqueous system for compatibility with the LC conditions. Many excipients are soluble in the organic system, but they will crash out during the final dilution. However, this approach is not commonly employed as a postextraction treatment as in many cases the API can be selectively extracted without dissolving all excipients or the dissolved excipients might not cause major interferences in the subsequent HPLC analysis.

Following the same concepts, if a drug–excipient interaction might be occurring, other extract treatments might be necessary to obtain acceptable recovery values. As discussed in Chap. 6, the addition of complexation agents or a change in pH might be necessary prior to chromatographic analysis.

9.4.2 Solid-Phase Extraction

In solid-phase extraction (SPE), pharmaceutical samples are subjected to an adsorbent phase for either retention of the analytes of interest to increase their concentration or to remove matrix interferences. The latter has been the focus of many SPE procedures in the bioanalytical field, for the rapid cleaning of blood and tissue extracts (James 2008). The concepts and the theory behind SPE extractions have been described in detail in a book chapter by Poole (2002) and summarized in Chap. 4 of this book.

Without question (as reflected by the number of applications published), the most frequent use of SPE in pharmaceutical determinations is to clean-up extracts prior to analysis, especially for bioanalytical samples or parenteral preparations. Krailler and coauthors described a simple procedure for the analysis of doretinel in a topical gel by cleaning the extracts from the high molecular weight thickening agent using a C18 cartridge (Krailler et al. 1991). It is important to note that to accomplish a sample clean-up, the analyte does not necessarily have to be retained in the adsorbent phase. As an example, in the author's laboratory, an extract containing a proprietary hydrophobic nonionizable compound was successfully cleaned from a coagulant polyethylene oxide matrix by using a silica SPE cartridge. These extracts were highly viscous due to the swelling of this polymeric matrix during the

Table 9.2 Case study results for cleanup of an osmotic tablet extract by SPE

Sample	%Recovery	Viscosity (cP)
Original	N/A	5.5
SPE Si (500 mg)	98	2.5
SPE Si (1,000 mg)	100	1.9
SPE Si (2,000 mg)	98	1.4

extraction of an osmotic controlled release formulation. The extraction consisted of a step-wise addition of solvents performed on a horizontal shaker at 280 opm. The first step included 3 h of shaking using acetonitrile followed by an additional 3 h after adding methanol. Samples were then diluted to volume with water for compatibility with the LC conditions. In this case, the analyte was not retained in the SPE cartridge, but the tight packaging of the silica (Si) particles acted as a barrier to retain the viscous components. The cleanup efficiency was tracked in this case by viscosity measurements of the final extracts and the % recovery of the active. The results for this application are summarized in Table 9.2. Note that the results are comparable to the MWCO filtration procedure mentioned in Sect. 9.3 in less time and with lower cost. In this case, the amount of silica in the cartridge was only limited by the ability to pass the viscous solution through the cartridge.

It is important to mention that SPE can be used for desalting extracts as well. A desalting step is necessary prior to the analysis of many parenteral formulations, particularly when using MS detection. For this purpose, it is usually common to retain the analyte in a C8 or C18 phase and wash out the cartridge with water to minimize the salt concentration. The analyte can then be eluted with acetonitrile or methanol for further analysis. This was described by Gilar et al. for the purification of a biopolymer prior to MS analysis (Gilar et al. 2001).

To maximize the benefit of a SPE procedure, most applications target both minimization of interferences and preconcentration of the active simultaneously. Hashem and Jira reported a simplified method for analysis of corticosteroids in tablets to improve the commonly used approach of solvent extraction (Hashem and Jira 2005). Tablets were ground to a fine powder, weighed, and loaded directly onto a prewashed C18 cartridge. These were then eluted with methanol into a volumetric flask and diluted to volume for quantitative assay. This simple procedure replaced a method where emulsions were problematic in a liquid–liquid extraction. This method was validated with LOD's close to 6 ng, recovery values >89%, and %RSD's of 6% or less.

The on-line modality of SPE has been more popular for pharmaceutical applications due to simplicity and potential efficiency gains. In on-line applications, the adsorbent phase is contained in the form of a precolumn directly connected to the flow path of a chromatographic system. Therefore, the loading and elution process is completely automated. This mode has been very popular for the on-line clean-up of pharmaceuticals from bioanalytical extracts, including bromazepam in plasma (Goncalves et al. 2005) and the simultaneous determination of lamivudine and zidovudine from serum (Estrela et al. 2004) prior to LC/MS analysis. Knochen and Giglio described a very simple and effective application of on-line SPE for the sensitive analysis of phenylephrine hydrochloride in a syrup formulation by FIA with UV detection (Knochen and Giglio 2004). A 2.6 cm column was packed with an anionic resin in a multiple valve system controlled by a program written in QuickBASIC. The column was washed with a buffered system at low pH and the sample loaded for 5 s to preconcentrate phenylephrine. The column was then rinsed to minimize the matrix interferences from the formulation prior to the elution with a stronger basic solution (0.1 M sodium hydroxide). To enhance UV detection, an additional valve introduced the necessary reactants to carry out the Emerson reaction with 4-aminoantipyrine and potassium hexacyanoferrate (III). Conley and Benjamin also utilized a C18 precolumn incorporated in the injection loop of a HPLC system for the analysis of a triple corticoid system and sulconazole in creams (Conley and Benjamin 1983). With this set-up, the nonpolar excipients were strongly retained in the precolumn loop, allowing the detection and quantitation of all actives without the need of additional off-line clean-up procedures.

Theoretically, immunosorbents are ideal candidates for successful preconcentration of an analyte and elimination of interferences due to their high selectivity. However, it is understandable that not many applications exist for their routine use in dosage form analysis due to the difficulties and cost associated with the preparation of antibodies for a specific analyte. Molecularly imprinted polymers (MIPs) present a sensible alternative and a few applications have been published utilizing them as platforms for SPE adsorbents. MIPs can be thought of as plastic antibodies (Poole 2002) prepared by molecular imprinting. This technique is used to prepare polymers with synthetic recognition sites having a predetermined selectivity for a specific analyte (or family) by the polymerization of functional and cross-linking monomers in the presence of a template molecule (the analyte). The adsorption and desorption of analytes from these phases are usually governed by hydrophobicity partition. Even though this is not a trivial procedure, several applications have been published where MIPs are used as SPE adsorbents for the successful extraction of pharmaceutical actives from dosage forms. Zander et al. described a procedure for the analysis of nicotine and its degradation products in nicotine chewing gum (Zander et al. 1998). For this purpose, MIPs were synthesized by polymerizing monomers of methacrylic acid (MAA), trimethylolpropane trimethacrylate (TRIM), and ethyleneglycol dimethacrylate (EDMA) in the presence of nicotine. The optimized procedure was able to preconcentrate the potential degradation products and minimize matrix interferences without utilizing liquid–liquid extraction steps with hazardous organic solvents. Hu and coauthors were able to synthesize MIPs for the selective extraction of trimethoprim from tablets (Hu et al. 2005). These MIPs were also based on MAA and EDMA monomers. Rezaei et al. also utilized MAA and EDMA monomers for the preparation of selective MIPs for the analysis of piroxicam from a capsule formulation containing vitamins B complex (Rezaei et al. 2008). The concentration of piroxicam in the eluants was determined by UV spectrophotometry. Because of the preconcentration and selectivity offered by the MIP phase, the detection limits were much lower than previously reported methods (0.10 ng/mL).

Table 9.3 Example excipient families used in ointment formulations (base/components)

Excipients		
Glycols	Lanolin	Glycerides
Petrolatum	Squalane	Vegetable oil
Camphor	Fatty acids/alcohols	Menthol

9.4.3 Liquid–Liquid Extraction (LLE)

As described in Chap. 4, liquid–liquid extraction (LLE) still has applicability to the extraction and preparation of dosage forms (some examples of USP monographs utilizing LLE were discussed in Sect. 9.4.1). LLE is highly effective for the cleanup of complex formulations, such as ointments and creams that contain a high amount of hydrophobic excipients and additives. Some of these excipient families are listed in Table 9.3. The addition of a LLE step will help reduce the presence of these excipients in the extracts and will increase extraction efficiencies.

As an example, Okamoto and coauthors described the analysis of seven active components in a commercial ointment by hydrophobic interaction electrokinetic chromatography (HIEKC) (Okamoto et al. 2001). To avoid excipient interferences, 0.5 g of the ointment was extracted with either hexanes or tetrahydrofuran (depending on the active being determined). After vigorous shaking to dissolve the excipients, the samples were subjected to either a mixture of methanol/formic acid or pure ethanol. An aliquot of the supernatant was then evaporated to dryness and reconstituted with a solvent system compatible with the HIEKC conditions. The LLE procedure was successful in minimizing interferences from the matrix and the cleanliness of the extracts allowed great reproducibility and quantitation limits lower than 50 μg/mL for all actives.

Miniaturization of the LLE procedure has increased the throughput and ease of these procedures. For a proprietary API, we were able to develop methodology involving LLE conducted directly in a HPLC vial and used to quantitate alkyl halide residues by GC. A volume (i.e., 1.00 mL) of API dissolved in an aqueous system (e.g., phosphate buffer pH 4.5 and acetonitrile) was transferred into a 2.8 mL standard chromatography vial. Another 1.00 mL of hexanes were added to the vial and agitated. After settling, the vial was placed on a GC autosampler. The needle of the autosampler was depth-programmed to pick up 1 μL only from the top organic phase. In this manner, interferences from the API were ionized and eliminated through the aqueous phase in an automated fashion. The small volumes used allowed the desired LOQ without the need of further transfers or steps. A few applications of "in-vial" LLE have been published in the literature, including the simultaneous determination of ropivacaine and bupivacaine in human plasma by GC/MS (Abdel-Rehim 2002). In this example, 400 μL of plasma (after adjusting pH and ionic strength) were mixed with 800 μL heptane containing 20% dichloromethane in a 2 mL vial. The vials were shaken for 10 min and centrifuged before injecting 50 μL of the organic phase into the GC. This version of LLE has made the technique much more attractive due to the improvements in sample handling (no transfers) and the small volumes used, eliminating the need for preconcentration steps prior to analysis.

9.4.4 Ionic Liquids

Ionic liquids are generally described as salts with a melting point below 100°C. Their general properties are very unique: often moderate to poor conductors of electricity, nonionizing (e.g., nonpolar), highly viscous, low vapor pressure, excellent thermal stability, and favorable solvating properties for a range of polar and nonpolar compounds. Therefore, many applications have been published in a wide range of fields including cellulose processing, solvents for many types of reactions, paint dispersants, transportation of highly reactive gases, solar energy, and waste recycling (Short 2006; Plechkova and Seddon 2008). Many applications of ionic liquids related to chromatography (as additives or stationary phases) have been published and, more recently, applications related to their use as extraction solvents have emerged. These ionic liquids can be used to prepare SPE sorbent materials as when combined with a silica substrate (Fontanals et al. 2009). The advantages of ionic liquids for headspace volatilc analysis in pharmaceuticals have been particularly recognized by Laus and by Andre et al. who have demonstrated their use for residual solvent analysis as well as for the analysis of volatile residual impurities (Andre et al. 2005; Laus et al. 2009). For these applications, ionic liquids offer the advantage of a negligible vapor pressure, chemical and physical stability, and are relatively inert. Because of these physical properties, ionic liquids have expanded the applicability of headspace analyses, as compounds with a little higher vapor pressure can now be analyzed due to the fact that increasing the temperature would not cause a substantial increase in vial internal pressure. In terms of pharmaceutical dosage forms, ionic liquids can be used to minimize matrix interferences when conducting headspace trace analysis, as demonstrated by Laus et al. for the analysis of sulfolane in Biocef tablets (Laus et al. 2009). The tablets were ground and dissolved in only 1.0 mL of the ionic liquid (1-*n*-butyl-3-methylimidazolium dimethyl phosphate) and analyzed directly by headspace GC/MS. It is important to note that one of the advantages of using ionic liquids for extraction is their ability to dissolve carbohydrate-derived excipients, increasing the chance of a successful extraction.

Furthermore, ionic liquids can be used to prepare aqueous two-phase systems (ATPS). These systems result from the mixing of two different polymers or by mixing one polymer with certain salts at high concentration. These are considered environmentally friendly (opposed to conventional LLE) and can be considered for simultaneously cleaning, extracting and enriching a sample. This is particularly true when most have the capacity of dissolving the complete pharmaceutical matrix. Li and coauthors described the preparation of an ATPS for the HPLC analysis of major opium alkaloids in *Pericarpium Papaveris* (Li et al. 2005). The complete extraction and preconcentration was achieved using less than 5 mL total volume, allowing sensitive analysis of codeine and papaverine with detection limits of 30 and 20 ppb, respectively, without any further processing.

9.4.5 Minimizing Interferences by Data Processing Methods

As previously discussed, minimizing the number of steps is always desirable to maximize efficiency in an extraction and sample preparation procedure. To this end, recent advances in data processing and chemometrics allow the analysis of target analytes even when a finite amount of interferences are present in the extracts. Furthermore, the analysis technique can be simplified from using chromatography to spectroscopy by adding these data treatment methods. Although additional model validation might need to be performed, these treatments can save consumable costs while increasing sample throughput once validated. Details of this approach are presented in a separate chapter of this book, but some applications are discussed below as they relate to the reduction of interferences in extracts.

Earlier multivariate applications involve the use of simple derivative UV spectrometry. Bonazzi et al. describe the determination of imidazole antimycotics in creams utilizing UV spectrometry after a supercritical fluid extraction (Bonazzi et al. 1998). Approximately 10 mg of the cream were subjected to four static equilibration cycles with one dynamic extraction cycle. The procedure utilized pure CO_2 followed by CO_2 modified with 10% methanol. The extract was then passed through ODS (hypersil) SPE material to trap the analytes for further elution. Even with the use of SFE and SPE, the extracts still contained residual amounts of excipients due to the complex matrix. Also, it was desired to achieve conditions that could provide sensitive quantitation of four different imidazole drugs simultaneously. Instead of following a more classical and potentially time consuming approach utilizing an additional SPE step followed by chromatography, the authors minimized all interferences by taking the second derivative of UV spectral data. All recoveries were >97% of label claim with calibration correlation coefficients >0.995 for all analytes. Toral et al. also utilized derivative spectrophotometry (first order) for the simultaneous analysis of dapsone and pyrimethamine in tablet formulations (Toral et al. 2003). In this case, the tablets were crushed and dispersed in acetonitrile. After centrifuging, the supernatant was analyzed directly by UV employing first derivative for minimizing baseline noise while allowing the simultaneous determination of the two active drugs.

More sophisticated methods of data processing have been applied for the simultaneous analysis of actives from pharmaceutical formulations without treatment of extracts or using chromatography. De Luca et al. describes a very comprehensive design of experiments to build a calibration set that can be used to determine $1-N$ components in different mixtures (De Luca et al. 2006). This design was used to analyze a drug system with 1 through 4 components in different combinations by multivariate UV processing methods involving principal component regression and partial least squares. The authors developed a computer algorithm that could calculate a calibration set matrix (minimizing the number of samples needed) based on user's input data: concentrations, number of components, and mixture type (binary, tertiary, etc.). Unlike the derivative UV methods previously described, these multivariate methods utilize the full UV spectra, maximizing the amount of analytical information used. Nonetheless, the selection of wavelength range is still critical to the construction of an accurate model.

It is important to note that these methods can yield very good recoveries and accuracy, but difficulties can be encountered for more complicated matrices and in cases where the drug loading is low, increasing the noise in the spectral data and therefore, the error in the prediction values from the models. As with any other analytical procedure, these models must be subjected to a validation process involving recoveries, precision, error in predictions (accuracy), and the model fit to the analytical data.

9.5 Improving Analyte Detection by Derivatization

In some instances, postextraction steps are necessary to be able to accurately detect a target analyte. Although there have been considerable advances in specialized detection systems for species lacking chromophores (i.e., light scattering detector [LSD], charged aerosol detector [CAD], MS, etc.), these types of detectors might not be readily available to use for commercial and/or compendial methods. Derivatization is an alternative to these systems, where a high yield, fast, and selective reaction is conducted to add a "tag" to the target analyte so that it can be sensitively seen by the selected detection system (e.g., chromophores, fluorophores, N or P containing groups, etc.). An added bonus to these derivatization reactions are the potential gains in sensitivity by a selective tag possessing special properties for detection (e.g., higher UV or visible absorption wavelengths). It is important to note that, derivatization does not always imply tagging but can also mean the conversion of the analyte of interest into a different derivative by rearrangements induced by exposure to light, acid, base, oxidizing agents, and so forth. Complexation reactions are also helpful, especially in spectrophotometric detection. One example of this is the determination of 4-aminophenol by derivatizing with sodium nitroprusside described by Bloomfield (2002). Table 9.4 presents some commonly used derivatizing agents (tags) for pharmaceutical and biomedical analysis. Very thorough reviews (Krull et al. 1994; Görög 1998) exist to help in the evaluation and selection of appropriate derivatizing agents. If the reactions are selected carefully, the analytical procedure can be conducted without any major hurdles. The method validation should include the evaluation of the completeness of this reaction as well as its irreversibility, selectivity, and reproducibility. The need for internal standards should be evaluated during method development.

Several methods in the US Pharmacopeia (USP) utilize derivatization procedures to improve detection of species lacking chromophores. Tobramycin, an aminoglycoside

Table 9.4 Some examples of derivatizing agents used in pharmaceutical applications for UV or fluorescence detection

Reagent	Target analytes	References
5-Dimethylaminonaphthalene-1-sulfonyl-chloride (Dans-Cl)	1° and 2° amines, phenols	Krull et al. (1994)
(9-Fluorenyl) methyl chloroformate (FMOC)	1° and 2° amines	Krull et al. (1994), Liu et al. (2009), and Narola et al. (2010)
7-Chloro-4-nitrobenzo-2-oxa-1, 3-diazole (NBD-Cl)	1° and 2° amines, phenols and thiols	Krull et al. (1994) and El-Elmam et al. (2004)
o-Phthaldialdehyde	Amines, alcohols	Krull et al. (1994)
Hexylchloroformate (HCF)	1° and 2° amines	Vanhoenacker et al. (2009)

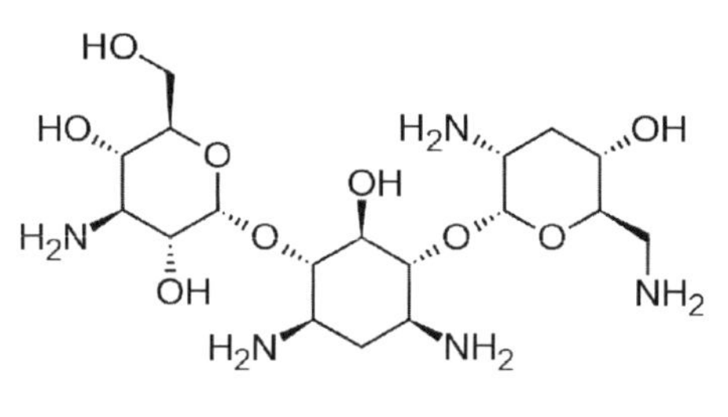

Derivatizing Agents

2,4 Dinitrofluorobenzene +

Tris(hydroxymethyl) aminomethane

O – phthaldialdehyde

Fluorescamine

Dansyl Chloride

Chloranil

Fig. 9.3 Chemical structure of Tobramycin and derivatizing agents that have been employed for its detection by UV

antibiotic, is an example. Figure 9.3 shows the chemical structure of Tobramycin. The USP Monograph for Tobramycin cites a derivatization procedure involving the reaction with 2,4-dinitrofluorobenzene and tris(hydroxymethyl)animomethane (USP Monograph for Tobramycin 2009). Other derivatization reagents have emerged for this compound (and related compounds) as discussed by Sampath and Robinson (1990).

Derivatization procedures can be conducted on-line or off-line. In on-line methods, the reaction used for derivatization occurs within the instrument being used for separation and detection. Some also include chromatographic vials in autosamplers in this classification. Xu et al. (2010) reported an on-line derivatization procedure for the analysis of nonsteroidal antiinflammatory drugs (NSAIDs) by GC. These families of compounds have a carboxylic acid moiety and can be ion-paired with tetrabutylammonium hydrogen sulfate. Once these pairs enter the high temperature environment in a GC injection port, they form the butyl ester of the corresponding acid. In this manner, they were able to be determined by the more conventional GC/MS and the higher mass also presented some sensitivity advantages. Since GC/MS is such a desirable technique, it is common to use derivatization to improve volatility and/or thermal stability so that GC analysis is feasible. On the one hand, the improved volatility results from the minimization of polar groups in the analyte moiety (e.g., –OH, –NH, –SH). On the other hand, decreased volatility could also be a reason for derivatizing,

Table 9.5 Some examples of derivatizing agents commonly used for GC analysis (Regis 2010)

Reagent	Type of reaction	Example compounds
Bistrimethylsilylacetamide (BSA)	Silylation	Alcohols, amides, carboxylic acids
Bistrimethylsilyltrifluoroacetamide (BSTFA)	Silylation	Alcohols, amides, amines, amino acids, carboxylic acids, sulfonamides
Trimethylsilyldiethylamine (TMS-DEA)	Silylation	Amino acids, carboxylic acids
Fluorinated anhydrides	Acylation	Alcohols, amines, nitrosamines, sulfonamides
Pentafluorobenzoyl chloride (PFBCl)	Acylation	Alcohols, 2° amines
Fluoracylimidazoles	Acylation	Alcohols, amines, carbohydrates/sugars, catecholamines
Dialkylacetals	Alkylation	Alcohols, amides, amines, carboxylic acids, sulfonamides
Tetrabutylammonium hydroxide (TBH)	Alkylation	Alcohols, amino acids, carboxylic acids
Pentafluorobenzyl bromide (PFBBr)	Alkylation	Alcohols, sulfonamides

to avoid losses while conducting the sample preparation. If MS is not an option, derivatization could also be used to improve detectability by electron capture detection (ECD) by adding halogenated groups. In summary, three major types of reactions are still employed for these purposes: acylation, alkylation, and silylation. Table 9.5 summarizes some examples of derivatizing agents used to enhance the feasibility of analysis by GC.

Another classification of derivatization procedures is related to its completion before or after the separation step. Precolumn derivatization is the most common mode, but it is also possible to conduct the derivatization procedure postcolumn just prior to detection. The precolumn mode offers the added advantage that the derivatization process itself might help in the resolution and selectivity of the analytes of interest. A classic example is the analysis of chiral compounds (e.g., amino acids) by achiral chromatography. As described by Görög, homochiral reagents (added to samples or to the mobile phase) can be employed to form covalently bonded diastereomers that could be separated by conventional achiral columns (Görög 1998). An example is the use of D or L-*O*-(4-nitrobenzyl)-tyrosine methyl ester to enable the separation of enantiomeric N-protected amino acids. It should be noted that these agents not only help in the separation of the species, but they usually improve sensitivity as the tags contain better absorbers or emitters.

The postcolumn derivatization mode is advantageous when the derivatives are not stable and have to be detected within a short timeframe from preparation, as it is the case for aminoglycosides (Fabre et al. 1989). *Ortho*-phthalaldehyde is the most popular reagent for derivatizing this family of compounds. Although the reaction can be carried out pre or postcolumn, the derivatives are unstable and a postcolumn arrangement is preferred to yield a robust method. Another advantage of the postcolumn approach is that the reaction is carried out for each analyte individually, since these have been "isolated" by the analytical column. This is thought to yield cleaner reactions and uniform conversions across analytes (Pickering 2007). For postcolumn derivatization, the reagents are introduced using a syringe or peristaltic pump after the analytical column. The reaction can be carried out inside a coiled loop and it can be configured to allow heating and/or exposure to UV light. A method for determining several biphosphonates simultaneously uses a postcolumn photoreactor set-up to degrade the analytes into orthophosphate (Perez-Ruiz et al. 2009), which in turn reacts with molybdate to form 12-MPA, a compound that oxidizes thiamine to thiochrome. Thiochrome can be readily detected by fluorescence.

9.6 Ensuring Extract Compatibility with Containers/Vessels

The compatibility of extracts with containers or vessels used in the sample preparation is not usually studied in depth. However, it is important to keep in mind several factors that could be useful when troubleshooting problems. Several examples of pharmaceutical analytes interacting with surfaces have been reported in the literature. These have been discussed in detail by Yahya et al. (1988) and revisited by Nickerson et al. (2009). The main effects of an analyte–container interaction are potency losses and highly variable results that are dependent on the type of solvent used. These interactions are more critical when the analyte is present in lower concentrations (e.g., analysis of low dose formulations).

Sample containers and vials are available in a wide array of materials including borosilicate glass, borosilicate amber glass, silanized glass, polypropylene, polymethylpentene, and many others. In the case of borosilicate glass, polar compounds (e.g., primary amines) could have a strong interaction with the polar silanol groups (Si–O–H) on the glass surface. These interactions are highly dependent on the selection of sample solvent, as they are related to the ionization state of the molecule and its hydrophobic/hydrophilic interactions with both the solvent and the glass surface. To minimize these interactions, silanized (or deactivated) glass is commercially available, where most of these silanol groups have been derivatized with more hydrophobic chains. However, hydrophobic compounds might have a stronger interaction with this type of glass. Hydrophobic compounds may also have a strong interaction with plastic materials. Some strategies to troubleshoot and/or minimize container effects are summarized in Table 9.6. Nickerson et al. discuss an interesting example where the potency of a proprietary compound decreased with time of contact of the extracts with glass surfaces (i.e., sample vials), in particular silanized glass (Nickerson et al. 2009). Variable potency results were obtained, as the first samples in the HPLC analysis run sequence had acceptable results but the results had a clear downward trend in potency as the samples waited in the autosampler to be analyzed. The interaction of this compound with plastic sample vials was even stronger. To minimize the interaction (and obtain constant potency results over time), the authors increased the % of methanol in the diluent to increase the affinity of the analyte to the solvent and disturb its interaction with the vial surface. One has

Table 9.6 Some factors to consider when troubleshooting container compatibility problems

Factor	Rationale
Analyte concentration	If the analyte concentration is increased, interactions with surfaces may become less significant
pH of diluent	Will alter interactions by changing the ionization state of the analyte and/or impurities
%organic	Will affect the partition between the analyte and the surface by increasing/decreasing its affinity for the solvent
Ionic strength	Will influence ionic interactions that might be occurring by competing for ionic sites on the surfaces or on the analyte
Surfactant	The addition of a surfactant (e.g., SDS) to the solvent could help minimize interactions with glass surfaces

to pay careful attention to the fact that these interactions apply to both the main component and related impurities/degradants.

Interactions are not the only factor related to sample containers affecting the integrity of an extract. As will be discussed in Sect. 9.7, one needs to also consider the photo reactivity of the analytes and related impurities to assess if using amber vials/glassware is necessary. Another factor to consider (and evaluate) is the presence of leachables and extractables due to sample containers. Many vendors now offer certified vials that have been prewashed and/or treated to minimize interferences.

9.7 Ensuring Extract Solution Stability

The stability of all solutions must be evaluated as part of the analytical method validation, as recommended by ICH Q2(R1). This evaluation is conducted as part of the method robustness, since instability issues may be a root cause for analytical variability. ICH Q2(R1) does not offer specific guidance related to acceptance criteria. The main focus is to confirm that the measured potency does not change significantly within an established time frame and that new or existing impurities are not growing above a certain established threshold. Usually, aliquots of solutions used during validation (API alone and/or API in the presence of excipients) are left under normal ambient conditions (room temperature unprotected from light). Separate aliquots are also stored under refrigeration and at ambient temperature protected from light. Samples are tested at specified time intervals (e.g., 24, 48, 72 h, etc.) and compared against freshly prepared standards. Several examples will be briefly discussed to highlight potential issues, points for consideration and potential solutions.

Compound A was an early development candidate for which the chromatographic LC method involved the use of 0.01% TFA (trifluoroacetic acid) as mobile phase A and 0.01% ACN (acetonitrile) as mobile phase B with a starting ratio of 70% A/30% B. As such, the initial extraction solvent for the IR tablet method was selected as 70:30 0.01% TFA/ACN based on matching the LC conditions, existing solubility data, and preliminary recovery experiments. When conducting the solution stability

Table 9.7 Solution stability data for Compound B at ambient temperature unprotected from light

	Initial		Day 3		Day 5	
Conditions	%Area API	%Area Deg	%Area API	%Area Deg	%Area API	%Area Deg
Water	98.6	0.11	93.0	5.9	89.7	9.2
Water/ACN (50/50)	98.7	0.28	98.6	0.33	98.5	0.5
0.1% Formic acid	98.3	0.34	93.3	5.3	90.2	8.8
10 mM NH_4Ac	98.1	0.40	92.5	6.5	88.4	10.6
0.1% NH_4OH	91.6	7.2	7.3	91.0	1.3	96.8

experiments, it was noted that Compound A underwent rapid degradation under normal laboratory conditions to a (M+18) species, up to 0.9% in 24 h and close to 5% in 5 days. Samples under ambient conditions protected from light or under refrigerated conditions showed no significant changes from initial. From this example, two factors are key: the selection of dissolving/extraction solvent and isolating the mechanisms of degradation. Selection of solvent is crucial (as discussed in the previous section), since in this case the presence of TFA significantly increased the rate of degradation of Compound A. It is key to study the effects of acid/basic solvents to the degradation kinetics. The solvent was subsequently changed to water/ACN and the degradation rate decreased by more than half. Placing the samples in different conditions as part of the validation exercise immediately helped identify that photosensitivity was the main cause of degradation. The method then specified the use of amber glassware at room temperature.

Compound B was a development candidate with extreme sensitivity to moisture, undergoing a hydrolysis degradation pathway that considerably reduced the calculated assay value. For the API validated method, ACN alone was used as the dissolving solvent to improve solution stability at ambient conditions. However, peak splitting in the LC was observed when the method was used by different laboratories. The starting LC conditions were 0.1% MSA (methanesulfonic acid)/ACN (95:5 v/v). Another major issue was that the IR tablets needed at least 40% aqueous for complete dispersion and extraction. In this case, the hydrolysis reaction could not be completely avoided, but refrigerated solutions were refrigerated and were stable for at least 5 days due to a considerable reduction in reaction rate at the lower temperature and no impurities changed or formed above 0.2%. As in the previous case, the acid was eliminated from the extraction solvent, although it is important to note that in this case acid catalysis did not occur and the reaction was catalyzed by a basic environment. Table 9.7 summarizes some solution stability experiments and the area % results for both the API and the main hydrolysis degradant.

For extracts of dosage forms, it is also important to consider the effects of the excipients present when studying solution stability. Therefore, it is recommended that solution stability experiments be conducted for both filtered/centrifuged and unfiltered/not centrifuged samples. Compound C was a candidate in development utilizing an osmotic dosage form. This dosage form contained high amounts of a swellable polymer and the final extracts were viscous in nature. Therefore, as time went on, the swelled polymer caused entrapment of the API reflected in the loss of potency during solution stability experiments. The potency changed from 98 to 92% in day 3 of the testing. In these cases, solutions needed to be filtered using a molecular weight filter (as described in Sect. 9.3) within 24 h of the extraction procedure. The filtrates were then stable up to 7 days at ambient temperature.

In conclusion, it is critical to assess the stability of extracts to have a robust method and avoid transfer issues. Some factors to consider are temperature, protection from light, nature of the dissolving/extraction solvent, and the presence of excipients. It is also important to take into account the analyte concentration in the extracts. Solution stability studies should be performed on both stock solutions and subsequent dilutions (if dilutions are necessary) as the stability issues might become more severe as the concentration of the analyte decreases.

9.8 Chapter Summary

The importance of evaluating if postextraction steps are required for the accurate quantitation of an analyte has been established throughout this chapter. As shown in Fig. 9.1, many factors should be considered prior to analysis due to the additional complexity caused by the presence of dissolved, partially dissolved, or undissolved excipients in extracts of pharmaceutical dosage forms. Other considerations regarding stability and compatibility with the equipment used are equally important.

Postextraction steps are geared toward two major goals: the sensitive detection of the analyte of interest and the reduction of the matrix background. Strategies around these two goals were summarized, including ways of increasing analyte concentration, clean-up procedures for extracts utilizing various analytical techniques, improving detection and troubleshooting stability issues. These postextraction steps should be evaluated during method validation and risks for potential method transfer or robustness issues should be identified in a proactive manner.

References

Abdel-Rehim M (2002) Determination of ropivacaine and bupivacaine in human plasma by programmed temperature vaporiser-fast gas chromatography-mass spectrometry (PTV/fast GC/MS) utilizing in-vial liquid-liquid extraction. J Sep Sci 25:252–254

Alsante KM, Baertschi SW, Coutant M, Marquez BL, Sharp TR, Zelesky TC (2011) Degradation and impurity analysis for pharmaceutical drug candidates. In: Ahuja S (ed) Handbook of pharmaceutical analysis, 2nd edn. Elsevier, San Diego

Andre M, Loidl J, Laus G, Schottenberger H, Bentivoglio G, Wurst K, Ongania K-H (2005) Ionic liquids as advantageous solvents for headspace gas chromatography of compounds with low vapor pressure. Anal Chem 77:702–705

Bloomfield MS (2002) A sensitive and rapid assay for 4-aminophenol in paracetamol drug and tablet formulation by flow injection analysis with spectrophotometric detection. Talanta 580:1301–1310

Bonazzi D, Cavrini V, Gatti R, Boselli E, Caboni M (1998) Determination of imidazole antimycotics in creams by supercritical fluid extraction and derivative UV spectroscopy. J Pharm Biomed Anal 18:235–240

Conley DL, Benjamin EJ (1983) Automated high-performance liquid chromatographic column switching technique for the on-line clean-up and analysis of drugs in topical cream formulations. J Chromatogr 20:337–344

De Luca M, Ioele G, Risoli A, Ragno G (2006) Improvement of multivariate calibration techniques applied to 1-to-N component mixtures through an optimized experimental design. Microchem J 83:24–34

El-Emam AA, Hansen SH, Moustafa MA, El-Ashry SM, El-Sherbiny DT (2004) Determination of lisinopril in dosage forms and spiked human plasma through derivatization with 7-chloro-4-nitrobenzo-2-oxa-1,3-diazole (NBD-Cl) followed by spectrophotometry or HPLC with fluorimetric detection. J Pharm Biomed Anal 34:35–44

Estrela Rita De Cassia E, Salvadori MC, Suarez-Kurtz G (2004) A rapid and sensitive method for simultaneous determination of lamivudine and zidovudine in human serum by on-line solid-phase extraction coupled to liquid chromatography/tandem mass spectrometry detection. Rapid Commun Mass Spectrom 18:1147–1155

Fabre H, Sekkat M, Blanchin MD, Mandrou B (1989) Determination of aminoglycosides in pharmaceutical formulations – II. High-performance liquid chromatography. J Pharm Biomed Anal 7:1711–1718

Fontanals N, Ronka S, Borrull F, Trochimczuk AW, Marce RM (2009) Supported imidazolium ionic liquid phases: a new material for solid-phase extraction. Talanta 80:250–256

Gilar M, Belenky A, Wang BH (2001) High-throughput biopolymer desalting by solid-phase extraction prior to mass spectrometric analysis. J Chromatogr A 921:3–13

Goncalves JCS, Monteiro TM, Neves CS, Gram KR, Volpato NM, Silva VA, Caminha R, Goncalves M, Santos FM, Estolano Da Silveira G, Noel F (2005) On-line solid-phase extraction coupled with high-performance liquid chromatography and tandem mass spectrometry (SPE-HPLC-MS-MS) for quantification of bromazepam in human plasma: an automated method for bioequivalence studies. Ther Drug Monit 27:601–607

Görög S (1998) The changing face of chemical derivatization in pharmaceutical and biomedical analysis. Fr J Anal Chem 362:4–8

Hashem H, Jira T (2005) Chromatographic applications on monolithic columns: determination of triamcinolone, prednisolone and dexamethasone in pharmaceutical tablet formulations using a solid phase extraction and a monolithic column. Chromatographia 61:133–136

Hu S-G, Li L, He X-W (2005) Comparison of trimethoprim molecularly imprinted polymers in bulk and in sphere as the sorbent for solid-phase extraction and extraction of trimethoprim from human urine and pharmaceutical tablet and their determination by high-performance liquid chromatography. Anal Chim Acta 537:215–222

ICH Q2(R1): Validation of analytical procedures: text and methodology, October 1994

James C (2008) Sample preparation. In: Venn RF (ed) Principles and practice of bioanalysis, 2nd edn. CRC Press, Boca Raton

Knochen M, Giglio J (2004) Flow-injection determination of phenylephrine hydrochloride in pharmaceutical dosage forms with on-line solid phase extraction and spectrophotometric detection. Talanta 64:1226–1232

Krailler RE, Adams PJ, Lane PA (1991) Quantitation of doretinel in a topical gel using high performance liquid chromatography with solid phase extraction sample clean up. J Liq Chromatogr 14:2383–2398

Krull IS, Deyl Z, Lingeman H (1994) General strategies and selection of derivatization reactions for LC. J Chromatogr B 659:1–17

Laus G, Andre M, Bentivoglio G, Schottenberger H (2009) Ionic liquids as superior solvents for headspace gas chromatography of residual solvents with very low vapor pressure, relevant for pharmaceutical final dosage forms. J Chromatogr A 1216:6020–6023

Li S, He C, Liu H, Li K, Liu F (2005) Ionic liquid – based aqueous two – phase system, a sample pretreatment procedure prior to high performance liquid chromatography of opium alkaloids. J Chromatogr B 826:58–62

Liu Y, Xu F, Zhang Z, Yang C, Song R, Tian Y (2009) Analysis of synephrine in Da-cheng-qi decoction by HPLC employing precolumn derivatization with 9-fluorenylmethyl chloroformate. J Chromatogr Sci 47:925–930

Narola B, Singh AS, Santhakumar PR, Chandrashekhar TG (2010) A validated stability-indicating reversed phase HPLC assay method for the determination of memantine hydrochloride drug substance with UV-detection using pre-column derivatization technique. Anal Chem Insights 5:37–45

Nickerson B, Joseph RM, Palmer C, Opio AM, Beresford GH (2009) Analytical method development: challenges and solutions for low-dose oral dosage forms. In: Zhang J (ed) Formulation and analytical development for low-dose oral drug products, 1st edn. Wiley, New York

Okamoto H, Uetake A, Tamaya R, Nakajima T, Sagara K, Ito Y (2001) Simultaneous determination of ingredients in an ointment by hydrophobic interaction electrokinetic chromatography. J Chromatogr A 929:133–141

Opio AM, Nickerson B, Xue G, Warzeka J, Norris K (2011) Expanding the application of the tablet processing workstation to support the sample preparation of powders for oral suspension. JALA 16:229–234

Pan C, Liu F, Ji Q, Wang W, Drinkwater D, Vivilecchia R (2006) The use of LC/MS, GC/MS, and LC/NMR hyphenated techniques to identify a drug degradation product in pharmaceutical development. J Pharm Biomed Anal 40:581–590

Perez-Ruiz T, Martinez-Lozano C, Garcia-Martinez MD (2009) A sensitive post-column photochemical derivatization/fluorimetric detection system for HPLC determination of bisphosphonates. J Chromatogr A 1216:1312–1318

Pickering M (2007) An overview of post column derivatization methods from a pharmaceutical applications perspective. www.pickeringlabs.com. Accessed Sep 2010

Plechkova NV, Seddon KR (2008) Applications of ionic liquids in the chemical industry. Chem Soc Rev 37:123–150

Polson C, Sarkar P, Incledon B, Raguvaran V, Grant R (2003) Optimization of protein precipitation based upon effectiveness of protein removal and ionization effect in liquid chromatography-tandem mass spectrometry. J Chromatogr B 785:263–275

Poole CF (2002) Principles and practice of solid-phase extraction. In: Pawliszyn J (ed) Sampling and sample preparation for field and laboratory, 1st edn. Elsevier, Amsterdam

Prabhu SV, Ballard JM, Reamer RA, Ellison DK (1992) Isolation and identification of impurities in L-696,229 drug substance. Talanta 40:989–994

Regis (2010) www.registech.com. Accessed Sep 2010

Rezaei B, Mallakpour S, Rahmanian O (2008) A selective solid-phase extraction and preconcentration method using molecularly imprinted polymer for piroxicam in pharmaceutical sample. Anal Lett 41:1818–1831

Sampath SS, Robinson DH (1990) Comparison of new and existing spectrophotometric methods for the analysis of tobramycin and other aminoglycosides. J Pharm Sci 79:428–431

Sarangi B, Chatterjee SK, Dutta K, Das K (1985) Thin layer chromatographic identification and quantitation of niacin and niacinamide in pharmaceutical preparations. J AOAC 68:547–548

Sherma J, Ervin M (1986) Quantification of niacin and niacinamide in vitamin preparations by densitometric thin layer chromatography. J Liq Chromatogr 9:3423–3431

Short PL (2006) Out of the Ivory Tower. Chem Eng News 84(17):15–21

Toral MI, Tassara A, Soto C, Richter P (2003) Simultaneous determination of dapsone and pyrimethamine by derivative spectrophotometry in pharmaceutical formulations. J AOAC Int 86:241–245

USP Monograph for Butabarbital Oral Solution (2009) USP32–NF27, 1729
USP Monograph for Erythromycin Estolate Oral Suspension (2009) USP32–NF27, 2289
USP Monograph for Tobramyacin (2009) USP32–NF27, 3756
Vanhoenacker G, Dumont E, David F, Baker A, Sandra P (2009) Determination of arylamines and aminopyridines in pharmaceutical products using in-situ derivatization and liquid chromatography – mass spectrometry. J Chromatogr A 1216:3563–3570
Xu L, Jiang M, Li G (2010) Injection port derivatization following sonication-assisted ion-pair liquid-liquid extraction of nonsteroidal anti-inflammatory drugs. Anal Chim Acta 666:45–50
Yahya AM, McElnay JC, D'Arcy PF (1988) Drug sorption to glass and plastics. Drug Metabol Drug Interact 6:1–45
Zacharis CK, Verdoukas A, Tzanavaras PD, Themelis DG (2009) Automated sample preparation coupled to sequential injection chromatography: on-line filtration and dilution protocols prior to separation. J Pharm Biomed Anal 49:726–732
Zander A, Findlay P, Renner T, Swietlow A, Sellergren B (1998) Analysis of nicotine and its oxidation products in nicotine chewing gum by a molecularly imprinted solid-phase extraction. Anal Chem 70:3304–3314

Chapter 10
Sample Preparation Method Validation

Edmund J. Bishop, Dawen Kou, Gerald Manius, and Hitesh P. Chokshi

Abstract The analytical procedure describes all the steps and details associated with performing an analysis. The validation process establishes the performance characteristics of the analytical procedure to meet the output requirements for the intended analytical application. During most validation processes, little attention is spent on sample preparation conditions and their effect on the overall analytical method. Potential approaches and practices commonly used for the sample preparation component of the analytical procedure during the validation process are discussed.

10.1 Introduction

Analytical method development and sample analysis consume a considerable amount of time and effort during the pharmaceutical development process. Throughout this process, dependent on the phase of development, the analytical method(s) may undergo redevelopment, validation and revalidation several times. Chemical characterization, formulation development, toxicological studies, human clinical studies, and final registration stability studies all have finite considerations to take into account in order to develop and obtain a validatable analytical method. Sample preparation is by definition a key component of the analysis process, and central to this capability assessment is obtaining the compound in analytically friendly solutions so that meaningful results can be obtained to ensure the safety, efficacy, and purity of the final dosage forms.

E.J. Bishop (✉)
Quality Laboratory Operations, Celgene Corporation,
86 Morris Avenue, Summit, NJ 07901, USA
e-mail: ebishop@celgene.com

B. Nickerson (ed.), *Sample Preparation of Pharmaceutical Dosage Forms: Challenges and Strategies for Sample Preparation and Extraction*, DOI 10.1007/978-1-4419-9631-2_10,

The International Conference of Harmonization (ICH) as well as the United States Pharmacopeia (USP) provide guidelines addressing expectations on the validation parameters for analytical methods. Although sample preparation procedures can be found in individual compendial methods, stand-alone guidelines are not available for validation or demonstration of acceptability of sample preparation. Therefore, sample preparation has traditionally relied on pharmaceutical analysts to determine the appropriate set of conditions and sample treatment prior to the analytical instrument for separation, detection, and assay. The sample handling process for preparation of analytical detection is often marginalized by such simple phrases as "grind-and-find" and "dilute-and-shoot." This chapter will discuss the role of sample preparation in and its impact on method validation, as well as the terminology, common practices, and potential approaches for validation of the sample preparation component of pharmaceutical analysis, with the main focus on dosage forms and API. Sample preparation and validation for bioanalytical methods (e.g., for blood, plasma, tissue, or urine samples) will not be covered here.

10.2 Analytical Performance

The overall scope of validation depends on the purpose and information that one is attempting to obtain from the chosen analytical procedure, e.g., identification tests, quantitative active-drug tests (potency or assay), quantitative tests for synthetic impurities and related substances, or limit tests for impurities/degradants. Other tests that play a formative role in the development and quality control during a drug product's life-span are particle size determination (drug substance) and dissolution (drug product).

The purpose of method validation is to demonstrate that the method meets predefined acceptance criteria in analytical performance. These are defined by specificity, accuracy, precision, linearity, as well as detection and quantitation limits, with HPLC being the main tool for pharmaceutical dosage form analysis. To understand these parameters it is necessary to discuss the underlying principles that govern the performance and expectations surrounding the analytical method particularly in relation to the sample preparation process. Several excellent texts and reviews have been written on analytical validation and the underlying principles (Hokanson 1994a; Hokanson 1994b; Bakshi and Singh 2002; Shabir 2003; Ermer and Miller 2005). It is not the purpose here to discuss each parameter in depth, but to discuss each concept in the context of sample preparation and its role in the validation process.

10.2.1 Specificity

Specificity refers to the ability of the method to separate the analyte(s) of interest from the other components in the sample, including impurities, degradants, sample matrix, and others. It is critically important to be able to ensure that the active drug is differentiated from any potential interferences. Typically, a diluent blank is used to determine if it introduces any peak in the chromatogram; a placebo to demonstrate no interference from the sample matrix; and a reference standard solution to show that the impurities and degradants are resolved from the main peak. Resolution criteria can be applied, and photodiode array detection at varying wavelengths can be used to confirm peak purity.

In some cases, interferences can be removed by sample preparation intervention. For example, depending on the sample preparation method used for solid oral dosage forms such as capsules, the capsule shell may or may not need to be included in the placebo. If a sample preparation method specifies that a capsule needs to be opened to remove the contents for testing, testing the capsule shell as an excipient material of the sample matrix during method validation is then not necessary. This step is usually more common when the capsule material has exhibited a known coelution with the analytes of interest or if simplification of the sample preparation is desired.

10.2.2 Accuracy

By ICH definition, the accuracy of an analytical procedure expresses the closeness of agreement between the value which is accepted either as a conventional true value or an accepted reference value and the found value (ICH 2005). This is sometimes termed as measurement trueness. Further, since it is the deviation from the true represented value, accuracy is a measurement of systematic error and can be estimated as the deviation of the mean measured value from the true representative value.

In the analytical method validation process, accuracy determination is usually demonstrated by "spike-and-recovery" or value comparison using a second, well-characterized procedure whose accuracy has been previously defined and/or established. Recovery studies in the case of drug substance are simple and straightforward, as long as the analytical procedures or solvents used in the solubilization process do not degrade the drug moiety or cause artificial contamination of the drug as it is measured. For validation of drug dosage forms, spiking the drug substances of known purity and quantity (within a predefined quantifiable range of the analytical method) into the excipient matrix, usually made from placebo, is used to simulate the drug dosage form. Care should be taken to ensure that spiking truly mimics the analytical procedures for the actual samples. For example, spiking drug substance into ground placebo tablets without further grinding them together may produce perfect recovery during validation. However, in the analysis of real active tablets, low recovery can occur if during grinding some drug substance sticks to mortar/pestle surfaces and gets locally concentrated. Therefore, grinding after spiking (instead of spiking after grinding) is more simulative of reality, and can uncover the problem during validation. In this case, grinding may not be the right technique for preparation of such tablets.

Accuracy can also be determined against an accepted reference method (a compendial method or a previously validated method) using a sample of known or unknown value. If the methods produce statistically the same results, then the new method would be considered equivalent to the reference method and thus, be deemed validated. This approach can be used to validate automated sample preparation methods where a manual sample preparation method already exists. In general, its use for pharmaceutical dosage forms is less common than in other fields such as environmental analysis where samples are usually of unknown value and/or in sample matrices that are difficult to simulate.

The assessment for accuracy can also be applied to quantitation of impurities, both degradation and manufacturing process related. In some cases, it is not possible to have availability of all impurities (particularly degradants) but meaningful accuracy can be obtained by comparison with other analytical determinative methods.

Although complete recovery is obviously preferred, in certain cases such a goal is difficult or impossible to achieve, e.g., to fully recover trace amounts from equipment surfaces in cleaning validation methods (used to verify manufacturing equipment having been adequately cleaned). If the recovery is low (say less than 85%) but consistent, a correction factor in the final calculation step can be used to compensate for the low recovery.

10.2.3 Precision

Precision, or reproducibility of a measurement, is influenced by random error. Here the ICH definition for precision expresses the closeness of agreement (degree of scatter) between a series of measurements of the same homogeneous sample under prescribed conditions. It is usually expressed as the variance, standard deviation, or coefficient of variation of a series of measurements (ICH 2005). Repeatability expresses the precision under the same operating conditions over a short interval of time. All measurements have associated random error to some degree and a single measurement is almost universally not accepted as the true representative value. Both precision and accuracy are defined by the boundaries that a true value may reside within. Only by reproducibility of a measurement can the true value be approximated.

Precision is greatly affected by sample preparation errors and the variability can also have an impact on the accuracy (Mitra 2003). In other words, results of high variability or low precision are less likely to achieve high accuracy. It has been discussed in a previously published treatise that analytical reproducibility decreases disproportionately with decreasing concentrations (Horowitz et al. 1980). It is clear that when one is at or close to the limit of quantitation (LOQ)/detection (LOD), precision variability is affected more by the complexity of the sample preparation and interferences from the sample matrix. For this reason, the precision acceptance criteria for assay and impurities should be different. For example, while precision for potency assay generally has to be no more than 2% RSD, 10% RSD is usually considered acceptable for trace-level residual solvents.

Sample preparation, still largely a manual process in pharmaceutical analysis, can be expected to add the majority of the variability to the total analytical procedure. Common sources of errors and variations attributable to analytical preparation are: incorrect weighing, general or specific sample mishandling, incorrect glassware usage–volumetric dilutional errors or low/high meniscus on pipette transfers, incorrect pipette volumes, and incomplete mixing, shaking, extraction, and filtration. Sample preparation procedures should be designed to minimize handling errors. For example, the minimum sample weight should be consistent with the precision of the balance. Small sample weights (less than 10 mg) call for the use of micro-balances instead of regular analytical balances. When measuring a small amount of liquid, using diluent weight (e.g., gravimetric dispensing) instead of diluent volume (e.g., volumetric dispensing) will improve precision. Excessive sub-dilution steps should also be avoided, as each step increases the chance of dilution error/variation.

10.2.4 Linearity/Sensitivity

While method sensitivity is not a parameter that is a requisite of the validation process, it is a valuable component of the overall analytical technique. Sensitivity is the capability of detecting small differences in concentration of an analyte-of-interest. Extractability and recovery during sample preparation is central to and has a large influence on the analytical sensitivity. Greater sample analyte extraction capability yields higher recovery and analytical sensitivity. Sensitivity of an analytical method is most easily obtainable and quantified from a linearity curve. The linearity curve represents the linear dynamic range of sample concentrations that the method can be used to quantify the analyte. Linearity can be further defined as the ability to obtain test results (within a defined range) that are directly proportional to the analyte concentration in the sample. The range is the interval between the upper and lower concentrations that acceptable levels of accuracy and precision have been demonstrated.

10.2.5 Detection and Quantitation Limits

The lowest concentrations that can be measured with a degree of confidence are the lower limits defined by the linearity curve and these apply to both the active drug and impurities. The limit of detection (LOD) is the lowest detectable signal distinct from the measurement background noise (baseline). This has traditionally been defined as a signal-to-noise ratio of approximately 3:1. There are some other ways to determine the LOD based on the standard deviation of the response and the slope of the calibration curve, but these approaches are not as straightforward and visual as the signal-to-noise method (ICH 2005). Closely related, but distinct, is the LOQ. This is a measurement at a concentration usually defined as approximately 10:1, signal-to-noise.

Impurities and/or degradants are usually at the extreme lower range of analytical methods, close to LOD or LOQ. In pharmaceutical analysis, the LOQ should not be higher than the level of impurities at the ICH reporting threshold of 0.05% of the drug substance (or 0.03% for daily intake exceeding 2 g/day) (ICH 2006). The sample matrix can contribute to baseline noises and interferences, which have a large impact on detectability and quantifiability at or close to these limits. An example is cleaning validation methods where interferences from the sampling procedure and preparation must be carefully controlled. Blanks from swab materials must be carefully monitored and analyzed to ensure that the analyte of interest is not consumed by the noises and interferences from the swabbing material or that the drug is not degraded or lost during the sampling procedure.

10.3 Validation with Regard to Sample Preparation

Sample preparation procedures need to take into consideration the nature of the analytes, the complexity of the sample matrix, the purpose of the analysis, and the requirements of the analytical instrument/technique, in order to provide analytes in an appropriate medium for the analytical method. Improper sample preparation procedures can render a method unable to meet acceptable method validation criteria as discussed in the previous section. Most importantly, a method cannot be considered properly validated if the sample preparation used in method validation does not truly mimic the sample preparation procedures for real samples, or if all aspects of sample preparation are not included in the method validation process and in the final defined method parameters, such as the correct diluent, glassware, crushing or grinding (if used), extraction conditions (e.g., time and temperature), dissolution, sonication, mixing, shaking, or vortexing procedures, pipetting and dilution scheme, filtration steps, applicable concentration range, etc. Being an integral part of the overall analytical method, sample preparation can only be validated together with the method instead of as a separate component alone.

Sample preparation can be considered to be routine when the analyte is soluble, stable, detectable and quantifiable without the application of additional steps other than dissolving in a suitable solvent. Nonroutine sample preparation encompasses a range of technical procedures from the less labor intensive (minor modifications), to those that are more labor intensive (major modifications). Nonroutine sample preparation involves isolating, recovering, preserving, and sometimes chemically altering the analyte.

10.3.1 In Cases of Routine Sample Preparation

Drug products, API and intermediates that generally fall into this category are those liquids and powders that are readily and completely solubilized, do not show indications of coming out of solution with changing conditions and are observed to be

Table 10.1 Analytical recovery from low concentration of peptide in saline and sub-diluted solutions in diluent A

Peptide concentration	88 μg/mL in saline	44 μg/mL (88 μg/mL in saline, 1:2 dilution with diluent A)	17.6 μg/mL (88 μg/mL in saline, 1:5 dilution with diluent A)	8.8 μg/mL (88 μg/mL in saline, 1:10 dilution with diluent A)
Analytical recovery (%)	33.1	29.8	44.1	48.2

chemically stable. It should be understood that sample solubilization may need the assistance of extensive shaking or agitation, ultrasonication, varying applications of heat, or a combination of these. Where no additional or complicated multistep extraction procedures need to be undertaken, the validation of the sample preparation parallels that of the general validation parameters of the method as previously discussed.

The diluent used to dissolve the sample is an important factor in routine sample preparation, and needs to be chosen carefully. It is common that cosolvents produce better solubilization and result in more complete extraction than a single solvent acting alone. A typical diluent for HPLC applications is a mixture of water and some organic solvent, acetonitrile or methanol being the most common, with some additives/modifiers if necessary. An improper diluent can cause many problems for method validation. Diluent that reacts with the main compound or excipients should be avoided. For example, if a formulation contains acidic excipients, such as fumaric acid, a diluent containing methanol will react with the acid to form an ester. This will create an extraneous peak in the chromatogram that grows over time during and after sample preparation, and in turn affects the specificity and recovery of the analytical method. On the other hand, a diluent optimized for sample stability sometimes is not the best for chromatography. For example, 100% acetonitrile as the diluent for a compound subject to hydrolysis in water provides excellent solution stability but may result in significant distortion in peak shape, which can lead to higher variability in precision. In this case, the percentage of water in the diluent needs to be carefully chosen to achieve reasonable solution stability while maintaining acceptable peak shapes in chromatography.

The effect of diluent on analytical recovery for certain types of compounds can be significant and has the potential to skew the accuracy of the obtained results, especially at low concentrations. For example, in the analysis of a peptide by ultra high performance liquid chromatography (UPLC), the nominal concentration of the external reference standard prepared in diluent A (water:acetonitrile:TFA; 800:200:1, v/v/v) was 200 μg/mL. The analytical recovery from a 45-μg/mL sample prepared in the same diluent was 93.4%, while the recovery of a sample prepared in a different diluent (normal saline) at the same concentration was dramatically decreased to merely 8.3% (Kou et al. 2010). Furthermore, when the peptide samples came in saline solution at low concentrations, further dilution with the diluent A could only improve the UPLC recovery to a limited degree. Table 10.1 shows the recovery of the same peptide at 88 μg/mL in saline and at sequentially lower concentrations diluted in diluent A. Even with a 1:10 dilution the improved recovery was still less than 50%, far from satisfactory for method validation purposes (Kou et al. 2010). Therefore, special attention should be paid to low concentration drug product samples that are at or below the nominal analytical standard concentration and do not require dilution before analysis. These samples are in a different matrix than the diluent used to prepare reference standard solutions. Method validation using only the diluent for standard preparation would fail to uncover any potential recovery issues associated with the samples in a different solution medium. Method validation should therefore include use of the sample preparation diluent.

Labware is routinely used in sample preparation and can be critical for certain types of analytes. For example, peptides and proteins have a tendency to adsorb to contact surfaces, leading to the loss of analyte and, in turn, lower analytical recovery. Labware and instrumentation parts and exposed surfaces can be made or coated with materials such as polymers or polyimides to prevent or minimize analyte adsorption (Chen et al. 2004; Liu et al. 2007). If a method requires labware made of specific passivated material or surface-treatments, it should be clearly indicated in the method and included as part of method validation. It is worth mentioning that the recovery of the peptide discussed in the preceding paragraph was significantly improved by using a special UPLC injection needle kit made from PEEKsil material to minimize peptide adsorption (Kou et al. 2010).

Complications can arise when samples that do solubilize begin to chemically change or precipitate out of solution with time. It may become necessary to modify the sample preparation environment. For example, certain types of samples may need the addition of antioxidants and chelating agents to prevent artifacts due to oxidative or free-radical degradations. The solubility and stability of the active drug, impurities, and degradants should all be considered. Light, humidity, and temperature often become factors for consideration and control. For these sample preparation procedures, validation may have to include a variety of stability profiles specifically designed for that system. One of the most common is measuring the stability of the analyte in solution as it resides in its autosampler vial over the course of varying lengths of time. In all cases, extensive experimentation is usually required to define the parameters of the sample preparation steps including those of time in the process.

The following are some validation examples by routine sample preparation procedures. Fig. 10.1 shows that the API in different tablet formulations is not interfered by the placebo excipients and that the active molecule retains its integrity. Furthermore, the identity of the main peak has been confirmed by peak elution time, and by spectroscopy, and no other anomalous peaks have been generated. Recovery studies are conducted to demonstrate that the API can be accurately assayed in the formulation across a range, such as ±20% of its targeted value. Recovery studies must include levels above and below the nominal or target sample preparation concentration and yield a satisfactory linearity plot of the data across this range. Typically, multiple samples are prepared at each concentration and analyzed. Table 10.2 demonstrates one such successful recovery study. The recovery data can be plotted to evaluate its linearity. For one of the formulations, the generated plot and calculated parameters can be observed in Fig. 10.2.

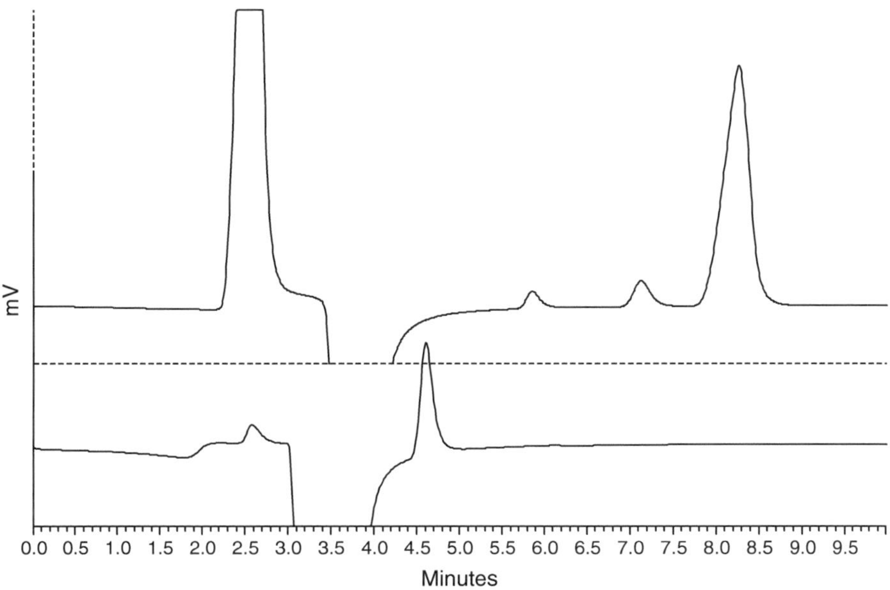

Fig. 10.1 Overlaid chromatograms of a test mixture of an organic acid with addition of 0.1 N HCl eluting at 8 min and two synthetic precursors eluting at 5.75 and 7.25 min, respectively (*top*) and the virtually unretained peak of the sodium salt of the acid (*bottom*). Sodium chloride elutes in the solvent front of the *top* chromatogram. The major excipient precipitates out and is not a factor

Table 10.2 Recovery data for weighed additions of API to placebo that mimics the actual experimental tablets using formulations that contain two different sets of excipients

Tablet formulation #1			Tablet formulation #2		
API Conc. mg/mL	% of nominal	Percent recovery	API Conc. mg/mL	% of nominal	Percent recovery
0.3245	81.1	100.16	0.4599	115.0	98.87
0.4107	102.7	100.31	0.3715	92.9	100.48
0.3672	91.8	100.50	0.3145	78.6	100.60
0.4576	114.4	99.96	0.3767	94.2	100.88
0.4869	121.7	99.97	0.3528	88.2	100.98
0.3895	97.4	100.44	0.3927	98.2	100.35
0.3916	97.9	100.22	0.4759	119.0	100.37
0.3300	82.5	100.55	0.3168	79.2	100.75
0.4913	122.8	99.96	0.4147	103.7	100.84
0.4317	107.9	100.09	0.4263	106.6	100.65

Nominal = 0.4 mg/mL

Std. precision: 5 Std. A: % RSD = 0.43; 3 Std. B: % RSD = 0.35; % difference = 0.12; 5 Std. A + interspersed Std. A: % RSD = 0.44; 0.40; 0.39; 0.39

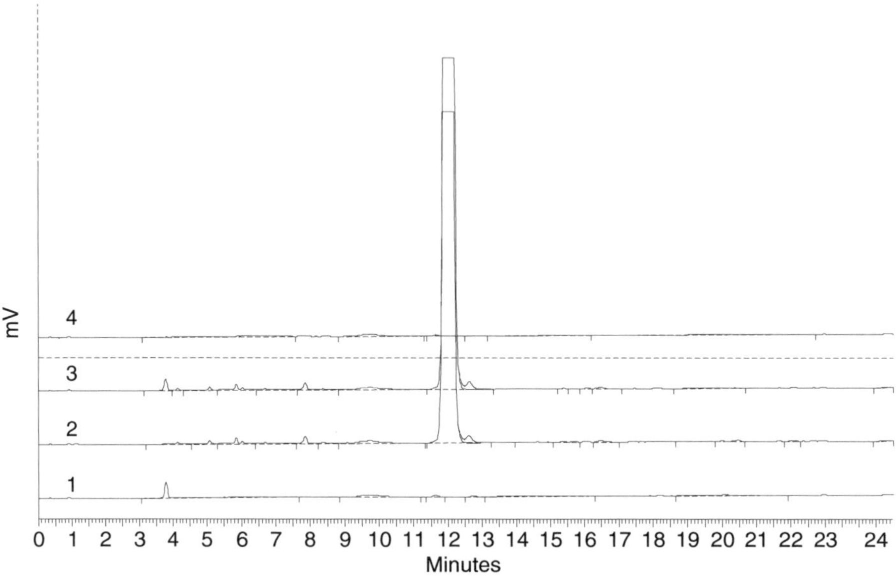

Fig. 10.2 Overlaid chromatograms of two experimental tablet formulations of the same active ingredient (numbers 2 and 3) and their respective placebos (numbers 1 and 4)

10.3.2 In Cases of Nonroutine Sample Preparation

Questions of solubility arise where excipients do not dissolve completely in the extracting solvents. If preliminary validation experiments demonstrate that the test analyte is unaffected then filtration, such as using a suitable syringe filter, is a simple and effective additional step. Table 10.3 shows calculated validation data for a powder that was formulated at two different strengths with all the excipients remaining proportionally the same. Except for filtration, no additional sample treatment was necessary.

If there is interaction between analyte and excipients that prevents complete solubilization, the addition of an additive such as a surfactant, emulsifier, or chelating agent may be warranted. Formulations that are suspensions, by definition, will have incomplete or nonsolubility. It would be necessary to weigh aliquots of suspensions that are undergoing spin-bar agitation for use in sample preparation.

The use of extensive sample preparation is usually necessitated by formulations containing complex matrices, e.g., lotions, creams, fermentation broths, etc. For oral dosage forms the advances in polymers and processes (such as hot melt extrusion or HME), to improve bio-availability or to achieve desirable timed-release profiles, have led to novel formulations that are more complex and challenging for sample preparation. For example, some polymers are difficult to dissolve, and once dissolved and injected onto HPLC, are strongly retained by the HPLC column, causing column fouling and peak broadening over time. In this case, additional sample preparation steps may be necessary to remove the polymer(s) from the sample solution prior to HPLC analysis.

Table 10.3 Validation data for a powder at two formulation strengths

Validation parameter	Conditions	Results				Criteria (pass/fail)
Specificity	Evaluated diluent (blank), placebo and reference standard	Diluent (blank) and placebo do not interfere with analyte peak in chromatograms				No interference from diluent or placebo PASS
Linearity	Range studied = 0.053–0.2674 mg/mL	$R = 0.99999$				$R \geq 0.99$ PASS
		$m = 5.78 \times 10^7$				
		$b = 3{,}836$ (line passes through origin)				$b = \pm 12965$ PASS
		The plot shows data points randomly distributed around the line				PASS
Linearity at trace levels	Range studied = 0.019–3.744 µg/mL	$R = 0.99997$				$R \geq 0.99$ PASS
		$m = 6.09 \times 10^4$				
		$b = -122.7$ (line passes through origin)				$b = \pm 449.4$ PASS
Detection and Quantitation Limit	Calculated by VALIDAT	LD = 0.028 µg/mL LQ = 0.107 µg/mL				LQ = 0.05% of a nominal sample concentration of 0.2 mg/mL PASS
Precision	Injection repeatability	% RSD = 0.74, $n = 5$				RSD ≤ 2.0% PASS
	Analysis repeatability	% RSD = 0.72, $n = 6$				RSD ≤ 2.0% PASS
Recovery	Three sets of synthetic formulations of API + placebo were prepared at the 70, 100, and 130% levels for both the low formulation (23%) and the high formulation (30%)	For 23% formulation				98.0 – 102.0% recovery PASS
		Level (%)	Avg %	%RSD	Range (%)	
		70	100.79	0.90	99.94–101.75	
		100	100.32	0.74	99.62–101.10	
		130	99.95	0.23	99.69–100.09	
		For 30% formulation				98.0 – 102.0% recovery PASS
		Level (%)	Avg %	%RSD	Range (%)	
		70	99.53	0.35	99.32–99.93	
		100	99.98	0.15	99.85–100.14	
		130	99.78	0.62	99.12–100.35	
Solution Stability	A solution of the reference std was stored at ambient in a clear flask under normal lab light	The solution was assayed periodically and was found not to have produced new impurity peaks, increased existing impurity peaks, or significantly changed its response factor after 3.5 months				PASS (Acceptable for use within this time frame)

Note: no exceptional techniques were applied other than by syringe filtration into auto-sampler vials

Major sample preparation often involves some form of extraction of the sample components. One widely used technique over the many years of chemical analysis has been liquid–liquid extractions (LLE) (Gupta et al. 1996; Behymer et al. 1992; Eerkes et al. 2003). Extensively used and laborious, this technique employs immiscible solvents, usually one aqueous and one organic that selectively solubilize the components. The aqueous solvent can be distilled deionized water or a buffered version of it. The solution pH may, and usually does, have a significant impact on the extraction and initial solubility. LLE is discussed in more detail in Chap. 4. In supercritical fluid extractions (SFE) (Abbas et al. 2008) carbon dioxide, in most applications, replaces the organic solvents, which can be advantageous in terms of lower cost, reduced flammability, and toxicity, greater dissolving abilities for certain types of molecules and less likelihood to destabilize the chemical equilibrium of the solubilized components. SFE is discussed in more detail in Chap. 5.

Solid phase extractions (SPE) (Lai et al. 1997; Christie 1992; Hilton and Thomas 2003) can be used in sample clean up and also in removing soluble but interfering species. These are often characterized as mini HPLC columns, reversed phase or reversed phase containing ionic exchange groups, or are simply adsorbents. One of the oldest techniques is open-column chromatography in which a sorbent such as silica is used to trap extraneous excipient material, or selectively adsorb the analyte of interest. A simple and extensively used variant of this is thin layer chromatography, particularly in identification tests. What might be termed closed-column chromatography is sample pretreatment that could couple two or more different types of columns, such as size exclusion, ion-exchange, and reversed phase in order to remove interfering peaks. For biologicals, there is the use of dialysis membranes to separate the proteins of interest from the media such as in peptide mapping of protein/peptide APIs. Additional information on SPE is provided in Chap. 4.

Another approach is modification of the molecular nature of the analyte, as for example by free acid or base formation, by derivatization (Manius and Viswanathan 1977; Manius et al. 1979, 1993) or by some other type of chemical transformation. An example of this is the conversion of the sodium salt of a drug molecule (an organic acid), which is virtually unretained on a reversed phase column, to its free acid form simply by the addition of 0.1 N HCl to the sample solutions. This can be seen in the multiplots via refractive index detection in Fig. 10.3. The free acid is retained and resolved, in this test mixture, from two synthetic precursors. An additional bonus in this sample preparation was that the major excipient precipitated out of solution at the low pH and did not interfere or remain to contaminate the column or the resulting analytical test solution.

The introduction of additional preparation measures (e.g., heating of the sample solution, longer sonication times, etc), some of which are harsh and even destructive, could present more challenges for validation. The following questions should be asked when assessing the sample preparation method:

- Has the compound of interest undergone any change or chemical modification?
- Has sample preparation produced new interferences or increased existing ones?
- How close to 100% is its calculated recovery?

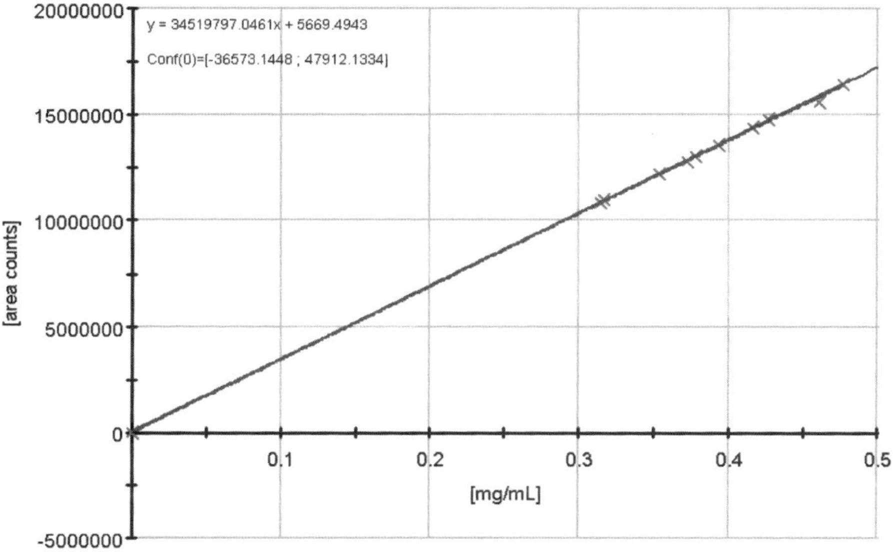

Statistical Data
Reference value for relative errors (yRef) = 483282828 area counts, Slope (b) = 34519797.04606 area counts/mg/mL, Intercept (a) = 5669.49428 area counts, Intercept (a) relative to 100% (= 4.83283e+008 area counts) = 0.00%, r = 0.99996, SDa = 20105.96815 area counts (=354.63%), Confidence interval of a = 5669.49428 ± 42242.63908 area counts (5%-two-sided), SDb = 72232.89786 (=0.21%), Confidence interval of b = 34519797.04606 ± 151761.31841 (5%-two-sided), Method SD = 0.00186 mg/mL, Method RSD = 0.95235%
Lack of fit
Repeatability RSD srel = 0.01%

Fig. 10.3 Linear regression plot and tabulated parameters of the recovery data

Validation of methods with extensive nonroutine sample preparation procedures can be facilitated by a robust sample preparation method development and prevalidation strategy before full scale validation starts. Critical sample preparation parameters should be optimized from well-designed experiments instead of being picked randomly or arbitrarily. After adequate method development and prevalidation work, the analyst should have a high degree of confidence that the method will meet the full scale validation requirements. One should not get big surprises in the final validation stage, such as discovering that the method cannot be validated (e.g., the recovery is less than 70% from the NINE samples prepared!). Extractions requiring multiple steps with varying sample treatment also need to be defined and documented in the final method for future analysts to obtain comparable results.

The complexity of the formulations and corresponding sample preparation will most likely affect the overall analytical method performance. It is sometimes unrealistic to set the same validation acceptance criteria for complex sample preparation as for simple routine sample preparation. For example, the precision criteria for drug substance are generally tighter than for drug products. The recovery criteria for dosage forms with very low drug loading (<1%) can have a wider range than for high drug-loading ones. Higher tolerances are given to validation criteria for inhalation products (e.g., dry powder inhaler) due to the inherent difficulty to extract and recover drug from the surfaces of inhalation sampling apparatus. The specific acceptance criteria should be consistent with the types of products, formulations, and the actual sample preparation techniques and procedures used.

10.4 Revalidation

A compendial method, including its sample preparation procedures, can be applied to drug substance analysis directly without the need for further validation. For a drug dosage form, however, a compendial method needs to be verified to be suitable for the specific matrix, although a full validation is generally not required.

A validated method should be revalidated if the original scope of the method (e.g., the concentration range) has changed or the sample preparation procedures have changed. In fact, a change in sample preparation is a common cause for revalidation. For example, a method initially validated for a tablet form of the drug product should not be used for a reformulated encapsulated dosage form without revalidation due to the change in sample matrix. In this case, the original sample diluent may not be adequate to recover the API from the formulation. Gelatin capsule material may require a combination of aqueous solution and sonic energy to break up the capsule sufficiently to release the API-excipient powder. Too high an organic concentration in the initial solution and the capsule will not release the interior material. Once the capsule is breached, the API may then need to have a higher concentration of organic solvent (depending on the API physicochemical solubility dynamics) to dissolve in the final extracting solution. Another example is that a sample preparation method originally developed and validated for an immediate release formulation often does not work as well for controlled (or modified, sustained) release formulations of the same molecule. Controlled release formulations may need an organic solvent (such as methanol or acetonitrile) to first dissolve the polymers often used in such formulations in order to release the drug within a short time. After the matrix is dissolved and the drug is extracted, some aqueous solution is added or a sub-dilution is performed in a suitable diluent to make it compatible with the HPLC mobile phase. Again, due to the change in sample preparation, the analytical method needs to be revalidated.

10.5 Analytical Method Transfer and Validation

When an analytical method is transferred from one laboratory to another, some form of verification or validation is usually required. Four approaches are commonly used in analytical method transfers: comparative testing, revalidation, covalidation, and validation waiver. Revalidation is a complete or partial validation of a method that has been previously validated at the originating laboratory. Covalidation usually consists of each of the two (or more) laboratories performing part of the validation, in some cases with experimental overlap to ensure that the results are comparable between the groups. A validation waiver can be acceptable when the receiving lab is deemed fully qualified to run the specific validated method without further verification, typically based on the laboratory's previous experience with similar methods. Whatever the approach, a method transfer protocol usually needs to be established with the acceptance criteria clearly defined and delineated prior to the actual experimental performance. The results of the transfer from all participating groups and laboratories must be properly documented (Swartz and Krull 2006). When comparative testing is used in method transfer, a statistical method such as a *t*-test or *f*-test can be employed to determine equivalency of testing results. Table 10.4 shows some examples of comparative testing in experimental design and acceptance criteria for analytical method transfer (Swartz and Krull 2006).

Validation issues could occur during method transfer, such as low recoveries or highly variable results, if the sample preparation method is not robust or rugged, or the method is not clearly written or missing critical information/details. It is very important that sample preparation procedures are clearly described, with critical steps delineated in sufficient detail. For example, if the sample requires vigorous mixing or sonication for a certain length of time to dissolve completely, it should be included in the instructions. If such information is left out, the receiving lab may perform this procedure without adequate time for dissolution, resulting in low recoveries in validation. Furthermore, certain procedures, such as tablet grinding in general and hand grinding in particular, are more prone to operator errors, high variability and low recovery; they present a challenge for revalidation during method transfer, if the method can be initially validated at all.

10.6 Conclusions

There will always be a need for sample preparation in analytical procedures. Most analysts strive for the most simple, quickest and least resource consuming manner of sample solubilization in order to avoid lengthy and labor-intensive extraction steps. Many analysts will sacrifice some method performance in an attempt to avoid laborious sample preparation. Sample preparation is often underrated and not fully utilized to augment and advance throughput, reproducibility, and trace-concentration analysis.

Combining sample preparation steps into a seamless, less labor-intensive integration with analytical instrument determination is the ideal. The validation process is an attempt to standardize the analytical process by addressing parameters that all methods have in common and by addressing the analytical procedure in toto. Sample preparation procedures must be properly designed and adequately developed first so that they are validatable. The discussions in this chapter and the other chapters in this book should help analysts achieve this goal. Validation of sample preparation can be a simple process if the API and its formulated excipients

Table 10.4 Examples of comparative testing in experimental design and acceptance criteria for analytical method transfer (reproduced from Swartz and Krull (2006) with permission from Advanstar Communications)

Type of Method	# Analysts	# Lots or Units	Acceptance Criteria	Notes
Assay	2	Three lots in triplicate	A two one-sided *t*-test with intersite differences of ≤ 2% at 95% CI	Each analyst should use different instrumentation and columns if available, and independently prepare all solutions. All applicable system suitability criteria must be met
Content uniformity	2	One lot	Include a direct comparison of the mean, ±3% and variability of the results, (% RSD), i.e., a two one-sided *t*-test with intersite differences of 3% at 95% CI	If the method for content uniformity is equivalent (same standard and sample concentrations, LC conditions, and system suitability criteria) to the assay method, then a separate AMT is not required
Impurities, degradation products	2	Three lots in duplicate (triplicate if done together with the assay)	For high levels, a two one-sided *t*-test with intersite differences of 10% at 95% CI. For low levels, criteria are based on the absolute difference of the means ±25%	All applicable system suitability criteria should be met. The LOQ should be confirmed in the receiving laboratory, and chromatograms should be compared for the impurity profile. All samples should be similar with respect to age, homogeneity, packaging, and storage. If samples do not contain impurities above the reporting limit, then spiked samples are recommended
Dissolution	NA	Six units for Immediate release, 12 units for extended	Meet dissolution specifications in both laboratories, and the two profiles should be comparable, or based on the absolute difference of the means, ±5%	A statistical comparison of the profiles (e.g. F2) or the data at the Q time points(s) similar to that performed for the assay may be performed

ID	One unit	Chromatography: confirm retention time. Spectral identification and chemical testing can also be used, assuming operators are sufficiently trained and the instrumentation can provide equivalent results		
Cleaning validation		Two spiked samples, one above, one below spec	Spiked levels should not deviate from the spec by an amount 3× the validated standard deviation of the method, or 10% of the spec, whichever is greater	Essentially a limit test. Low and high samples to confirm both positive and negative outcomes are required

are readily soluble; if there are no interferences from the excipients; if there are no chemical or hydrogen bonding interactions and if stability is not an issue during the analysis. Validation can then follow the usual method validation guidelines. Whenever modifications need to be made to isolate the drug (whether it is API or formulated drug product), to enhance its recovery, to improve its detectability, and to assure a reliable and rugged assay result, additional stepwise validation work may need to be done. The preparation procedure also needs to be carefully described to ensure that it is easy to follow, key to the process, and reproducible. A quasi in-process approach may be followed depending, to some degree, on the discretion of the analyst. Ultimately, the regular method validation procedures are applied.

References

Abbas KA, Mohamed A, Abdulamir AS, Abas HA (2008) A review on super-critical fluid extraction as new analytical method. Am J Biochem Biotechnol 4:345–353

Bakshi M, Singh H (2002) Development of validated stability-indicating assay methods – critical review. J Pharm Biomed Anal 28:1011–1040

Behymer TD, Bellar TA, Ho J, Budde WL (1992) Determination of benzidines and nitrogen-containing pesticides in water by liquid-liquid extraction or liquid-solid extraction and reverse phase high performance liquid chromatography/particle beam/mass spectrometry. EPA Method 553 Revision 1.1

Chen H, Brook MA, Sheardown H (2004) Silicone elastomers for reduced protein adsorption. Biomaterials 25:2273–2282

Christie WW (1992) Solid-phase extraction columns in the analysis of lipids. Adv Lipid Methodol 1:1–17

Eerkes A, Shou WZ, Naidong W (2003) Liquid/liquid extraction using 96-well plate format in conjunction with hydrophilic interaction liquid chromatography-tandem mass spectrometry method for the analysis of fluconazole in human plasma. J Pharm Biomed Anal 31:917–928

Ermer J, Miller JHM (2005) Method validation in pharmaceutical analysis: a guide to best practice. Wiley-VCH, Weinheim

Gupta R, Mauri R, Shinnar R (1996) Liquid-liquid extraction using the composition-induced phase separation process. Ind Eng Chem Res 35:2360–2368

Hilton MJ, Thomas KV (2003) Determination of selected human pharmaceutical compounds in effluent and surface water samples by high-performance liquid chromatography-electrospray tandem mass spectrometry. J Chromatogr A 1015:129–141

Hokanson GC (1994a) A life cycle approach to the validation of analytical methods during pharmaceutical product development, part 1. Pharm Technol 18:92–100

Hokanson GC (1994b) A life cycle approach to the validation of analytical methods during pharmaceutical product development, part II. Pharm Technol 18:118–130

Horowitz W, Kamps L, Boyer K (1980) Quality assurance in the analysis of food and trace constituents. J Assoc Off Anal Chem 63:1344–1354

ICH (2005) International Conference on Harmonization Technical Requirements for the Registration of Pharmaceuticals for Human Use, Q2(R1): Validation of Analytical Procedures: Text and Methodology. www.ich.org

ICH (2006) International Conference on Harmonization Technical Requirements for the Registration of Pharmaceuticals for Human Use, Q3A(R2): Impurities in New Drug Substances. www.ich.org

Kou D, Zhan S, Chokshi HP (2010) unpublished results

Lai C-K, Lee T, Au K-M, Chan A (1997) Uniform solid-phase extraction procedure for toxicological drug screening in serum and urine by HPLC with photodiode array detection. Clin Chem 43:312–325

Liu J, Sun X, Lee ML (2007) Adsorption-resistant acrylic copolymer for prototyping of microfluidic devices for proteins and peptides. Anal Chem 79:1926–1931

Manius G, Viswanathan V (1977) GLC determination of lasalocid and its bromo-analog as their silyl derivatives. J Pharm Sci 66:769–772

Manius GJ, Liu TMY, Li Wen L-F (1979) The quantitation of sugars from differential culture media by gas-liquid chromatography. Anal Biochem 99:365–371

Manius GJ, Li Wen L-F, Palling D (1993) Three approaches to the analysis of trace formaldehyde in bulk and dosage form pharmaceuticals. Pharm Res 10:449–453

Mitra S (2003) Sample preparation techniques in analytical chemistry. Wiley, Hoboken

Shabir GA (2003) Validation of high-performance liquid chromatography methods for pharmaceutical analysis: understanding the differences and similarities between validation requirements of the US Food and Drug Administration, the US Pharmacopeia and the International Conference on Harmonization. J Chromatogr A 987:57–66

Swartz ME, Krull I (2006) Analytical method transfer. LCGC North Am 24:1204–1214

Chapter 11
Application of Quality by Design Principles for Sample Preparation

Jackson D. Pellett, Beverly Nickerson, and Ivelisse Colón

Abstract Sample preparation methods must be rugged and robust to ensure accurate results. Pharmaceutical development teams rely on these results to make critical decisions about formulations, manufacturing process conditions, and packaging. Once decisions are made, the manufacturing conditions and associated analytical methods are transferred to clinical and commercial manufacturing facilities where they will be used over the course of many years. At this stage, ensuring extraction robustness becomes increasingly important with the potential to impact patient safety, product efficacy, and business efficiency. This chapter describes a Quality by Design approach that can increase the probability that sample preparation conditions remain rugged, robust, and suitable for use throughout a product lifecycle.

11.1 Introduction

11.1.1 Importance of Understanding and Controlling Extraction Conditions

Previous chapters have covered many of the challenges encountered during the development of sample preparation and extraction procedures. A variety of different dosage forms have been highlighted (i.e., solids, nonsolids), all with unique considerations. Regardless, the objective is always the same: to develop a method that will enable transfer of the compound(s) of interest *from* the dosage form into a matrix that is compatible with the downstream analytical technique (e.g., HPLC,

J.D. Pellett (✉)
Small Molecule Pharmaceutical Sciences, Genentech, Inc., 1 DNA Way,
South San Francisco, CA 94080, USA
e-mail: pellett.jackson@gene.com

B. Nickerson (ed.), *Sample Preparation of Pharmaceutical Dosage Forms: Challenges and Strategies for Sample Preparation and Extraction*, DOI 10.1007/978-1-4419-9631-2_11,

LCMS, IR). For quantitative methods, this transfer must also be done in a manner that will enable accurate and reproducible assessment of the amount of drug or impurity in the original dosage from.

As highlighted in Chap. 7, there are a number of variables that can impact sample preparation. These can be associated with the following:

- Manufacturing conditions used to produce the product (e.g., blending time, dryer temperature, compression force)
- Material properties of the dosage form or components of the dosage form (e.g., API particle size distribution, film-coat thickness, tablet hardness)
- Equipment used for sample preparation (e.g., shakers, sonicators, flasks)
- Settings on equipment or process steps (e.g., shaker speed, sonication time, number of inversions)
- People (e.g., degree of training, experience with a specific technique, clarity of written procedures)
- Laboratory environment (e.g., light, temperature, humidity)

Many of these are variables that are readily controllable. By following a systematic approach such as the one recommended in Chap. 7 for solid oral dosage forms, an analytical scientist is likely to select conditions that produce reliable results. The physicochemical properties of the drug substance and impurities will be collected and reviewed. A sample diluent that provides high solubility will be selected. Dispersion conditions will be chosen that increase access of diluent to the compounds of interest. Sample preparation conditions along with other aspects of the method will be validated. Method conditions will be documented as a set of test procedures to be followed by a limited number of analytical scientists under a narrow set of operating conditions. However, as the method continues to be transferred and used throughout the product lifecycle, situations that were not originally anticipated are increasingly likely to be encountered.

Consider the following:

1. Laboratory A developed and validated an extraction procedure for IR tablets using a reciprocating shaker. Development studies showed that 30 min shaking at 200 oscillations per minute (opm) was sufficient to disintegrate the tablet and completely extract the drug from the resulting particulates. The method is to be transferred to Lab B for release of clinical trial material. However, Lab B only has orbital shakers. *Is 30 mins shaking likely to still be adequate?*
2. Laboratory C does have a reciprocating shaker. However, it has not been well maintained. When set at 200 opm, the shaker is actually only operating at 150 opm (although the actual speed is not routinely verified and therefore Lab C is unaware of the bias). *Is Lab C likely to achieve complete extraction in 30 mins?*
3. A manufacturing site has been producing osmotic controlled release tablets. There is concern that the current supplier of one of the excipients critical to controlling the drug release rate of the tablets, hydroxypropyl methylcellulose, may not continue to be able to supply a sufficient amount of the material.

An alternative supplier is being considered, and process development studies are being planned. *Is the current assay method still acceptable for the evaluation of process development samples?*

4. The quality group at a commercial manufacturing site recently performed a process capability assessment. Results indicate that they can expect approximately 1% of all lots to fail the assay specification of 95.0–105.0% LC. The site would like to reduce this to <0.1%. *Prior to evaluating the manufacturing process, the site would like to know how much error is associated with the measurement system.*

Often these questions are addressed in a *reactionary* manner. For example, in the first two cases the laboratories might choose to go ahead with an analytical method transfer exercise (AMTE). If Laboratories B and C meet predefined acceptance criteria for equivalency with Laboratory A, then they will be considered qualified to use the method. If equivalency is not demonstrated, then the cause(s) for the failure will be investigated. There are risks with this approach. First, a failed AMTE will result in a delay in laboratory qualification and may also delay testing and release of clinical or commercial product. Meanwhile, laboratory resources will be consumed attempting to identify what went wrong in the study. If something with the original method is found to be problematic, additional method development and validation work may need to occur. Second, a successful AMTE offers little guarantee of successful long-term method performance. The exercise generally takes place over a short period of time using two to three lots of material. Often, the participants chosen for the study are among the more experienced analysts. After transfer, the method may not continue to perform as intended, especially if the method is operating at conditions in which small changes have significant impact on performance. The level of risk increases if the scientist:

(a) Is not aware of the relationship between specific operating conditions and method performance.
(b) Does not know how close the operating conditions are to conditions that will result in unacceptable method performance.
(c) Is not able to establish suitable controls to ensure that the method will continue to provide acceptable performance.

11.1.2 Benefits of Applying Quality by Design Concepts to Analytical Methods

There are several reasons why a company may consider implementing Quality by Design (QbD) for measurement systems:

First, if a company is already practicing QbD for the development and control of manufacturing processes, applying these concepts to methods is a natural extension. The additional knowledge gained about method capability and risk could be valuable in determining whether a method is suitable for supporting the process or product control strategy. For example, if a process chemist knows that water content must be below 1.0% to avoid formation of an impurity, the analytical chemist knows that he or she must develop a method that is very capable around this limit. High variability or bias could lead to production of API that fails impurity specifications. In this case, the method(s) developed to quantify impurities must also be accurate and precise. If not, it will be difficult to develop a true understanding of the functional relationship between water content and impurity formation. A poor method will also increase the probability of "misclassifying" clinical or commercial lots. Material that should fail specification will be tested as "passing," resulting in a potential risk to patients. Other material will be tested as "failing" specification when it should pass. Although this does not pose a risk to patients, it could result in acceptable material getting destroyed at significant cost to the producer (Chatfield and Borman 2009).

Second, a better understanding of method capability could lead to better process understanding. Understanding relationships between manufacturing process parameters and product quality attributes is at the heart of QbD. Testing of material (in process or final product) is necessary to understand these relationships. Therefore, the uncertainty of the measurement systems becomes important. This is illustrated in (11.1):

$$\sigma^2_{\text{measured}} = \sigma^2_{\text{process}} + \sigma^2_{\text{method}} \tag{11.1}$$

The total variability associated with the number that is generated (σ_{measured}) is a function of the variability associated with the process as well as the method. As σ_{method} is reduced through method improvements, σ_{measured} will become smaller and more representative of process variability. A good rule of thumb is that σ_{method} should contribute no more than 30% of the total variability, and ideally less than 10% (Wells 2010; Rasmussen et al. 2005).

Third, strict application of QbD principles will ensure that suitable test methods are used throughout the product lifecycle. A key component of QbD is lifecycle management. Method performance can be regularly monitored, and action should be taken if results indicate a drop in method capability or failure to meet performance requirements. By monitoring method performance, action can be taken *proactively* to reduce risk to both patients and producers.

Fourth, QbD provides a framework for evaluating and implementing new technology. If a laboratory is considering moving from manual to automated sample preparation equipment, the knowledge gained during the development of the original method is very valuable. For example, during method understanding studies a laboratory might find that low pH is needed for adequate drug solubility. This aspect would be important to maintain in any automated method.

Most analytical laboratories are already practicing some elements of QbD. Knowledge gained from the development of methods for similar products is leveraged and improved upon. Multivariate approaches are routinely used to identify robust method conditions, especially for chromatographic methods (Molnár et al. 2010; Cole et al. 2004; Ye et al. 2000). Method control strategies such as system suitability and bracketing standard agreement are implemented to reduce the risk of poor method performance. Taking full advantage of the QbD benefits, however, requires a systematic approach to method development beginning with predefined objectives (the analytical target profile [ATP]), which emphasizes method understanding (risk-based experimental studies) as well as method control and knowledge management.

11.1.3 Overview of Analytical Quality by Design Concepts

There are a variety of different approaches that can be taken to demonstrate method robustness and ruggedness, some of which are introduced in ICH and USP guidance documents (ICH Q2(R1) 2006; USP 33-NF 27 Chapter<1225> 2010). The key is a well designed study, encompassing variables that have the highest risk of impacting method results. These same considerations hold true for the development and understanding of a manufacturing process. In fact, the importance of this in pharmaceutical manufacturing has gained enough recognition that formal guidance documents have been created (ICH Q9 2005; ICH Q8(R2) 2008; ICH Q10 2008).

QbD is defined in ICH Q8(R2) as "a systematic approach to development that begins with predefined objectives and emphasizes product and process understanding and process control." A very similar definition could be applied to the development of analytical methods, including sample preparation. In 2010, a working group representing the Pharmaceutical Research and Manufactures of America (PhRMA) and the European Federation of Pharmaceutical Industries and Associations (EFPIA) formally introduced two concepts to improve method robustness and ruggedness and to facilitate continuous improvement (Schweitzer et al. 2010). The first concept is that the steps, tools, and approaches in ICH Q8(R2), Q9, and Q10 can be applied to analytical methods. The second concept is the ATP, a mechanism for formally describing measurement system requirements. Figure 11.1 shows an overview of how these concepts are applied.

11.1.3.1 Analytical Target Profile

The purpose of any analytical method is to provide reliably accurate measurements of one or more critical quality attributes (CQAs) of a product or intermediate. Note: a CQA is defined in ICH Q8R(2) as "a physical, chemical, biological, or microbiological property or characteristic that should be within an appropriate limit, range, or distribution to ensure the desired product quality."

The ATP provides detail about how accurate and precise a measurement should be to achieve that goal. The ATP is derived with consideration for process understanding requirements and is aligned with process and product control strategies. More restrictive accuracy and/or precision limits may be required for an ATP supporting a product with a 95.0–105.0% assay specification compared with a product with a 90.0–110.0% specification. Sensitivity requirements should be in-line with regulatory reporting

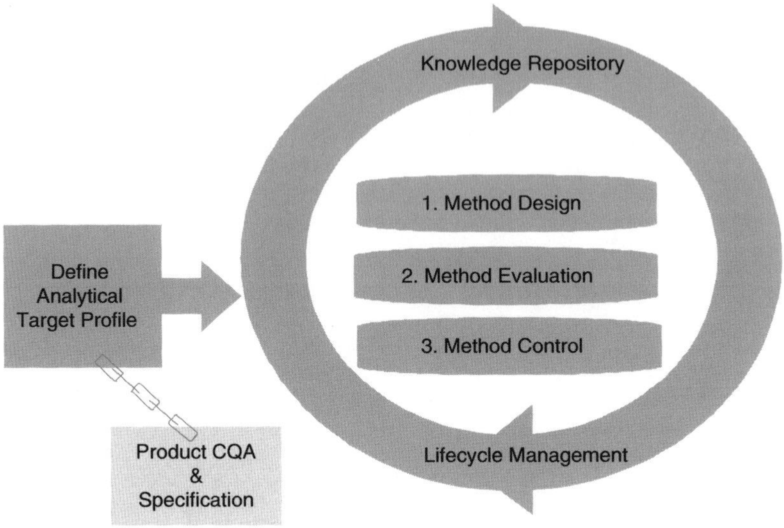

Fig. 11.1 Quality by design components as applied to analytical methods. Adapted from Schweitzer et al. (2010)

limits (e.g., ICH Q3A(R2) 2006; ICH Q3B(R2) 2006; ICH Q3C(R4) 2009), but may also depend on requirements for process understanding or stability modeling.

An ATP for an Assay method might be written as:

> *Assay.* The procedure must be able to accurately quantitate *Drug X* in film-coated tablets from 80 to 120% of the intended nominal concentration range with accuracy and precision such that measurements fall within ±3.0% of the true content with 95% probability.

The accuracy and precision requirements for this ATP can be described by the operating characteristic curve (OCC) shown in Fig. 11.2. A very similar approach was recently presented for comment in a USP stimulus paper for performance based monographs (Williams et al. 2009).

11.1.3.2 Method Design

Once an ATP has been defined, QbD concepts can be applied to design a suitable method. The first step is selection of an analytical technique. It should be noted that the above ATP does not mention anything about sample preparation or sample analysis technique. This is intentional, as there are likely to be multiple approaches that can be used to fulfill these requirements. It is up to the development scientist to select the one that makes the most sense at that time. In addition to the requirements mentioned in the ATP, other business drivers may influence technique selection. These could include the following:

- Equipment availability at testing laboratories
- Solvent cost (including disposal)

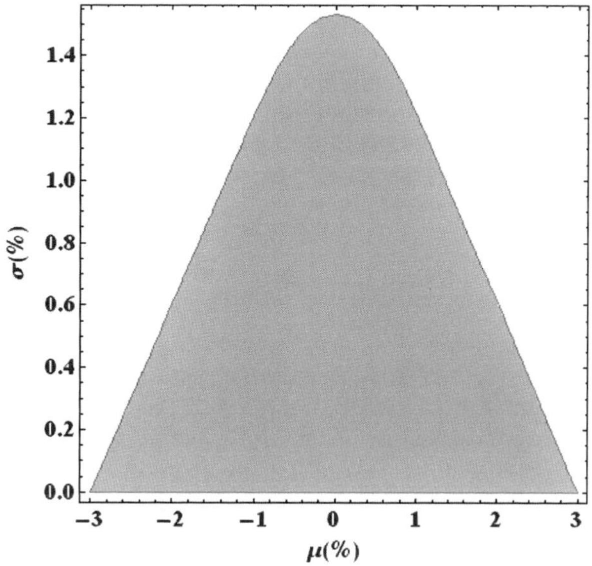

Fig. 11.2 Operating characteristic curve for method precision (σ) and bias (μ) constructed from the requirements in the example assay ATP

- Consumables cost (e.g., filters)
- "Green" considerations (see Chap. 14).
- Sample preparation time
- Capability for at-line or on-line use

It should be recognized, however, that throughout the product lifecycle, business considerations are likely to change. The cost of an extraction solvent may rise, or sample preparation time might become challenging as product volume increases. Advances in equipment and instrument technology will also occur. In many cases, the new technology will be more robust and provide an opportunity to improve method accuracy and precision. Sect. 11.4 describes how these future considerations can be managed within the framework of a QbD process.

After a technique is selected, method conditions are identified that have a high probability of meeting ATP criteria. Analytical scientists can rely on experience with similar products in selecting conditions, or may utilize platform approaches. At this point, a risk assessment is conducted to identify and prioritize method factors that have the potential to impact the performance of the test procedure. A variety of quality risk management (QRM) tools can be applied and several examples are provided in Sect. 11.3. If a company is already practicing QbD for product development, the same QRM tools and approaches can often be adapted for risk management of measurement systems.

11.1.3.3 Method Evaluation

One output of the risk assessment is a list of method factors that warrant further evaluation. These are input variables that have a reasonably high probability of influencing method performance in either a univariate or multivariate manner. Execution of appropriately designed studies will provide a more rigorous understanding of the relationship between method factors and method performance, and can also give a good indication of method ruggedness. In some cases, the results of the studies can be used to establish acceptable ranges of factors or a "method operable design region" (MODR).

11.1.3.4 Method Control

A method control strategy is developed from the knowledge gained during design, evaluation, and routine use of a method. These controls may include operation within the MODR as well as method performance verification such as system suitability. The goal of the control strategy is to ensure that optimal performance is achieved each and every time the method is run. It should be recognized that the control strategy established just after development may not be appropriate several years after product launch. As method understanding accumulates, the control strategy should be reassessed and modified as needed.

11.1.3.5 Lifecycle Management

Lifecycle management is a comprehensive approach to ensure that measurement systems continue to operate optimally and support process control needs. Procedures should be in place to manage information collected during development and throughout the commercial life of the product. This information should be readily accessible to those in quality organizations as well as those in development laboratories. How this information is acted upon is just as important. There should be clear direction for the kinds of events or new information that would warrant changes to the measurement systems. Changes could be as minor as clarifying a set of instructions in a test procedure. Or they could be potentially more significant – for example, changing from a chromatographic method for monitoring product assay to a spectroscopic technique such as NIR.

11.1.3.6 Knowledge Management

Effective knowledge management is required to leverage the extensive information that will be generated throughout the QbD analytical method development process. A knowledge management system provides "a systematic approach to acquiring, analyzing, storing, and disseminating information" (ICH Q10). Such a system will facilitate decision making and encourage continual improvement of measurement systems along with manufacturing process control strategies.

11.2 Method Design for Sample Preparation

The objective of the method design phase is to identify an initial set of extraction and sample preparation conditions that are likely to fulfill the requirements of the ATP. These conditions will serve as a baseline for subsequent steps in the QbD process, and a starting point for the collection of method understanding. It is quite possible that the conditions identified during the design phase may not end up as the final, commercial method conditions.

Chap. 7 describes a six step, systematic approach that works well for the identification of initial sample preparation conditions for solid oral dosage forms. As the authors note, the amount of time and effort spent on each step can vary depending on the purpose of the method and stage of development. A brief summary of each step is presented here:

11.2.1 Step 1: Review of Available API and DP Information

Information about the API solubility and solution stability will be helpful for selecting (or eliminating) potential extraction/dissolving solvents for screening studies. Information about excipients used in the dosage form and manufacturing conditions can be useful for selecting dispersion mechanisms (e.g., grinding, shaking, stirring, homogenization, etc.). Often, much of this information can be gathered from chemical synthesis or formulation development reports. Some physicochemical properties can also be estimated *in silico*.

11.2.2 Step 2: Evaluate and Select Diluent to Extract and Dissolve API

The purpose of this step is to develop knowledge about which solvent (or solvent combinations) will provide adequate solubility and recovery of drug and impurities. The most influential factors are typically the level of organic solvent, the diluent pH, and the type of organic solvent. A three staged diluent screening study, as described in Chap. 7, can be performed using API, excipient blends, and active drug product. The results of this study will provide considerable information about API solubility, chemical or physical interactions between excipients and the drug, and dosage form disintegration and dispersion. If designed in a multivariate manner, the screening study can also shed light on interaction effects, for example pH and organic level on API solubility. This knowledge can help focus method robustness studies later in the QbD process (see Sect. 11.3.2).

11.2.3 Step 3: Evaluate and Select Appropriate Conditions to Disintegrate/Disperse the Dosage Form

Once an extraction diluent has been selected, a mechanism for disintegrating and dispersing the drug product must be chosen. This will be largely dependent on the type of dosage form (e.g., immediate release tablets, delayed release tablets, orodispersible tablets, hard gelatin capsules, soft gelatin capsules); however, consideration should also be given to the type of equipment available in any laboratory that may test the product. Examples of disintegration/dispersion mechanisms for a variety of products are provided in Sect. 7.4.

11.2.4 Step 4: Evaluate and Select Agitation Parameters

Agitation is used to help disperse the dosage form and facilitate dissolution of drug and impurities. The most common means of agitation are as follows: mechanical shaking, manual shaking, stirring, vortexing, and sonication. Along with each approach are a number of factors that may need to be controlled to ensure adequate dispersion and dissolution (see Table 7.5). For example, if mechanical shaking is chosen, one would need to select what type of shaker to use (e.g., orbital, reciprocating, wrist action), what speed to set the shaker to, how long to shake, what container to use, and how that container should be oriented. If an appropriate diluent has been selected (Step 3), the influence of these factors on dissolution should be minimized. Nevertheless, they are important to evaluate during the design phase and a greater understanding can translate into shorter, more efficient sample preparation conditions.

11.2.5 Step 5: Evaluate and Select Appropriate Means to Remove Insoluble Components

Once the compounds of interest are in solution, any insoluble excipients can be removed by filtration, centrifugation, or solid phase extraction. This is done to avoid issues with downstream separation (column lifetime) and detection (scattering). As with the other steps in method design, the choice in filtration and centrifugation conditions can have an impact on method reliability. Drug and/or drug related impurities can be retained on the filters, leading to artificially low results. In some cases, especially methods with low wavelength detection, nondrug-related impurities extracted from the filters can make quantitation and interpretation difficult. For centrifugation, appropriate relative centrifugal force and time settings must be selected. Additional information on sample clarification can be found in Chap. 9.

11.2.6 Step 6: Confirm Sample Preparation Method Works

The final step in the method design process is to confirm that the combination of conditions selected in the previous five steps result in a method that works. Drug product containing known amounts of active ingredient and impurities should be tested. At this stage, it may also be prudent to challenge the method with samples that could be encountered during routine use. Results from stressed stability studies or "aberrant" manufacturing conditions can provide an indication of method sensitivity toward the physical properties of the dosage form.

11.3 Method Evaluation for Sample Preparation

During the method design phase, it will become clear that there are a number of different factors that can influence sample preparation and extraction. Selection of conditions will have been influenced by some previous knowledge of the relationship between these factors and method performance. The goal of the method evaluation phase is to build upon that knowledge and establish a more complete understanding of these relationships. Additional consideration should be given to long-term use of the method; what challenges might be encountered 1, 5, or 10 years from now? Will the method be used in more than one laboratory? Are the physical properties of the dosage form likely to change over the course of the product lifecycle?

The list of factors that could influence method performance is often quite extensive and it may not be practical to study every possible relationship. Project teams often choose to use decision making and prioritization tools to help identify which relationships warrant further evaluation and which do not (Borman et al. 2007; Graul et al. 2010). There is little value in spending resources to study factors that are unlikely to be encountered or are expected to have a minimal effect on extraction. Section 11.3.1 describes an approach to help identify risks associated with sample extraction and some of the experimental strategies that can be used to better understand the importance of these risks.

11.3.1 Risk Assessment

Many of the tools and approaches used for identifying and managing risks associated with a pharmaceutical manufacturing process can also be applied to analytical methods (ICH Q9 2005). A formal or informal risk assessment will help to identify hazards and evaluate the risks associated with those hazards. A risk assessment can be broken down into three stages: risk identification, risk analysis, and risk evaluation.

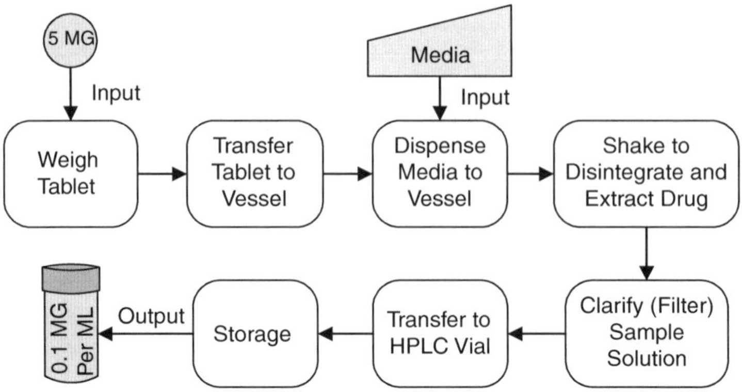

Fig. 11.3 Process flow diagram for steps associated with preparation of an immediate release tablet for HPLC analysis

11.3.1.1 Risk Identification

For sample preparation, the goal of this stage is to create a comprehensive list of any method factors that could affect extraction. As with the other stages in a risk assessment, it is a good idea to include scientists from multiple laboratories in this exercise. It is particularly valuable to solicit input from those who have had *hands-on* experience with the method, such as personnel from development, quality control, stability, and contract laboratories. These scientists will have had different experiences with the method (and similar methods) and will bring different perspectives on risk probability. To minimize the chances of missing important factors, the team should go through the method step-by-step and discuss in detail how each operation is performed. It is a good idea to map the method process as discrete unit operations in a process flow diagram (PFD). An example PFD for the preparation of an immediate release tablet is shown in Fig. 11.3.

Once the process has been mapped, method factors associated with each unit operation are listed. In addition to the process steps, it is important to consider properties of all ingoing materials, as these could also influence method performance. In the example shown in Fig. 11.3, there are two ingoing materials: the tablets and the media used to disintegrate the tablets. Tablet properties that could affect disintegration and solubilization could include the following:

- Hardness
- Age
- Storage conditions
- Shape
- Film-coat thickness
- Excipient vendor
- Particle size distribution of drug substance

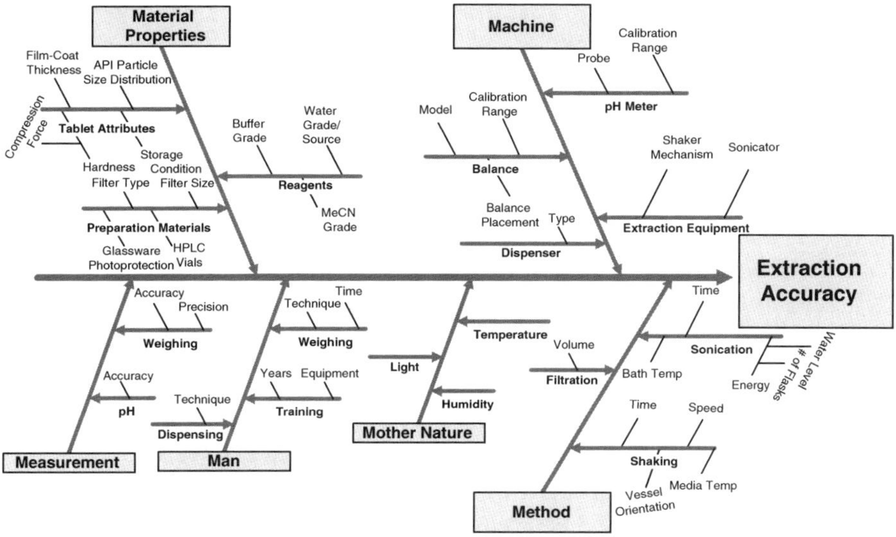

Fig. 11.4 Ishikawa diagram of method factors that could potentially impact extraction accuracy

Media properties that could be important include the following:

- pH
- Buffer and organic composition accuracy
- Ionic strength
- Diluent Temperature
- Diluent Purity (for an impurity method)

After the ingoing material properties have been identified, similar lists can be generated for each of the unit operations. Consider the shaking step as an example:

- Shaker motion (orbital, reciprocating, wrist action)
- Shake speed
- Shake time
- Vessel orientation (upright, sideways)
- Vessel size
- Vessel shape (volumetric flask, Erlenmeyer flask, glass bottle)
- Laboratory light (*potentially important for photo labile compounds*)

Risk assessment teams may wish to organize and present method factors as Ishikawa diagrams (Ishikawa 1985) sometimes called "fishbone diagrams." This is an excellent way to visually describe the relationships between method factors and a single method attribute. An example Ishikawa diagram for extraction accuracy is provided in Fig. 11.4.

11.3.1.2 Risk Analysis

Once all method factors have been identified, a risk analysis can be performed. This is an exercise to qualitatively or quantitatively estimate the influence of each method factor on method performance. Again, this exercise benefits from the collective knowledge of scientists who have had prior experience with the specific method or with very similar methods. Additionally, it is common practice to indentify someone from outside the group who is unfamiliar with the method to facilitate the risk analysis. The facilitator is responsible for keeping discussions on track, ensuring that opinions from all participants are heard, and encouraging the team to come to reasonable consensus on the risk level (or influence) of each method factor. It is also important that discussion and rationale supporting risk scores are captured, especially when not obvious. Teams should identify one or two people responsible for capturing the risk analysis output including supplementary discussion notes. Two of the more commonly used risk analysis tools are the cause & effect (C&E) matrix (e.g., Table 11.1) and failure mode effects analysis (FMEA) (e.g., Table 11.3) (Dept of Defense 1980; Van Leeuwen et al. 2009).

A C&E matrix can be thought of as a semiquantitative extension of an Ishikawa diagram. The first step is to identify the attributes of the method (or a subsection of the method) that are important for method performance. For the sample preparation part of the method, there could be several attributes: concentration accuracy of the final sample solution, precision of the final sample solution, and stability of the sample solution. The attribute is then scored based on the expected impact that it would have on the final measurement system result. For a measurement system used to determine the purity of a tablet, sample solution stability might score a "10" (based on scoring system in Table 11.2). Slight degradation of the drug substance in solution could result in large changes in degradation levels (levels that are not reflective of levels present in the tablet). The score may not be as high for a method used to determine tablet potency since in this case slight degradation in solution would have only a minor impact on the calculated amount of drug in the tablet.

After method attributes have been scored, each *method factor–method attribute* pair is scored based on the expected strength of the relationship. If a method factor is known or highly likely to have a strong relationship with an attribute, a high score will be assigned. For example, based on experience with sample preparation for other similar tablet formulations, the team may assign a "10" for the "shaker speed – concentration accuracy" pair (based on criteria in Table 11.2). An example of a scored sample preparation C&E matrix for an assay/purity method is shown in Table 11.1. The "Total Score" for each method factor is the sum of the expected relationship between that factor and each of the method attributes. The higher the "Total Score," the more influence (or risk) the method factor might have on method performance. A relative risk of all method factors is easily obtained by sorting by total score (highest to lowest).

Table 11.1 Scored cause & effect (C&E) matrix for preparation of an IR tablet

Unit operation	Factor name	Sample solution stability	Concentration accuracy	Concentration precision	Final score	Factor type	Experimental strategy
		10	5	5			
Shaking and extraction	Media temperature	9	9	9	180	XD	DOE 1
Shaking and extraction	Vessel orientation	9	9	9	180	F	Fixed
Shaking and extraction	Shake time	5	10	10	150	XD	DOE 1
Clarify (filter)	Filter membrane type	5	10	10	150	XC	OFAT 1
Tablet attributes	Product storage humidity	9	1	5	120	XC	DOE 1
Shaking and extraction	Shake speed	1	10	10	110	XD	DOE 1
Transfer to HPLC vial	Vial type	5	5	5	100	XC	Ruggedness 1
Clarify (filter)	Filter size	1	9	9	100	XC	OFAT 1
Shaking and extraction	Shaker mechanism	1	9	9	100	F	Fixed
Transfer to HPLC vial	Vial vendor	5	5	5	100	N	Ruggedness 1
Dispense media	Dispenser type	5	5	5	100	XC	Ruggedness 1
Clarify (filter)	Filtrate volume	1	9	9	100	XD	OFAT 1
Tablet attributes	Hardness	1	5	5	60	N	Ruggedness 1
Weigh tablet	Transfer time	1	1	1	20		Not evaluated
Weigh tablet	Balance accuracy	1	1	1	20		Not evaluatied
Weigh tablet	Balance precision	1	1	1	20		Not evaluated
Transfer to HPLC vial	Pipette transfer type	1	1	1	20		Not evaluated

Note: it is impractical to attempt to study each potential relationship. Teams often look for natural break points in the final scores and group method factors into 3–4 groups. Experimental studies will focus primarily on the two highest scoring groups. In this example the groups are color coded to clearly differentiate the groups

Table 11.2 Example scoring system for completing a cause and effect matrix

	Score	Criteria
Method attribute	10	Known to impact measurement system result (have evidence)
	7	Likely to impact measurement system result (may not have direct evidence)
	5	Unlikely to impact measurement system result (no evidence, though)
	1	No impact on measurement system result (have evidence)
Relationship between method factor and attribute	10	Known strong relationship between factor and attribute (based on data in hand or experience)
	9	Expected strong relationship between factor and attribute (no data to support)
	5	Medium relationship or uncertain
	1	No relationship between factor and attribute (based on data in hand or experience)

11.3.1.3 Risk Evaluation

The objective of the evaluation phase is to compare the identified risk against preestablished risk criteria (i.e., what level of risk is acceptable). After going through an initial risk analysis and performing some experimental studies, a team will begin to develop some understanding about factor/attribute relationships. At this stage, the team should consider the strength (or uncertainty) in understanding and whether current control systems adequately address risk. An FMEA is an excellent tool for this exercise.

In an FMEA, potential failure modes for the method are classified by their severity and likelihood of occurrence. An example failure mode for sample preparation might be a shaker table that is operating at 50 opm below the set-point (see Table 11.3). The team would consider how operating at 150 opm rather than 200 opm might affect disintegration and extraction efficiency. If it is likely that the tablet would not disintegrate in the time specified in the method, a high number (9 or 10) might be assigned for "severity." Next, the team would assign a number for likelihood of occurrence. Shaker tables that are properly maintained typically do not show such large changes in speed. If the laboratory has observed unreliable shaker performance a couple of times in the last year, they might assign a medium score (5) for "occurrence." Finally, the team would consider whether the product quality issue resulting from the failure mode is likely to be detected before the product reaches the consumer. If the answer is yes, then a low score (1) may be assigned for "detection." The three scores are multiplied to obtain a risk priority number (RPN). Failure modes with high RPN scores should be addressed either through further experimentation and/or through improvements to control systems. In the shaker table example, the laboratory might choose to do additional laboratory work to better understand

Table 11.3 Scored failure mode effects analysis (FMEA) for preparation of an IR tablet

Method factor	Potential failure mode	Potential effects of failure mode	SEV	Severity justification	Potential cause(s) for failure	OCC	Current controls	DET	RPN
	What could happen?	What are the potential effects on the method?	How severe is the effect on the method	What is the impact to method performance?	What are the potential causes for the failure?	How often does the Failure occur?	What are the existing controls that prevent either the failure itself or its cause?	How well can the Failure be detected?	SEV, OCC, DET
Shaker speed	Shaker operating below set point	Incomplete extraction in specified time	10	Low assay value for tablet, which could result in lot failure Super-potent tablets being released	Mechanical issues	5	None. Shaker speed is not routinely verified	1	50
Light exposure	Sample solutions exposed to light for excessive amount of time	Formation of a photo degradation product	5	Unexpected degradation product would result in laboratory investigation	Sample solutions left near windows. actinic glassware not used	7	Procedure states that *care should be taken to protect solutions from excessive light exposure*	1	35
Vessel orientation	Bottles shaken up-right rather than vertically	Incomplete extraction in specified time	9	Low assay value for tablet, which could result in lot failure Super-potent tablets being released	Operator does not follow test procedure as written	1	New operators must be trained prior to running method. Orientation is clearly indicated in procedure	1	9
Very hard tablets	Tablets exposed to high temperature and high humidity	Tablets do not disintegrate	9	Very low assay values for tablet, which could result in lot failure	Integrity of packaging is compromised. Impact will be observed on 40°C/75%RH stored samples	1	Packaging is robust. Effect would be observed in disintegration results	1	9

the relationship between shaker speed, shake time, and amount of drug extracted. Perhaps by increasing the specified shake time in the method, the extraction efficiency would be less sensitive to shake speed. Alternatively, the team may feel that it would be more prudent to focus on a preventative maintenance program to reduce the risk of the failure mode (e.g., shaker operating below set point).

11.3.2 Experimental Studies

11.3.2.1 Pre-experimental Planning

Prior to diving into laboratory studies, it is well worth the effort to devote some time to pre-experimental planning. Having a statistician involved in the planning is also recommended and can increase the probability that study results will lead to valid conclusions.

The first step in pre-experimental planning is to clearly state the objectives of the experiment(s) and to identify potential constraints (e.g., material, equipment, or resource limitations). The next step is to identify method factors and levels (or ranges) of interest. Information from previous experiments can be used to estimate experimental error, which in turn can help to determine factor ranges. Often, teams find it useful to classify method factors before selecting experimental designs. One classification system is described in Table 11.4. Note that the C&E matrix in Table 11.3 includes a column to designate method factor type.

The third step in the pre-experimental planning is to identify which response variables (or method attributes) are of interest. For an assay method, one might want to determine both the average and standard deviation of the concentration of active drug in sample solutions. Additionally, the concentration after a specified storage time may be of interest if solution stability behavior is not already well understood. Similar response variables could be envisioned for a purity method – but for impurity and degradant levels rather than active drug.

11.3.2.2 Selection of Experimental Designs

There are a variety of options when it comes to choosing an experimental design. The stage of development often influences design selection. For example, early in development, one may choose to perform a screening study with broad ranges to get a general sense of the sample extraction "space." Later in development, optimization designs with narrower ranges may be selected to more fully understand relationships between specific method factors and attributes. Several design options will be highlighted in this section, but others (or variations of these designs) may be more appropriate to meet the objectives of the study. Some useful references on statistical designs include (Burdick et al. 2003, 2005; Montgomery 2000; Vander Heyden et al. 2001).

Table 11.4 Example method factor classification system used for preexperimental planning

Category	Definition (examples)
Experimental (XD – discrete) (XC – categorical)	*Discrete* – method factors that can be numerically varied and are selected for experimentation Examples of "XD" factors for sample preparation include shaker speed, shake time *Categorical* – method factors that are nonnumerical, but can be set at multiple levels for the purpose of experimentation Examples of "XC" factors for sample preparation include shaker type (orbital, reciprocating, wrist-action®), shaker model and vessel orientation (sideways, upright)
Fixed (F)	Method factors that are likely to have an effect on a method attribute, but for the purpose of experimentation are of no interest and will be held at a specific (fixed) level An example of an "F" factor could be "orbital shaker" if there is no interest in evaluating other shaker types Note: in this case it would be important to state in the test procedure that an orbital shaker must be used
Nuisance (N)	Method factors that cannot be controlled or are allowed to vary randomly from a specified population Examples of "N" factors for sample preparation could include the humidity of thc laboratory used for storage and testing of the tablets or the operator performing the testing Note: while the humidity may not be controllable, the humidity could likely be measured during experimental studies and treated as a covariate in data analysis. Since "N" factors cannot be controlled, the objective is to find regions for "X" and "F" factors that minimize the influence of "N" factors

One commonly used approach is known as *one-factor at a time (OFAT)*. A single set of conditions is chosen as the baseline level. Each factor of interest is then varied successively while the other factors are held constant at baseline levels. An example of the output of an OFAT study for sample preparation is shown in Fig. 11.5.

From this study, one might conclude that over the ranges studied, shaker speed and shake time show no influence on the amount of drug recovered from the tablet. Additionally, both the orbital and reciprocating shakers show complete recovery while the wrist action® shaker does not. The results of this study also suggest that the method could be optimized by reducing the shake time from 15 to 5 min, and reducing the shaker speed from 200 to 150 opm (potentially reducing wear on the shaker).

The major disadvantage of the OFAT approach is that interactions between factors are not considered. This omission can easily lead to false conclusions about appropriate method conditions. In the example above, if the shake time study is run at 200 opm, there is no effect on the amount of drug recovered. However, if the study were run

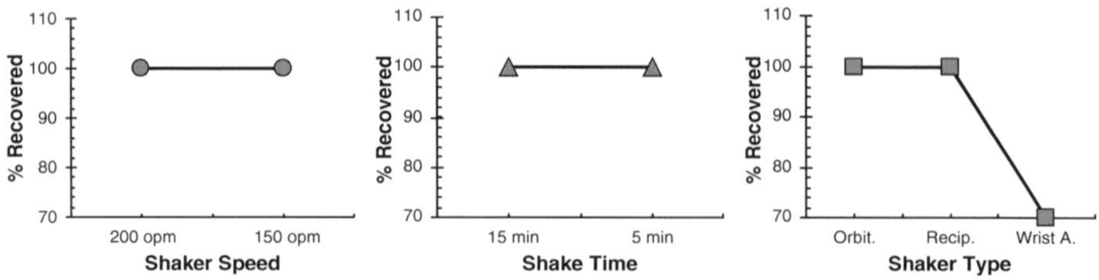

Fig. 11.5 Output from OFAT studies to understand influence of shaker speed, shake time and shaker type on recovery of drug from a tablet. Baseline factors and levels are speed (200 rpm), time (15 min), type (orbital)

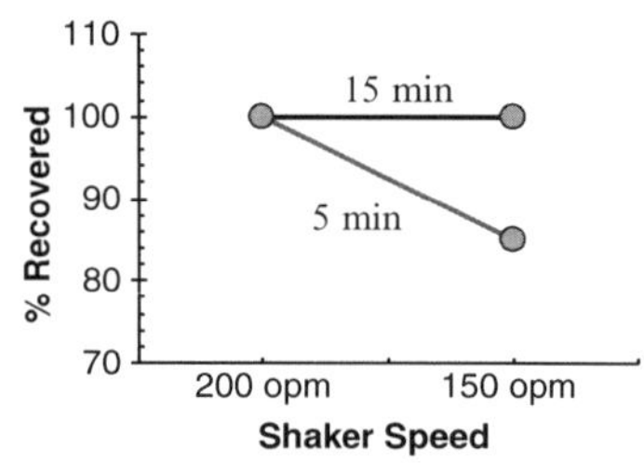

Fig. 11.6 Illustration of interaction effects between shaker speed and shake time

at 150 opm (results from the shaker speed study suggested this was OK), complete extraction after 5 min may not be obtained (Fig. 11.6).

Interaction effects are quite common for method factors associated with sample preparation. OFAT studies should only be considered if the factor of interest is known not to have an interactive effect with any of the other factors being studied.

When multiple factors are being studied, and it is not clear if there are interaction effects, *factorial designs* are more appropriate. In these designs, multiple factors are varied together rather than individually. In the shaker example, both speed and time might be varied simultaneously in a 2^2 factorial design. This design allows a better understanding of the individual effects of each factor (main effects) as well as any interaction effects between the 2 factors. To obtain additional confidence in the effects of each of the two method factors, the scientist may wish to repeat the study twice for a total of eight experiments (Table 11.5).

However, for sample preparation, there are typically more that two factors identified during risk analysis as potentially impacting method performance. A 2^3 factorial design (eight runs) can be used to study three factors at two levels, a 2^4 factorial design (16 runs) for four factors at two levels, and so on. As the number of factors of increases by one, the number of runs required for a full factorial study doubles. It is not uncommon to have seven+factors of interest which could require 128 (or more) runs!

Table 11.5 Example of a 2^2 factorial design to study effects of shaker speed and time

Exp. no	Shaker speed	Shake time
1	150 opm (low)	5 min (low)
2	150 opm (low)	15 min (high)
3	200 opm (high)	5 min (low)
4	200 opm (high)	15 min (high)

As one can imagine, execution of all of 128 runs might consume a great deal of time and material. When studying four or more factors, teams should consider *fractional factorial designs*. These designs can greatly reduce the number of experiments needed to identify factors that have large effects on extraction efficiency, and are often used for screening experiments. More detail about the application of fractional factorial designs can be found in references (Montgomery 2000; Vander Heyden et al. 2001). Alternatively, method factors can be grouped and evaluated in several separate smaller studies.

Once method conditions have been defined, *gauge repeatability and reproducibility (Gauge R&R)* studies are commonly conducted to determine how much variation is associated with the measurement system and what factors are the largest contributors to that variation. For methods designed and evaluated using QbD principles, the biggest contributors may already be known and controlled. Nevertheless, Gauge R&Rs can be an excellent means of confirming whether the amount of variation introduced by a method is small relative to the variation introduced by the manufacturing process. If there is a desire to be able to improve a manufacturing process, as a general rule, the measurement system should be contributing <30% (and ideally <10%) to the total variation (Rasmussen et al. 2005; Wells 2010).

The repeatability part of a Gauge R&R represents the variability coming from one person measuring the same material several times in a short interval (e.g., same day). Reproducibility represents the variability associated with multiple people across multiple laboratories testing the same material. It is important that the material selected for the study is representative of material that is likely to be encountered during routine testing. For example, if tablet hardness was likely to affect disintegration rate (and therefore extraction efficiency), one would want to make sure that material selected for the Gauge R&R included tablets across the typical hardness range. Likewise, one would also want to make sure that skill levels of laboratory scientists selected for the study are representative of those expected to routinely use the method. Although it can be tempting to use the most experienced scientists for the studies, this practice should be explicitly avoided to reduce the risk of *underestimating* measurement system variability. On the contrary, unrealistic sources of variability should not be incorporated into a Gauge R&R. For example, an SOP may state that a scientist cannot test production material until he or she has been formally trained on the method by an experienced scientist. One would not want an "untrained" scientist participating as there is a risk of *overestimating* measurement system variability.

11.3.2.3 Data Analysis

Once experiments have been completed, statistical methods can be used to analyze the results. Assuming that proper attention was paid to pre-experimental planning, this step should be relatively straightforward. Software packages used for experimental design can often also be used for data analysis. Most packages allow results to be presented graphically, which is particularly helpful for illustrating the relative importance of different method factors and the influence of multifactor interactions. As with the experimental planning, it is always a good idea to involve a trained statistician involved with the data analysis and interpretation of results.

11.3.2.4 Multivariate Study Example for Controlled Release Formulation

Examples provided to this point have been for relatively simple immediate release dosage forms. The formulations are designed to release drug relatively quickly, so extraction for analytical purposes is not typically too challenging. Rapid and reproducible extraction from controlled release formulations, on the contrary, can be quite difficult. Typical extraction times for an osmotic formulation such as swellable core technology (SCT) tablets (Thombre et al. 2004) can be 12 h or more due to the presence of polymeric excipients in the sweller layer as well as in the tablet film-coat. In addition, SCT methods often require stepwise addition of solvents and the use of costly molecular weight cut off (MWCO) centrifuge filters. This next example illustrates how multivariate studies can be used to evaluate extraction methods for more complicated dosage forms, and how the results can be graphically presented to provide a great deal of information about method robustness.

In this example, a project team was interested in developing a more efficient extraction procedure for SCTs that was also less complicated (fewer steps) and less expensive (elimination of MWCO filters) than the current 12 h shaking procedure. A 30-min, five step process was developed using methanol as an extraction solvent and a Polytron® homogenization probe to mechanically break apart the tablet (Fig. 11.7).

After a fractional factorial screening study was conducted to establish initial homogenization conditions, a follow-on 2^3 full factorial study was designed to optimize the method. The three factors evaluated were the number of homogenizer pulses (2–8), the duration for each pulse (20–60 s), and the pulse speed (8k–12k rpm) (Fig. 11.8). The results of the study indicated that over the ranges studied, pulse speed did not have a significant effect on extraction. However, pulse duration and the number of pulses were both shown to impact the amount of drug extracted (Fig. 11.9). Furthermore, the curvature in the contour plot shows the interaction effect between these two method factors. Better extraction is obtained with more pulses that are each longer in duration. Although this interaction effect is not unexpected, the contour lines clearly show the sensitivity of extraction toward these two factors. In this case, a method with a minimum of 8 pulses, 60 s each, is necessary to achieve complete and robust extraction.

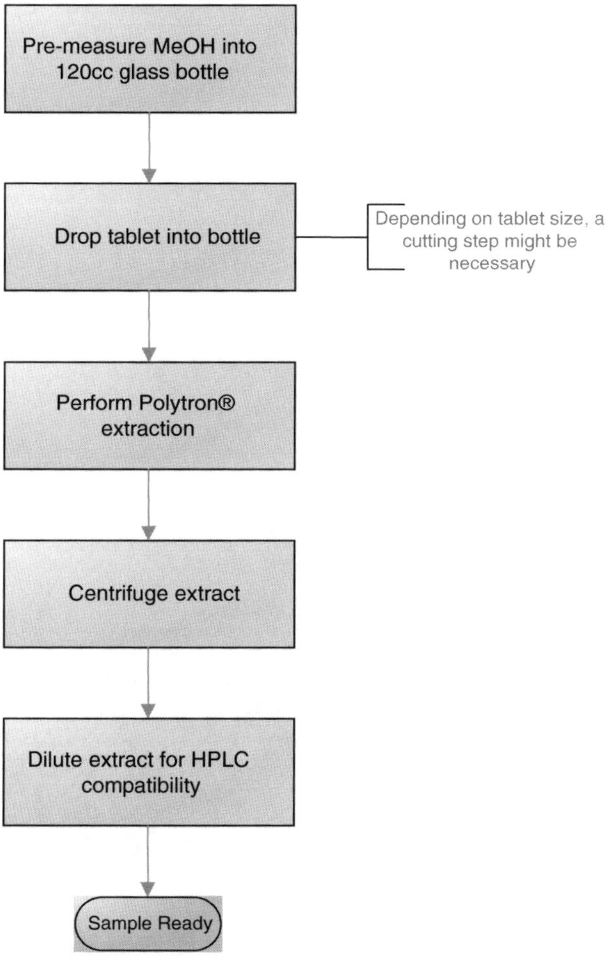

Fig. 11.7 PFD for the optimized SCT method using Polytron® homogenization

11.4 Method Control Strategy and Lifecycle Management

Once sufficient method understanding has been obtained, an appropriate control strategy can be established to ensure that the method will be suitable for routine use. Results from experimental studies can be used to establish optimal settings for each method factor. In some cases ranges for these factors can be supported. For example, photostability studies may show that light protective glassware is required to prevent unacceptable levels of degradation. Part of the method control strategy could include a statement in the test procedure indicating that actinic glassware must be used to prepare and store sample solutions. But then, extraction studies might show that

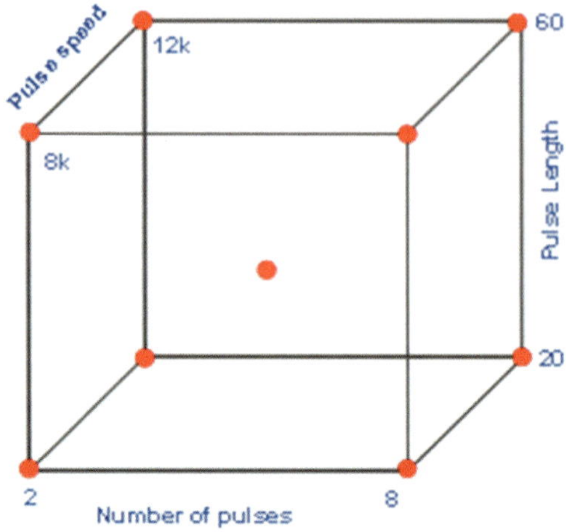

Fig. 11.8 Graphical representation of a 2^3 full factorial design (with center point) for optimizing homogenization conditions

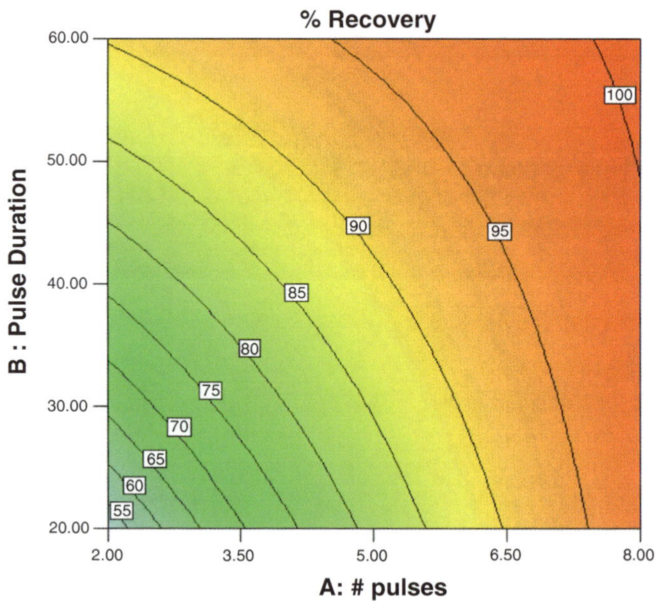

Fig. 11.9 Results of the 2^3 full factorial optimization design for SCT extraction using homogenization

shaking times between 5 and 20 min are acceptable, as long an orbital shaker is used at 200 opm. Since times less than 5 min were not evaluated, the scientist may feel more comfortable setting a minimum shake time slightly higher than the lowest level evaluated. A range of 10–20 min might be established. An example of how

Table 11.6 Method factor set points and ranges for the preparation of 5 mg drug A tablet solutions for assay and purity testing by HPLC

Method factor	Level
Glassware	Actinic, or foil covered 100 mL bottles
Bottle shape	Round or square
Diluent	80–90% 0.1 N HCl
	10–20% ACN
Diluent volume	50 mL
Bottle orientation	Sideways
Shaker type	Orbital
Shaker speed	200 opm
Shake time	10–20 min

sample preparation set points and ranges may be presented in a test procedure is shown in Table 11.6.

So, one element of the method control strategy is assurance that testing is performed within ranges specified in the test procedure. A second element is use of performance checks. System suitability requirements are often part of HPLC, NIR, and UV test procedures. The purpose is to verify that the instrument is performing properly close to the time of use (typically performed just prior to sample analysis). However, most of these checks focus on instrument/detector performance and ignore the sample preparation component of the method.

Theoretically, performance checks could extend into sample preparation. For example, a previously characterized lot of tablets could be tested along side the tablets of interest. A suitability requirement might be that the potency for the reference lot must be within 1 or 2% of the originally determined potency. However, before implementing such a control check, one should keep in mind that it will require additional steps in the procedure and could also require maintenance of a reference lot of drug product under conditions whereby physical and chemical properties remain unchanged. Visual observations could also be considered "performance checks." For example, one might include a statement in the method that "After completing the shaking step, carefully examine the bottle to ensure that the dosage form has completely disintegrated. If it has not completely disintegrated, continue shaking for an additional 5 minutes. If after an additional 5 minutes of shaking the dosage form is still not completely disintegrated, record the observations and total extraction time in your notebook and notify a lab manager."

With proper application of QbD principles, the importance of performance checks is minimized. During the method design and evaluation phases, sample preparation steps where failure could lead to poor method performance are identified and controlled.

Other potential elements of a method control strategy are shown in Fig. 11.10. One can develop a very robust method but then encounter frequent issues if the operators using the method have not been adequately trained. *Laboratory qualification* including familiarization with sample preparation equipment and participation

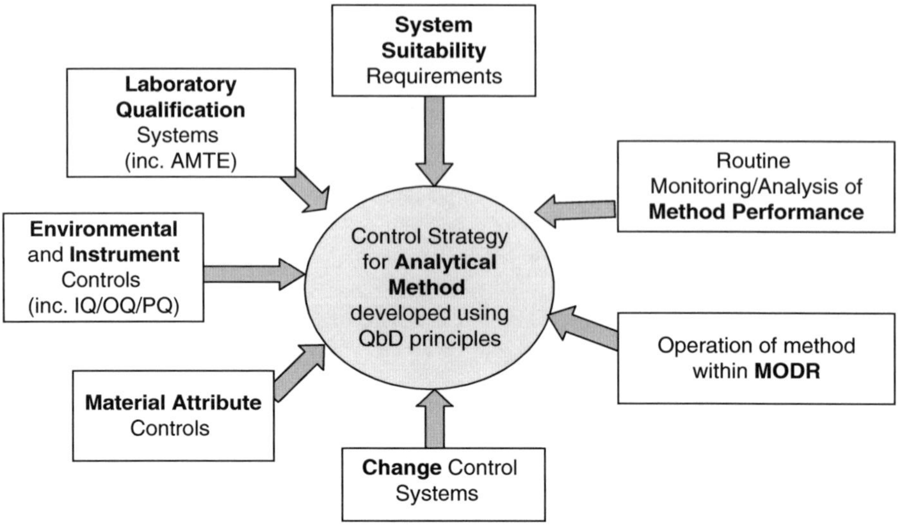

Fig. 11.10 Potential elements of a method control strategy

in method specific AMTEs can reduce the frequency of failures and laboratory investigations (Nethercote et al. 2010). The knowledge collected during method development and evaluation should be shared with receiving laboratories. AMTEs should also be designed to ensure that the new laboratory will be able to operate anywhere within the originally established MODR.

Environmental and *instrument control* systems are also very important. If equipment are not properly maintained (or correctly installed), one can expect issues with method performance. For samples that are thermal, humidity, or photo-sensitive, maintaining adequate laboratory controls can be critical. Control of ingoing *material attributes* can be accomplished by specifying reagent grades in test procedures. This can be particularly important for purity methods where impurities introduced through water systems or low grade solvents can bias results or otherwise complicate data analysis.

The ultimate goal of a method control strategy is to ensure that the method will continue to meet the ATP criteria over the lifecycle of the product. As such, one must be confident that the uncertainty in the measurement system is not changing (or at least increasing) over time. Using QbD approaches to define robust operating conditions and appropriate method performance checks will go a long way toward facilitating consistent method performance. However, there will be situations that were unforeseen during method design that will occur months, years, or decades after method implementation. A major strength of the QbD approach is that it provides a framework for evaluating risk associated with these new events. Indicators of method performance can be monitored and trended in a similar manner used to assess variation in manufacturing processes.

11.5 Knowledge Management

A knowledge management system is critical to leveraging the extensive knowledge that will be generated throughout the QbD analytical method development process. A knowledge management system provides “a systematic approach to acquiring, analyzing, storing, and disseminating information” (ICH Q10 2008). Knowledge needs to be captured from across the organization (e.g., from R&D as well as QC laboratories). As shown in Fig. 11.11, this knowledge is not merely documented and archived, but is instead captured in a system where the knowledge can be readily accessed. Throughout the lifecycle of the product, this knowledge is referenced (e.g., for use in future risk assessments when changes are made) and transferred (e.g., to other sites or projects).

In the QbD analytical method development process, knowledge is generated during method design, during method evaluation (e.g., outcomes of analytical method risk assessments, outcomes of experimental studies) and during method transfer activities. Various tools (e.g., PFDs, C&E matrices, fishbone diagrams) are available to aid in capturing information during these processes. Knowledge may also be available from other products within the same dosage form platform or from other products for the given compound. Additional knowledge about method performance will continue to accumulate every time commercial product is tested. The outcome of these activities can be captured in individual reports, which can then be referenced or linked within an overarching report such as a QbD Analytical Method Understanding Report. This report along with information from QbD manufacturing process understanding studies can be accessed and shared as needed. Both analytical method and process understanding reports should be continually updated as new learning is gained across the organization. Knowledge management is an iterative process. Therefore, there should be a mechanism to trigger an evaluation of whether the method or the design space needs to be reevaluated as new information is obtained, or when changes are made in the formulation or manufacturing process.

Once the knowledge for a given project has been captured in the knowledge repository, the potential impact of any proposed analytical method changes can be evaluated. These proposed changes may result from the desire to make method improvements, or to address changes in the formulation or manufacturing process. A risk assessment should be performed evaluating the proposed analytical method changes based on the knowledge that has been accumulated up to this point in time.

Lessons learned are valuable to share across projects as well. This may be particularly true for different products that use the same formulation platform as there may be common themes or challenges with respect to sample preparation challenges (e.g., known extraction issues due to gelling of a specific polymer used in an osmotic tablet formulation platform). It is therefore important to have a knowledge repository system that allows sharing of information across projects. With this central repository of knowledge, a given project can assess potential impact of proposed future changes to the analytical methodology.

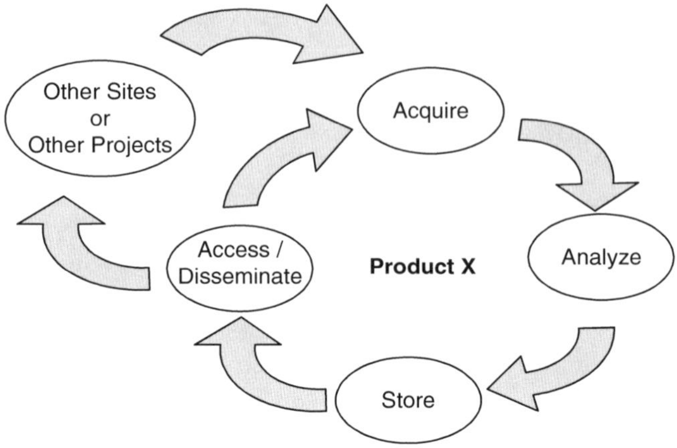

Fig. 11.11 Schematic of knowledge management process

11.6 Conclusions

Quality is not an attribute that can be "tested" into a product at the end of pharmaceutical development. It is a guiding principle that should drive all aspects of development, beginning with the identification of product requirements and continuing throughout the life-time of the product. The same holds true for the measurement systems that support these products. One cannot rely solely on a single validation exercise at the end of method development to ensure long-term quality performance. Instead, quality must be designed into measurement systems from the start. This is particularly important for sample preparation aspects of a method, since "time of use" performance checks (i.e., system suitability) are set up to flag chromatographic or detection problems rather than extraction issues. Application of QbD principles throughout a method lifecycle – including sound design approaches as described in Chap. 7 – should increase method understanding, reliability, and time spent on activities other than troubleshooting (Chap. 13).

Acknowledgments The authors thank the following individuals for insightful discussion and generous contributions to this chapter: Garry Scrivens, Kimber Barnett, Tim Graul, Melissa Hanna-Brown, Peter Jones, Loren Wrisley, Gregory Steeno, Mike Rouns, and Kelly Canter.

References

Borman PJ, Nethercote P, Chatfield MJ, Thompson D, Truman K (2007) The application of quality by design concepts to analytical methods. Pharm Technol 31(12):142–152

Burdick RK, Borror CM, Montgomery DC (2003) A review of methods for measurement systems capability analysis. J Qual Technol 35(4):342–354

Burdick RK, Borror CM, Montgomery DC (2005) Design and analysis of gauge R&R studies: making decisions with confidence intervals in random and mixed ANOVA models. ASA-SIAM Series on Statistics and Applied Probability, SIAM, Philadelphia, ASA, Alexandria, VA

Chatfield MJ, Borman PJ (2009) Acceptance criteria for method equivalency. Anal Chem 81(24):9841–9848

Cole DC, Pagano N, Kelly MF, Ellingboe J (2004) Use of design of experiments to optimize high-throughput semipreparative LC and LC/MS methods. J Comb Chem 6(1):78–82

Dept of Defense (1980) Procedures for performing a failure mode, effects, and criticality analysis. MIL-STD-1629A. United States of America Department of Defense Military Standard

Graul TW, Barnett KL, Bale SJ, Gill I, Hanna-Brown M (2010) Quality by design for analytical methods. In: am Ende DJ (ed) Chemical engineering in the pharmaceutical industry: R&D to manufacturing. Wiley, New York

ICH Guideline Q10 (2008) Pharmaceutical quality system. ICH Online. http://www.ich.org. Accessed Jun 2008

ICH Guideline Q2 (R1) (2006) Validation of analytical procedures: text and methodology. ICH Online. http://www.ich.org. Accessed Nov 2006

ICH Guideline Q3A (R2) (2006) Impurities in new drug substances. ICH Online. http://www.ich.org. Accessed Oct 2006

ICH Guideline Q3B (R2) (2006) Impurities in new drug products. ICH Online. http://www.ich.org. Accessed Jun 2006

ICH Guideline Q3C (R4) (2009) Impurities: guidelines for residual solvents. ICH Online. http://www.ich.org. Accessed Feb 2009

ICH Guideline Q8 (R2) (2008) Pharmaceutical development. ICH Online. http://www.ich.org. Accessed Nov 2008

ICH Guideline Q9 (2005) Quality risk management. ICH Onlinc. http://www.ich.org. Accessed Nov 2005

Ishikawa K (1985) What is total quality control? The Japanese way. Prentice-Hall, Englewood Cliffs, pp 63–64

Molnár I, Rieger H-J, Monks KE (2010) Aspects of the "design space" in high pressure liquid chromatography method development. J Chrom A 1217(19):3193–3200

Montgomery DC (2000) Design and analysis of experiments, 5th edn. Wiley, New York

Nethercote P, Borman PJ, Bennett T, Martin G, McGregor P (2010) QbD for better method validation and transfer. PharmaManufacturing.com. http://www.pharmamanufacturing.com/articles/2010/060/html. Accessed October 2010

Rasmussen HT, Weiyong L, Redlich D, Jimidar MI (2005) HPLC method development. In: Ahuja S, Dong MW (eds) Handbook of pharmaceutical analysis by HPLC, vol 6. Elsevier, San Diego

Schweitzer M, Pohl M, Hanna-Brown M, Nethercote P, Borman P, Hanson G, Smith K, Larew J (2010) Implications and opportunities of applying QbD principles to analytical measurements. Position Pap QbD Analytics Pharm Technol Eur 34(2):52–59

Thombre AG, Appel LE, Chidlaw MB, Daugherity PD, Dumont F, Evans LA, Sutton SC (2004) Osmotic drug delivery using swellable-core technology. J Control Release 94(1):75–89

USP 33-NF 27, Chapter<1225> (2010) Validation of compendial practices. USP-NF Online. http://www.uspnf.com. Accessed October 2010

Van Leeuwen JF, Nauta MJ, de Kaste D, Odekerken-Rombouts YMCF, Oldenhof MT, Vredenbregt MJ, Barends DM (2009) Risk analysis by FMEA as an element of analytical validation. J Pharm Biomed Anal 50(5):1085–1087

Vander Heyden Y, Nijhuis A, Smeyers-Verbeke J, Vandeginste BGM, Massart DL (2001) Guidance for robustness/ruggedness tests in method validation. J Pharm Biomed Anal 24(5–6):723–753

Wells J (2010) Gauge R&R for transactional six sigma projects – measuring the unmeasurable. QualityDigest.com. http://www.qualitydigest.com/inside/quality-insider-article/measuring-unmeasurable.html. Accessed October 2010

Williams RL, Abernethy DR, Koch WF, Hauck WW, Cecil TL (2009) Performance Based Monographs. Stimuli to the revision process. Pharmacopeial Forum 35(3):765–771

Ye C, Liu J, Ren F, Okafo N (2000) Design of experiment and data analysis by JMP (SAS Institute) in analytical method validation. J Pharm Biomed Anal 23(2–3):581–589

Section D
Additional Sample Preparation Topics

Chapter 12
Automation and Sample Preparation

Gang Xue and K. Rick Lung

Abstract Sample preparation for pharmaceutical dosage forms can be a challenging, time consuming, and labor intensive task. Long extraction times may be needed to completely extract the components of interest. In addition, multiple samples often need to be prepared and analyzed for such studies as stability, formulation development, manufacturing process optimization, etc. Automation can reduce the resources and/or time needed to prepare samples. This chapter will discuss different instrumentation available and strategies to automate sample preparation. Examples and case studies will also be presented.

12.1 Introduction to Laboratory Automation

For many years, laboratory automation has been widely applied to various aspects of the pharmaceutical industry, from as early as lead optimization to as late as commercial scale manufacture. It can be as simple as a sample injector (Polesello 1995), or as sophisticated as an adsorption, distribution, metabolism, and excretion (ADME) work station (Reichman and Gill 2009), as slow as a *single* cell manipulating micro operator (Zhang et al. 2009) or as fast as a subsecond sampling high throughput screening (HTS) system (Gloekler et al. 2010), as customized as a home-grown sample division system (Xue and Brown 2009) or as highly commercial as a Metrohm Karl Fischer titrator. The most obvious driver to automate is to reduce the required labor for laboratory operation, especially for labor intensive activities. This becomes particularly important at a time when the industry strives for productivity gain and

G. Xue (✉)
Analytical Development, Pfizer Global Research and Development, Pfizer Inc., 558 Eastern Point Road, Groton, CT 06340, USA
e-mail: gang.xue@pfizer.com

B. Nickerson (ed.), *Sample Preparation of Pharmaceutical Dosage Forms: Challenges and Strategies for Sample Preparation and Extraction*, DOI 10.1007/978-1-4419-9631-2_12,

minimized business cost. Another key benefit of automation is the improvement of experimental data quality. With robotic precision, the laboratory operation instruction can be repeated meticulously, and more importantly, each operation can be electronically tracked providing an accurate full audit trail. It eliminates or greatly reduces the chance of data error due to human error or mix-up. At the time of method transfer, the use of consistent lab automation equipment simplifies the transfer to downloading an electronic method from the corporate database. In addition, automation systems typically allow sampling more data points during process monitoring, resulting in more accurate process control and decision making. Finally, lab automation provides unmatched safety protection of laboratory personnel by minimizing the exposure of scientists to highly toxic or hazardous reagents.

For pharmaceutical sample preparation, all of the above-mentioned benefits are quite significant. As an example, Majors surveyed and summarized the chromatography analysis laboratory practices within the pharmaceutical industry during one of his recent presentations (Majors 2006). As much as 61% of the lab scientists' time was concluded to be spent on sample preparation, followed by 27% on data processing. With regard to sources of error generated during analysis, 30% trace back to sample preparation with an additional 19% originating from human error. Clearly, automation could play a big role in improving the sample preparation efficiency. In addition, active pharmaceutical ingredients (API) are generally most hazardous in solid form because of possible airborne inhalation risk. Once dissolved and diluted in solvents, the health risk can be significantly reduced and the sample is safer to handle in the lab.

The use of automated sample preparation technology for API and formulated drug assay and content uniformity testing is not new. Reports of automated sample preparation and analysis of pharmaceutical samples appeared as early as in the 1960s and 1970s (Beyer 1966; Hanna et al. 1976). In these early examples of automated tablet analysis, a homogenizer combined with the Technicon analysis technology was used. The Technicon technology used colorimetric reagents, peristaltic pumps, and mixing cells for on-line colorimetric analysis. As HPLC eventually became the method of choice for tablet analysis, it also replaced the Technicon analyzer in automated tablet analysis. In 1979, a fully automated sample preparation instrument that used a continuous flow extractor combined with HPLC analysis was reported (Huen and Thevenin 1979).

During the 1980s, robotics technology designed for the chemical laboratory became commercially available. Initially, these laboratory robots were designed to be a totally flexible system. Commercial instrument vendors, such as Zymark and Perkin-Elmer, supplied laboratory robot components, software, and user training that allowed the users to build their own laboratory systems. In principle, a robotic user in the 1980s could build an automated analysis system to perform any laboratory task from parts supplied by Zymark or Perkin Elmer (Cirillo 1986). A typical laboratory robot system then included a robot arm and a set of laboratory peripherals, each performing a specific laboratory unit operation (LUO). The primary function of the robotic arm was to pick up and place samples and laboratory glassware at designated locations. The function of each peripheral device within the laboratory robotic system was to perform a specific unit operation within the laboratory. For example, a laboratory robotic system might include a computer controlled solvent dispensing station. The unit operation for this solvent dispensing station was to dispense an accurate amount of reagent to a sample container. A general review of the available robotic arms and typical peripheral devices was published by Lochmuller et al. with many examples of laboratory applications in sample preparation (Lochmuller et al. 1985).

For these "first generation" laboratory automation systems, the user had to program the robotic arm and peripheral stations with a great amount of detail to accomplish a specific task. For example, the robot arm must be told to move to a sequence of coordinates that corresponded to picking up a beaker from a specific location at point A and placing it down at point B. Robot vendors typically supplied the user with the robotic programming platform with built-in drivers for unit operation such as the Zymark EASYLAB™ and Perkin Elmer Robotic Language (PERL™).

As the technology for the robotic applications became more specialized, dedicated automation systems began to emerge. Lab scientists also found programming each LUO too costly and time consuming. Throughout the next few years, vendors began to develop preconfigured robotic systems for a number of commonly used applications, which gradually shaped the modern sample preparation automation. This chapter discusses the available laboratory automation solutions for sample preparation of API, solid dosage forms and nonsolids in today's pharmaceutical laboratories.

12.2 Automated Sample Preparations for API and Powders

Compared with formulated drug products, sample preparations for API and reference standards are generally more straightforward since they only involve the dissolution of a single pure chemical entity. API solubility characteristics within various solvent systems should have been explored at a relatively early stage of drug development. Full recovery of the analyte in solution is usually not a problem with the proper selection of dissolving solvents or diluents. However, like drug products, the manual preparation of API and reference standards remains one of the most time consuming and laborious steps in the overall pharmaceutical analysis workflow.

Most API and reference standards are in powder form, and the typical manual sample preparation procedure involves accurate weighing of a specific amount of the powders and transferring it into a fixed volume volumetric flask, followed by dispensing of diluent, agitation, bringing solvents to scale and homogenization. Of all the steps, accurate weighing and diluent dispensing with volumetric flasks are the most labor intensive and error-prone, and scientists have been seeking help from lab automation to remediate these issues.

12.2.1 Automatic Powder Dispensing

Unlike liquids, powders are difficult for lab automation systems to handle. Each chemical has its unique physiochemical properties such as density, particle size, surface charge, cohesiveness, and hydroscopicity. As a result, while all liquids flow predictably

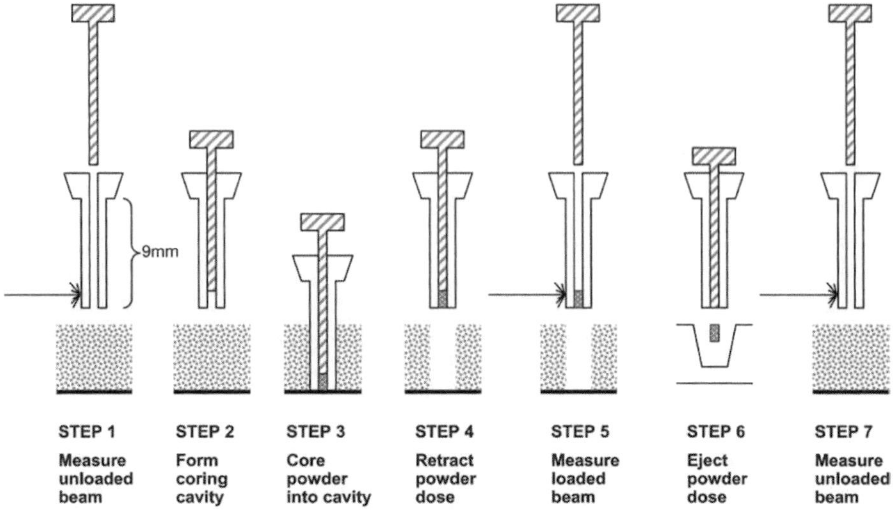

Fig. 12.1 Schematics of sampling process with a powder pipette with opto-mechanical mass sensor (reprinted from Cherng et al. 2004, with permission from Elsevier)

in response to pressure differences, the flow-ability of powders vary so dramatically that it is impossible to have one system automatically dispense all of the different API and reference standards. Actually, this is one of the main reasons for adding excipients into the drug formulations to improve their manufacturability. To date, a number of powder dispensing automation systems have been introduced with very distinct working mechanisms, each applicable to a range of powder categories.

12.2.1.1 Volumetric Powder Pipetting

The first type of powder dispensing automation is analogous to the widely used liquid micropipettes. As shown in Fig. 12.1, the powder pipettes consist of a high grade stainless steel tube (sampling probe) with a plunger in the middle. The plunger is adjustable within the tube to allow the setting of sampling volume (Step 2). When the sampling probe is dropped into the powder container, the powder is compressed into the coring cavity (Step 3) and extracted (Step 4). The dose of the powder is then ejected into the receiving container by actuating the plunger (Step 6). As illustrated, the powder pipette works very similarly to liquid pipettes except that the powder pipette relies on the powder cohesive force to keep the material within the sampling probe during transfer. Thus, this type of system is only applicable for medium to highly cohesive powders. Since the powder is dispensed via volumetric delivery, the actual mass is affected by the bulk density. Calibration is typically required for each powder.

The DisPo sampling probe made by BioDot falls into this category. Five different internal diameter sample probes (0.5, 0.8, 1.0, 2.0, and 3.0 mm) are available, offering a choice to best fit the target mass and cohesiveness of the powder. Corn starch was demonstrated to be consistently dispensed in the 0.1–5.5 mg range with 8.8% RSD and 20 different powders including ibuprofen, naproxen, acetaminophen, table salt, and grout were tested with success (BioDot Inc. 2009). Although manual pipette operation offers the most flexibility and simplicity for fast powder dosing, the probes could be mounted to a robotic arm to allow hands-free batch powder dispensing as well. The DisPo was reported to successfully dose 160 μg naproxen to 384-well plates automatically from a very limited amount of starting material (BioDot Inc. 2009).

Zinsser took a very similar approach in their powder pipette tip design (REDI) to "aspirate" and dispense granules and powder material. Additional functional modules are integrated with the powder pipette to improve system performance and reduce the carryover. Shaking or stirring is introduced to the source reservoir to even out the particle distribution. A special tool is designed to wipe off excessive powder outside the tip to improve the volume reproducibility and minimize spill. There is also a cleaning station to thoroughly clean the pipette tip between runs to prevent clogging and carryover. The available dispensing volume ranges from 1 to 300 μL, which is about 300 μg to 100 mg. The weight reproducibility depends upon the particle size distribution within the sample. Powders with narrow particle size distribution show better reproducibility. It has been reported that the REDI 2002 Plus achieves a reproducibility of ±5% for 1 mg, ±3% for 10 mg, ±1% for 50 mg, and ±1% for 150 mg with various powders (Zinsser 2010).

Simplicity, speed, and low wastage are the key advantages of the powder pipette approach. For example, the Zinsser REDI can fill a 96-well plate within 15 min and the DisPo is expected to be at a similar throughput. However, as a volumetric distribution approach, the mass accuracy depends heavily on the powder packing density, which is affected by many factors such as particle size distribution, source powder bed height, and compressing pressure during extraction. Although 5~10% RSDs were reported in well-controlled experiments with good behaving compounds, the mass variation is much more significant in practical routine lab operations. Gravimetric confirmation with micro or analytical balances is required to record the accurate mass when high precision analysis is required.

Recently, Cherng et al. reported a novel opto-mechanical sensor to accurately measure the mass of powder directly on the transferring pipette (Cherng et al. 2004). To do so, the conical sampling probe has to be made with minimal mass and high stiffness. A piezoelectric ceramic actuator applies a swept-sine signal to the base of the probe. Then the probe would respond to the excitation signal with a unique resonant frequency that is a function of its mass and stiffness. The shift in the resonant frequency thus reveals the mass difference before and after the powder is dispensed (Fig. 12.1 steps 5 and 7). In this work, the sensor demonstrated the capability of measuring as little as 30–70 μg of powder with less than 2.9% RSD in a few seconds. Although there are a couple of technical challenges to be resolved before the technology can be applied to routine use, it offers a fast and accurate remediation for the existing volumetric pipette products with precise powder weight measurements.

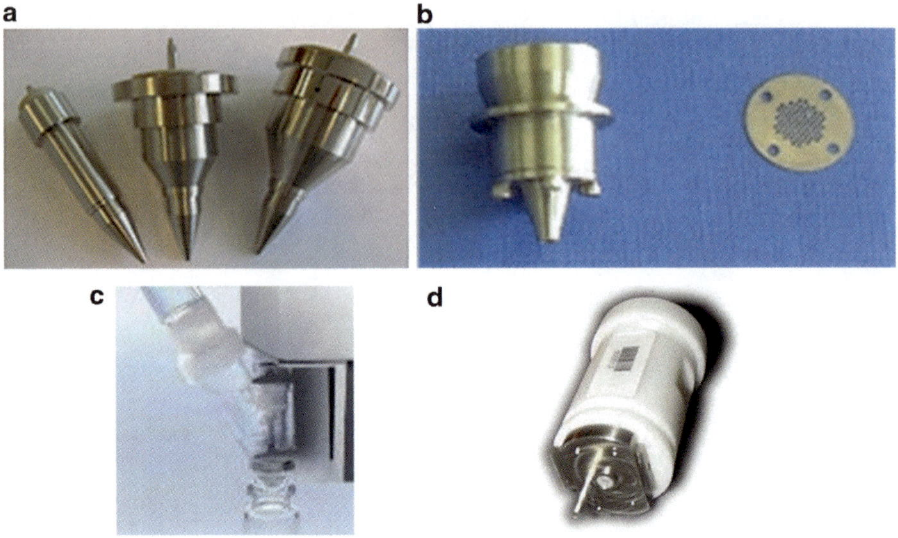

Fig. 12.2 Gravimetric powder dispensing heads (**a**) Powdernium; (**b**) Excelodose®; (**c**) Quantos; (**d**) adaptive precision powder dispenser. ((**a**) Reprinted with permission from FreeSlate. (**c**) Reprinted with permission from Mettler Toledo. (**d**) Reprinted with permission from Lab Automate Technologies)

12.2.1.2 Gravimetric Powder Dispensing

A second category of automated powder dispensing involves real time gravimetric feedback. It usually includes a reservoir container or hopper, a dispensing head, and a balance. The real time weight information from the balance is collected by embedded software to apply proportional-integral-derivative (PID) control to the dispensing head to optimize speed and reduce overshoot during dispensing.

AutoDose (now part of FreeSlate) is one of the early pioneers in gravimetric powder dispensing. Its powder doser, Powdernium as shown in Fig. 12.2a, is an integrated hopper and dosing head with three key components: the cylindrical or conical hopper (top), hopper plug (middle, hidden in the bottom of the hopper), and funnel. The hopper itself can be used for GMP storage. A hard stainless steel pin that runs through the middle of the hopper turns and agitates the powder up and down during dispensing to loosen up the powder in the hopper to assist flow. A piece of elastic stainless steel wire is attached to the pin and fits snuggly in contact with the wall of the hopper. As the pin rotates, the wire scrapes off any powder that sticks to the wall. The hopper plug is mounted to the end of the pin. The up and down agitation of the pin closes and opens the plug to control the powder flow. Finally, the funnel guides the flow into the opening of the receiving containers to minimize spills.

Compared with the volumetric powder pipettes mentioned above, the Powdernium is capable of handling a wider range of materials from low density to free-flowing, cohesive, micronized materials, and more. To accommodate the drastically different flow characteristics, a self learning adaptive dispensing optimization algorithm is included in the control software. Starting with very cautious dispensing by opening the hopper plug very narrowly and rotating the pin slowly, the algorithm will measure the powder flow rate and gradually adjust the two parameters up or down according to the difference between current and target weights. As the dispensed weight approaches the target, the dosing parameters are tuned down again to avoid overshooting the desired weight and eventually stopped when hitting the dosing specification. The parameters throughout the first dosing are recorded and then further optimized for the subsequent doses. In general, the dosing time drops significantly and the precision improves as more doses are completed. The optimal dosing profile is typically achieved after four or five doses and dynamically monitored and adjusted according to any variability in powder flow-ability during the dispensing process. Once optimized, each powder dispensing takes ~20–60 s with ~1–5% RSD while the first few learning doses could take several minutes. Unfortunately, the optimized dosing profile does not get stored after the dispensing job completes in the current software. When the same material is dosed again, the software has to repeat the learning process. Thus, the Powdernium is more suitable for one-to-many (i.e., dosing many times from one sample) dispensing applications (Xue and Brown 2009).

The Excelodose® precision powder micro-dosing system is developed by Meridica (now part of Capsugel) targeting accurate high speed capsule filling. It adopts metering technology with sophisticated predictive control software to achieve a good balance between precision and speed (Bryant et al. 2002). As shown in Fig. 12.2b, the metering dosing heads work similarly to a "pepper pot" principle. When tapped by a solenoid, the powder is released through the mesh at the bottom of the head. The dispensing weight is closely monitored by a micro balance and is fed back to the tapping action by the predictive control algorithm. As the weight approaches the target weight, the dispensing slows down, and eventually stops. Depending on the powder particle sizes and flow-ability, the dosing heads are available in a number of sizes, which differ in pore sizes and openings. Proper selection of a dosing head that suits the powder properties and target mass is important to achieve the optimal performance. Although not limited to this application, Excelodose® is currently marketed exclusively to capsule filling during exploratory formulation development (Hariharan et al. 2003; Mouro et al. 2006). Speed is one of the most striking merits of the system. With the complete capsule handling, which includes feeding, orientating, opening, dosing, closing, and measuring, the system can deliver a throughput of up to 600 capsules per hour. Excellent reproducibility of 0.6–0.8% RSD was reported for 50 mg dispensing of four different blends compared with typical 3–5% manual dosing variation (Fagan et al. 2006). In that application, the throughput was reported to be 330 capsules per hour. Although not at its maximum speed, it was still more than tenfold faster than hand filling capsules.

In 2008, Mettler Toledo debuted their analytical powder dosing device, Quantos (Laukart and Bensel 2008) (Fig. 12.2c). As a simple snap-on to their off-the-shelf XP series analytical balances, Quantos is the lowest cost gravimetric powder doser on the market today. Designed specifically for analytical sample preparation, it obsoletes the traditional powder weighing operation with spatula in most applications. The core of the dosing head is an indented stainless steel pin, which looks like a half-turn deep-throw auger. When retracted, the pin seals the opening at the bottom of the dispensing cone. When extruded out, the indent on the pin serves as the flowing channel for the powder. Similar to the Powdernium, a scraper is attached to the center pin to clean the cone inner surface and loosen the powder. The dosing operation is a combination of the turning and up-down agitation of the indented pin, along with side tapping from time to time. Again, a self-learning PID control algorithm is built in to optimize the dosing speed and precision. What is unique on the Quantos is the inclusion of a RFID chip on the dosing head, which not only records the powder information but also the dosing parameters. Thus, once optimized, the dosing software always automatically reads the dosing profile from the chip when the dosing head is loaded.

Simplicity is the most appealing feature of Quantos. One single head appears to cover a wide variety of the pharmaceutical compounds. A total of 12 powders were used in the early evaluation including very fine (alitame), electrostatic (acetaminophen), fluffy (HPMC-AS), free flowing (calcium carbonate), sticky, and clumping powders (Xue and Brown 2009). Quantos dispensed well 10 out of the 12 compounds. The two compounds that did not work were a very sticky smear (CAB-O-SIL) and a waxy clumpy drug, which would not be expected to be handled accurately by any automated system. The dispensing time ranged from 15 to 50 s for all samples up to 100 mg. The repeatability was better than 5% RSD all the way down to 5 mg.

Made of chemically inert material, the Quantos heads have very little risk of chemical alteration of powders. However, there was a concern with its mechanical impact on the physical properties of powder during the dispensing, especially the particle size and morphology. Polarized light microscopy was used to analyze the powders before and after dosing. Figure 12.3 shows the images of two compounds, one with large coarse crystals, and the other with very fine particles. No noticeable change in either particle size or morphology was observed.

Unlike the above three dosing heads that all rely on gravity to drive the powder down into the target container, the unique design invented by Mr. Rajesh Maheshwari of Lab Automate Technologies Inc called Adaptive Precision Powder Dispenser (Fig. 12.2d) applies pressurized nitrogen to the conical cavity of the dispenser to create a vortex that actively forces the powder out of the hopper. In addition, the positive pressure can be alternated with vacuum to push and pull the powder in the hopper to break up any conglomerate before the actual dispense. This active dispense not only further expands the spectrum of powders that can be dosed, but also allows direct dosing into capped containers sealed with septum. The limitation of this dispenser is that the adaptive software requires predetermination of the optimal dosing parameters for each individual chemical. It works for reference standards that would be reused for dosing again and again, but is tedious for samples that will be weighed only a few times.

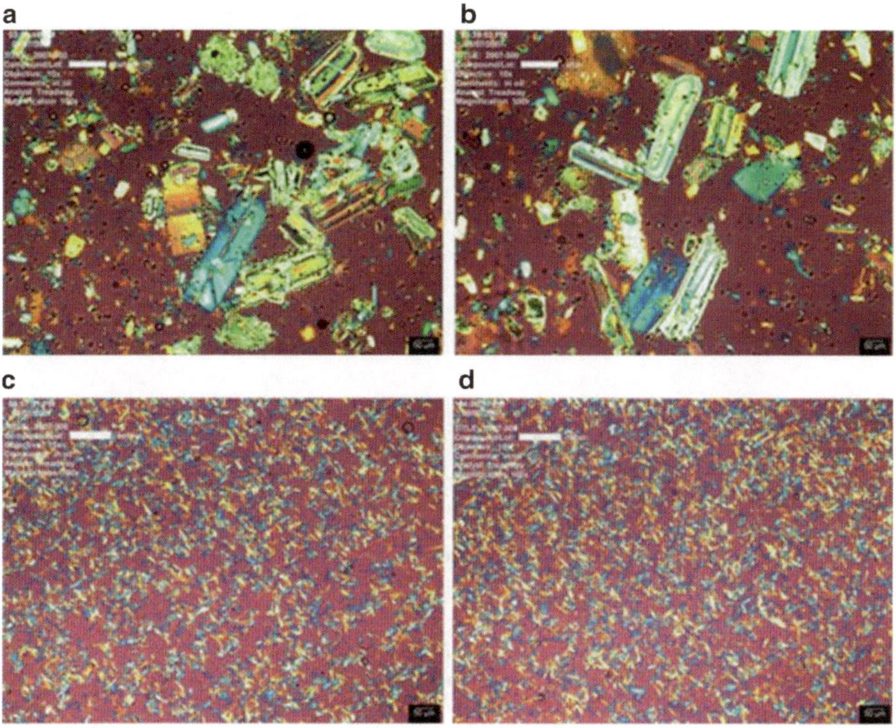

Fig. 12.3 Polarized light microscopy images of drug compound A (**a**) before dosing; (**b**) after dosing; and drug B (**c**) before dosing; (**d**) after dosing

12.2.2 Integrated Sample Preparation Workstation for Powder

Automated powder dosing devices address one of the two bottlenecks in API and reference standard sample preparation. The other laborious operation, volumetric dissolution and serial dilution, can be remediated by the numerous commercially available liquid handling automation systems such as Tecan, Hamilton, Thermo, and Gilson. Maximized benefit and efficiency, however, can be achieved only when the liquid handler is integrated with the powder dispenser.

The Sotax Zymark APW (Active Ingredients Processing Workstation, formerly called Prelude) is one of the first commercial instruments targeted to automate powder dissolution with analytical accuracy and precision. A picture of the latest model, APW3, is shown in Fig. 12.4a. Although the APW does not have a powder doser on board, its test tube rack design allows systems such as Powdernium to dispense the powder directly into the test tubes. The accurate weights can be transferred from Powdernium to the APW or the weights can be directly captured by the five-place balance on the APW by measuring an initial test tube tare weight and reweighing after powder dosing.

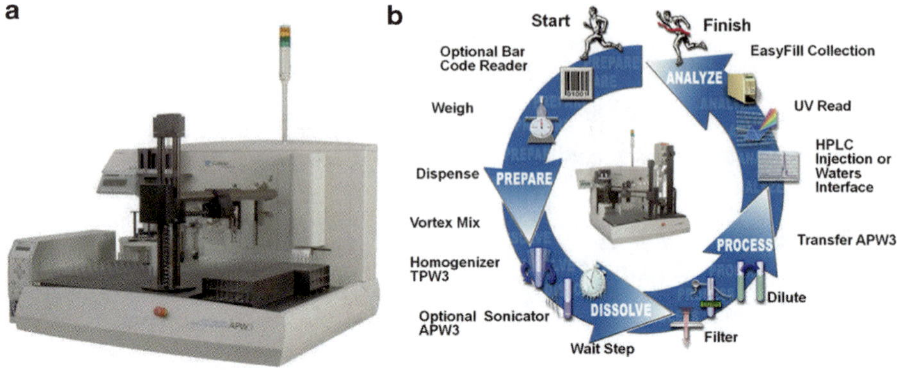

Fig. 12.4 (**a**) Sotax Zymark APW3; (**b**) APW3 operating protocol (reprinted with permission from Sotax)

After weighing, the APW would then dispense the diluent into the test tubes with syringe pumps. The delivered volume can be automatically adjusted according to the actual powder weight and preset target concentration, which is important when reproducible solid dispensing cannot be achieved for certain sample types. For most API and reference standards that have decent solubility, a simple vortex mixing is sufficient to dissolve the powder in the appropriate diluent. For those powders with poor kinetic solubility, the sonicating probe equipped on the APW3 can be very useful. Unlike a sonicating bath, the sonicating probe directly applies well controlled energy locally into the liquid within the test tube, which is far more reproducible and effective. Most sonication takes only seconds rather than tens of minutes. A temperature sensor is embedded within the probe not only as a gauge of applied energy but also to prevent degradation due to overheating.

Additional automated APW features include filtration, serial dilution, and direct injection for online HPLC or UV/Vis spectrometer analysis or collection into HPLC vials for offline analysis (using the EasyFill), which completes the preparation and analysis cycle as detailed in Fig. 12.4b. It is worth mentioning that each liquid dispensing step can be accompanied by gravimetric measurement by the onboard five-place balance, which assures the accurate capture of final sample concentration as specified by quantitative analysis.

Fermier et al. recently introduced an interesting customized cannula with pipette tips to serve as an onboard powder dispensing station for the commercial APW (Fermier et al. 2003). Different from the other volumetric dispensing devices such as the DisPo and REDI sampling probes, the cannula is simply connected to the house vacuum through a three-way solenoid valve. The vacuum flow rate and pipette tip volume can be adjusted to control the incremental powder mass (~0.5–40 mg) to be delivered in each transfer step. By gravimetric feedback, the cumulative transfer is capable of reproducibly delivering milligram quantities of powder into the test tubes. With this neat addition, the modified APW can offer complete automation of powder sample preparation from solid to solution ready for HPLC or UV/Vis analysis.

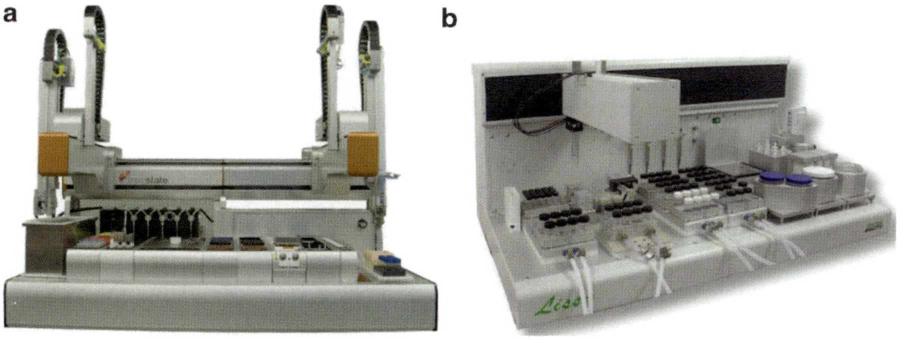

Fig. 12.5 (**a**) FreeSlate core module 3; (**b**) Zinsser Formulax. ((**a**) Reprinted with permission from FreeSlate. (**b**) Reprinted with permission from Zinsser)

Initially designed for solubility screening and crystallization, the Symyx Core Module 3 (now FreeSlate) and Zinsser Formula X are the first two commercial systems that fully integrate powder dosing, liquid dispensing, extraction, filtration, and serial dilution. These laboratory unit operations (LUOs) actually are the same as those for powder sample preparation. Not surprisingly, both systems have been applied to powder automation as well (Xue and Brown 2009). As shown in Fig. 12.5a, the FreeSlate Core Module 3 adopts the Powdernium powder doser and integrates it with seven syringe pumps for liquid dispensing. Heating, cooling, and magnetic stirring are available on deck to facilitate sample dissolution and dispersion. The available sample vial sizes range from 1 to 40 mL and a capping/uncapping option is available off deck. Similar to the APW, all volumetric liquid dispensing can be gravimetrically checked with the equipped five-place balance. Because of the long tubing connection between the syringe pump and dispensing tip, the weighing is actually critical for the accuracy and precision required by analytical sample preparation. Unfortunately, all vials have to be transported back and forth between the vial deck and balance for weighing, which significantly slows down the operation. The filtration block is a complicated multilayer sandwich design, which is not only very difficult to assemble, but is also only available in a 96-well titer plate format for 1 mL vials. The filtration is driven by the vacuum generated by syringe pump pull, which only works with a 1-μm and above pore size filter sheet. Most analytical samples that require 0.22 or 0.45 μm filtration have to be filtered off the deck.

The Zinsser Formula X includes its REDI powder doser, four syringe pump driven liquid dispensers, and one viscous media dispenser (Fig. 12.5b). All dispensing steps are gravimetrically controlled by the feedback from the integrated weighing cell, although the vials also have to be transported to the cell. In addition to vacuum, positive pressure can be applied to the filter plate as well, which greatly enhances the filtration performance. Unfortunately, only the 96-well format is available. On the extraction side, the available vortexing and overhead stirring options in addition to magnetic stirring offer some flexibility and the on-deck capping/uncapping block allows the use of screw-on caps to prevent evaporation. Compared with the Symyx Core Module 2, the Formula X does offer some nice additional features. Overall performance, however, of both systems are comparable for analytical powder sample preparation.

12.3 Automated Sample Preparations for Solid Oral Dosage Formulations

As discussed in Chap. 7, about two-thirds of the drugs are formulated in solid oral dosage forms such as tablets and capsules. Various excipients have to be blended and compressed with the API to control the pharmacodynamics of the medicine as well as enhance its manufacturability. At the time of analysis, the existence of excipients considerably complicates the sample preparation. Most solid sample preparation involves a multistep procedure consisting of such steps as weighing, diluent dispensing, dispersion, particle size reduction, solubilization, and filtration dilution. Any deficiency in any of the steps could cause an error in the analysis result. Chap. 3 summarized various agitation and particle size reduction techniques that can be applied to solid oral dosage forms to facilitate dispersion and extraction. A number of automation systems are built upon these techniques to streamline the entire sample preparation procedure.

12.3.1 Tablet Processing Workstation (TPW)

The first specialized automation system for solids sample preparation is the TPW made by Zymark (now Sotax) in collaboration with several pharmaceutical companies in the early 1990s (Fig. 12.6a). The TPW and its subsequent generations are probably the most widely used automation system for solids preparation. Developed on the Zymate robotic platform, the first generation of TPW is often referred to as Zymate-TPW. Unlike the early user self-built automation solutions, TPW integrated all necessary function modules along with the Zymate robot on a 3 foot by 5 foot bench. The entire system is controlled by EASYLAB, but preconfigured and fully programmed. In this way, the instrument is standardized and liberates the user from lengthy hardware integration and customized programming.

The core of Zymate-TPW is a high shear homogenizer (on the right of Fig. 12.6a), which is used to disintegrate the tablets and extract the active drug from the tablets. The homogenizer includes a set of rotating blades (rotor and stator) with speeds up to 40,000 rpm, combined with wet grinding, shredding, and shearing to break up the sample in the presence of a diluent. A peristaltic pump and several syringe pumps are included to dispense solvents and diluents. Unlike volumetric transfer in typical manual lab operation, all liquid dispensing is controlled gravimetrically over a balance (on the left of Fig. 12.6a) to achieve good accuracy and precision. The Zymate robotic arm (in the middle of Fig. 12.6a) is the system commander that handles sample transfer between homogenization, liquid dispensing, and staging stations.

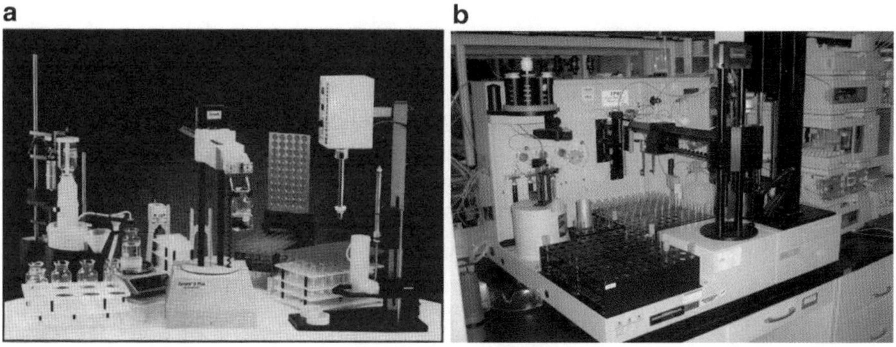

Fig. 12.6 (**a**) The first Zymate-TPW is based on the Zymate robot arm and is mounted on a 3 foot×5 foot bench. (**b**) The redesigned, more compact TPW II ((**a**) reprinted with permission from Sotax)

The LUO concept is first introduced with the Zymate-TPW because of the GMP requirement to validate computerized systems. Since the TPW is based on the user programmable Zymate system, an advanced user could alter the detailed operation protocol. Modular testing of the individual TPW unit operation against its intended function makes GMP validation much more practical. A typical operation protocol is listed below with five LUOs.

1. Pick up sample tube with tablets and pour tablets into the extraction vessel.
2. Dispense, with a high degree of accuracy, 100.00 mL of extraction solvent into the extraction vessel.
3. Place the extraction vessel under a homogenizer.
4. Run the homogenizer at 10,000 rpm for 5 min.
5. Pipette homogenate into a tube for storage.
6. Loop back and prepare the next sample.

Two important LUOs are missing from the first version of TPW solids preparation: filtration and serial dilution. To cover the gap, Zymark introduced a separate bench-top workstation called Benchmate™ to address these two LUOs and then later integrated it into the TPW named Benchmate-TPW.

In response to customers' demands for a more compact automation system, Zymark upgraded TPW to TPW Version II (TPW-II) in 1995 (Fig. 12.6b). The LUO concept was preserved in the redesign. The left side of the TPW II incorporated the Benchmate™ workstation functions of dilution, filtration, and mixing, while the right side contains the homogenization and solvent addition functions. The system, however, is very compactly integrated with a single vessel for solids extraction.

To accomplish sample preparation for a solid composite assay or content uniformity testing, sample preparation on a TPW II typically involves the following LUOs.

1. Dispense, with a high degree of accuracy, a specified volume of extraction solvent (e.g., 100.00 mL) into the extraction vessel.
2. Pick up sample tube with tablets and pour tablets into extraction vessel.

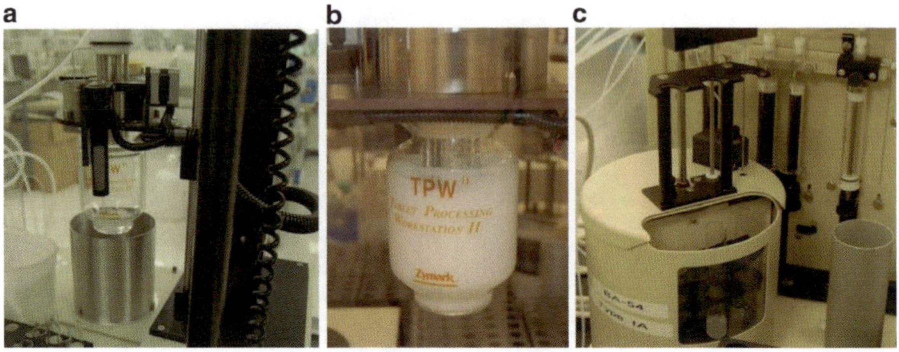

Fig. 12.7 (**a**) Solvent dispensing into the TPW-II extraction vessel. (**b**) Homogenization of the tablet sample in an extraction solvent. (**c**) The dilution station, syringe station and vortex station of the TPW-II are shown. The test tubes within the dilution station rests on a four-place (or five-place) balance

3. Soak the tablet and add a second extraction solvent if necessary.
4. Place the extraction vessel under a homogenizer and run the homogenizer at a specific RPM.
5. Pump sample through the filter station.
6. Dilute the sample gravimetrically.
7. Mix the sample at the vortex station.
8. Loop back and prepare the next sample.

In Step 1, a high degree of accuracy in solvent dispensing is accomplished by the gravimetric feedback from a three-place top loading balance. While the solvent is being pumped into the extraction vessel, the amount is precisely monitored by the three-place balance under the extraction vessel (Fig. 12.7a). The second and third steps are pick-and-place and a timing function that can be easily accomplished by the robotic equipment. In the fourth step, the extraction vessel is raised and the sample is homogenized for a specified amount of time (Fig. 12.7b). Similarly, accurate dilution in Step 6 is accomplished by the gravimetric measurement through the four-place or five-place balance. After homogenization for a specified amount of time, a second solvent can be added and the sample can be further homogenized. For HPLC and UV analysis, the sample must be clarified through a filter station. The dilution station shown in Fig. 12.7c could further dilute the prepared sample solution to the desired concentration and mix it using a vortex mixer. The TPW-II offers three options for the prepared final solution: directly inject the solution onto an HPLC, measure the solution using a UV spectrometer, or store the solution in test tubes or vials.

Another TPW system upgrade occurred in 2006. The general functionality and the appearance of the third generation of TPW (TPW3) are not very different from TPW-II as shown in Figs. 12.4b and 12.8. The electronics and the software have been significantly upgraded, which further improved the reliability and robustness of the instrument.

Fig. 12.8 Sotax/Zymark TPW3 (reprinted with permission from Sotax)

Ease of use and reliable extraction make TPW-II and TPW3 very popular both in analytical research and QC labs when compared with its predecessor. The new controlling software not only packages the low level instrument control, but also provides the users a simple graphic user interface for easy protocol design. It also enables data and communication connection with a number of external systems such as HPLC, UV spectrometers, and third party laboratory software packages.

A number of publications that report the successful use of the TPW for sample preparation have appeared since 1996. For example, Han et al. discussed the transfer of a manual test method of a capsule formulation to a TPW test method (Han and Munro 1999). Holler et al. discussed the extraction of multiple active ingredients in multivitamin tablets (Holler et al. 2003). Shamrock et al. discussed many of the basic cautions that must be made when a manual test method is converted to a TPW-II method and described typical validation procedures used for a late phase development compound (Shamrock et al. 2000).

Besides the testing of commercial products and late phase development compounds, the use of a TPW-II in an early phase development project was also reported. A feasibility study that examined a "direct to automation" approach was reported. The authors discussed a way of developing a TPW-II method without first developing a manual test method for the development compound (Lung et al. 2005).

In principle, it is not necessary to develop a manual test method first before an automated sample preparation method is developed. As described by Shamrock et al., steps must be taken to ensure that the API is soluble in the extraction solvent (Shamrock et al. 2000). By the time a tablet formulation is being developed, solubility data are usually available from previously completed solubility studies. However, proper cleanup procedures must be performed and verified by carry-over studies.

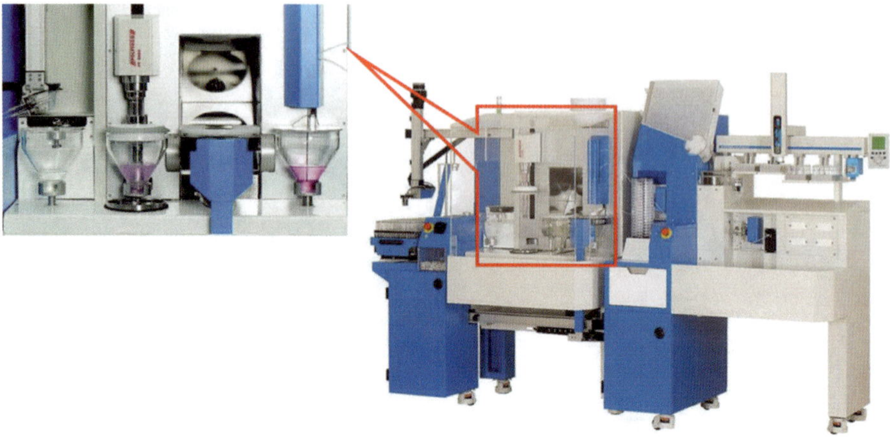

Fig. 12.9 The SOTAX CTS automated assay and content uniformity system (reprinted with permission from Sotax)

12.3.2 SOTAX Automated Content Testing System (CTS)

At about the same time that the TPW3 was released, SOTAX launched its content testing system (CTS) Automated Content Uniformity Testing System. Utilizing the same wet homogenization technology, the CTS works in a similar way as the TPW and offers almost the same LUOs.

The key unique features of the CTS are higher throughput through staggered sample processing and innovative vessel cleaning design, which significantly reduces solvent waste. As shown in the close-up view of the processing station in Fig. 12.9, three extraction vessels are juggled across four vessel positions (from left to right: liquid dispensing, homogenization, cleaning, and transfer) so that up to three samples can be processed simultaneously. The vessel washing step is accomplished in 30 s using a jet spray of water or solvent, followed by a 1-s ethanol rinse and air blow drying, which translates to using only about 25 mL of organic solvent. Compared with the TPW, this jet flow design effectively cuts the solvent waste by more than 60% and the parallel processing increases the instrument throughput by 75%. In addition, the unique conical shape of the vessel enables sample extractions in volumes ranging from 25 mL to 1 L and allows effective rinsing of the vessel neck during liquid dispensing.

A potential limitation of the system is its 9 feet by 3 feet footprint. It could be challenging to fit into R&D laboratories where space usually is at premium and sample load is relatively low. A suitable method conversion from the smaller instruments like TPW to CTS would allow successful method transfer from R&D to manufacture QC labs where the full advantage of the CTS throughput could realized.

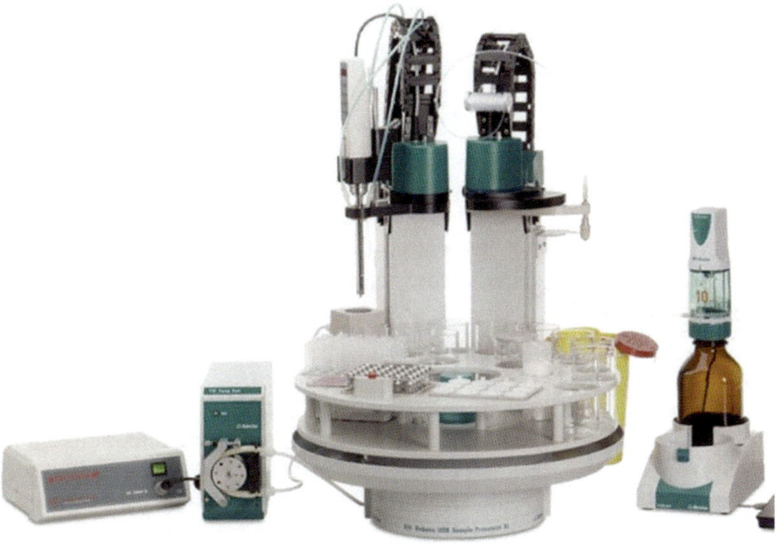

Fig. 12.10 Metrohm Robotic Soliprep system (reprinted with permission from Metrohm)

12.3.3 Metrohm Soliprep

Compared with the large CTS system, Metrohm took a different direction in designing its automated sample preparation system for solids. Another system built upon use of the Polytron® homogenizer, the Metrohm Soliprep system, shown in Fig. 12.10, is mostly integrated on a two feet in diameter turn table. With very few moving parts involved, this extremely compact benchtop instrument integrates the Polytron® with the proven Metrohm dosing technology and offers superb system robustness.

Simple yet flexible, the Soliprep still offers fully automated sample preparation for up to ten solid samples with the LUOs including liquid dispensing, homogenization, filtration, and dilution. Early evaluation results by the author's laboratory did reveal a few areas for improvement (Cometa 2009). First, the extraction vessels are flat bottom beakers about 2.5 in. in diameter while the Polytron® is only 0.5 in. in diameter. The rotating blade sometimes misses the tablet, which spins around the bottom of the vessel. Second, the vessel cap does not provide a sufficient seal to prevent evaporation. For extraction involving volatile solvents such as methanol and acetonitrile, significant bias is introduced to the sample concentration because of this problem. These issues, however, can be readily rectified through some minor design enhancements to make Soliprep a low cost tool for research and development analysis.

A few additional automation systems have just recently emerged, including the rapid tablet extraction (RTE) sample preparation system from PA Consulting (PA Consulting 2010), and the RTS automated PrepStation by RTS (RTS Life Science 2010). However, not much detailed information was available as this publication was prepared.

12.3.4 Volumetric vs. Gravimetric Solvent Dispensing

Gravimetric solvent dispensing is common in automated sample preparation systems, for example, as discussed with the TPW and CTS. The accurate weights captured by the balance ensure dispensing accuracy to match the requirement of Grade-A glassware for volumetric liquid transfer. Typically, the density of the extraction solvent can be measured and thus be used for gravimetric and volumetric correlation. Therefore, manual vs. automated sample preparation method conversion can be very straightforward. In most laboratories with well controlled temperatures (e.g., within 5°C), the solvent density usually does not need to be measured every day.

Manual vs. automated solvent dispensing becomes complicated in some cases that require multiple step liquid dispensing. For example, water or a 100% aqueous buffer sometimes needs to be dispensed first to facilitate the disintegration of a tablet. Then the organic component, such as ethanol, is added to ensure sufficient solubility of the API in the final extraction solvent. When water and ethanol are mixed, one must keep in mind that solutions of water and ethanol are thermodynamically nonideal solutions. For nonideal solutions, the volume of the final mixture is not equal to the sum of the individual volumes of the water and ethanol. For example, when one mole of water, 18.0 mL, and 1 mole of ethanol, 58.0 mL, are mixed, the final volume is not equal to 76.0 mL. Instead, the final volume is the sum of the partial molal volumes of the components. Therefore, when one mole of water is added to one mole of ethanol, we observe apparent volume shrinkage when we observe that the volume of the final mixture is 74.3 mL instead of 76.0 mL.

When there is more than one solvent dispensing step in a TPW-II procedure, one must be aware that the TPW-II dispenses solvent by mass and does not keep track of the thermodynamically nonideal solution behavior of solvents such as ethanol–water, methanol–water, or acetonitrile–water. The apparent volume shrinkage for nonideal solutions can be measured experimentally and be compensated in a TPW-II (or TPW3 or CTS) method.

The following TPW-II method example illustrates one of the many possible experimental approaches that can be used to compensate for volume shrinkage when two solvents are mixed. A manual sample preparation procedure calls for first soaking six tablets with 100 mL of water for 5 min in a 200-mL volumetric flask. The soaking action allows the disintegrant in the immediate release formulation to react with water and disintegrate the tablet. The next step involves adding 50 mL of methanol and sonicating for 30 min. After sonication, an additional volume of methanol is added until the total volume is at the 200-mL mark of the volumetric flask. Since the manual sample preparation method is volumetric, no compensation of volume shrinkage is necessary. When this method is translated to a TPW-II method, compensation for volume shrinkage can be made by adding 103.6 mL of water and 103.6 mL of methanol. The partial molal volumes of 103.6 mL of methanol and 103.6 mL water sum to an exact total volume of 200.0 mL. Details of the method are shown in Table 12.1.

Table 12.1 Comparison of a volumetric manual method and a TPW3 gravimetric method that compensates for volume shrinkage

Manual method	TPW3 method
• Add six 10-mg tablets	• Add 103.6 mL of water
• Add 100 mL of water and soak for 5 min	• Add six 10-mg tablets and soak for 5 min
• Add 50 mL of methanol and immerse in sonic bath for 30 min	• Add 103.6 mL of methanol
• Cool to room temperature	
• Dilute to 200 mL mark with additional methanol	• Homogenize for 8 min
• Filter through a 0.45-μm filter	• Filter using a 0.45-μm filter
• Fill HPLC vial/clean-up	• Fill HPLC vial/clean-up

In some cases, the amount of "extra" solvent that must be added by the TPW-II can be calculated from the partial molal volumes of common solvents. However, partial molal volume data are not available for all combinations of mixed solvents. In many cases, the volume shrinkage can be measured experimentally. For the above example, 100.0 mL of methanol and 100.0 mL of water can be pipetted into a 200-mL volumetric flask. After mixing and equilibration to room temperature, the additional volume of premixed (50:50) solvent that must be added to the volumetric flask to equal 200 mL can be precisely measured either volumetrically or gravimetrically to determine the volume shrinkage.

12.4 Automated Sample Preparations for Nonsolid Formulations

Other than homogenous liquid formulations such as intravenous (IV) solutions, nonsolid formulations typically pose more problems than solids for sample preparation due to high viscosity, poor flow-ability, or heterogeneity. These types of formulations include oral suspensions, gels, and many parenteral dosage forms including creams, lotions, pastes, and transdermal patches. The challenges of preparing these samples include, but are not limited to, dispensing accurate weights, performing quantitative transfer, achieving uniform dispersion, and performing filtration, all of which make sample processing prone to human error. Unfortunately, these same challenges make it difficult for robotic systems as well. Full automation of sample preparation for these nonsolid formulations is almost impossible. However, automation of some of the LUOs could prove valuable in improving sample preparation efficiency and consistency.

Accurate dispensing of semi-solids can be very tedious. Automated gravimetric measurement would require only approximate aliquots of samples. With the actual weight captured, the follow-up liquid dispensing could be scaled accordingly to assure the final target concentration is achieved. The dissolution and extraction of active ingredients can occur within the initial container to eliminate sample transfer in a heterogeneous state. Once in solution, the sample could be filtered, transferred, or diluted just like solid samples, i.e., share the same sample preparation automation system.

The TPW-II was recently applied to the automated sample preparation of a powder for oral suspension (POS) formulation (Opio et al. 2010). Because of the use of xanthan gum in the formulation, the reconstituted suspension is extremely viscous. The analysis requires an aliquot of the POS to be diluted 1:25 and filtered to determine the potency of the preservative and purity of the sample by HPLC, while a second aliquot is further diluted and filtered for the potency determination of the API. Because of the viscosity of the sample solution, the positive displacement pipette that is typically used for volumetric transfer is awkward and the manual filtration is very difficult to perform. In the TPW-II method approximately 320 μL of the suspension was manually transferred into a pretared test tube. From that point, the TPW-II automates the rest of the preparation including an accurate weighing of the actual suspension transferred; performing a 1:25 dilution followed by API extraction with 60 s vortexing; filtering with a PVDF syringe filter; performing another 1:25 dilution from the filtrate and vortexing for 10 s. Cleanup was executed between samples. The total preparation time was approximately 15 min per sample.

Compared with the up to 3% variation observed in manual sample preparation, the reproducibility of the TPW-II preparation ranged from 0.5 to 0.7% RSD ($n=10$) for the three lots of POS samples tested. These results clearly demonstrate how significantly human error contributes to the difficult sample preparation of the POS. More importantly, because the manual preparation is so laborious, typically only the beginning, middle, and end of the 4-h manufacturing process are sampled and analyzed. With the TPW-II, the scientists were able to sample every 5 min to monitor the process, which actually allowed them to promptly catch an API segregation problem during the process scale up optimization.

The matrix in POS is usually fairly easy to break up and vortexing with suitable diluent is sufficient for reproducible extraction of the API as demonstrated in the above study. For cream, lotion, paste, and gel formulations, more vigorous agitations are required to break up the gummy samples. In a recent analysis of triclosan and fluoride in toothpaste, VonBehren introduced twenty 5 mm glass beads to the test tube containing the toothpaste (VonBehren 2009). With the agitating beads, they thoroughly dispersed the toothpaste with 15 min of vortexing with the TPW, and less than 1% RSD ($n=5$) precision was achieved for both triclosan and fluoride. The sonicating probe on the APW 3 is another very effective extraction approach for cream and paste samples and moderate heating can also help them to become more fluidic and thus facilitate both solubility and dispersion.

The inventor of the ever-popular CTC PAL, Leap Technology, also developed a compact Semi Solid Workstation based on the versatile X, Y, Z robot as shown in Fig. 12.11. Following the "syringe only" simple automation concept, the core of the Semi Solid Workstation is the adoption of RANIN Pos-D™ technology. A uniquely designed positive displacement pipette, the Pos-D™ has a disposable piston sliding within the plastic pipette tip. The snug fit of the piston to the inner wall of the tip assures a positive wiping motion to completely dispense samples without droplets. The piston is in direct contact with the sample, which leaves no air gap to assure constant aspiration force to effectively draw the viscous semi solids (Leap Technologies 2010a).

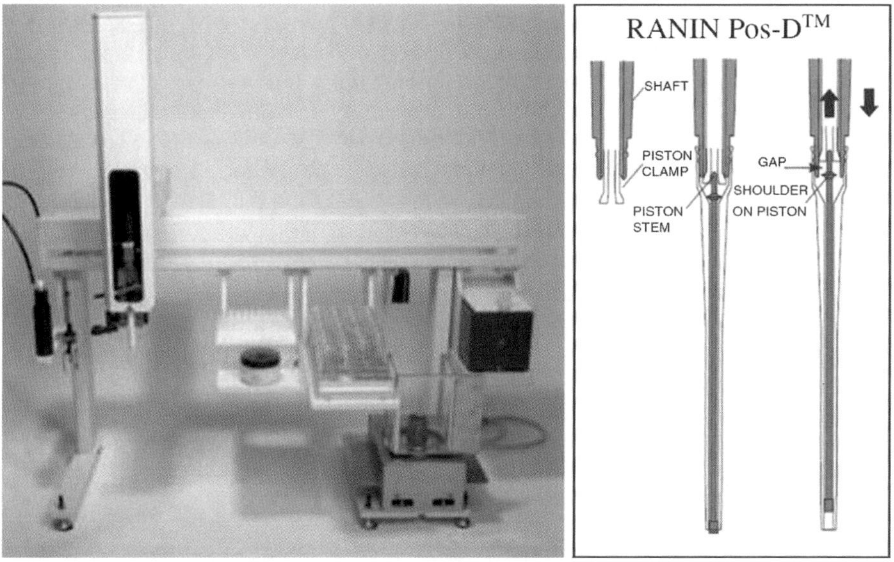

Fig. 12.11 The Leap semi solid workstation integrates the RANIN Pos-D™ positive displacement pipette, vial gripper, liquid dispenser, balance and vortex for viscous sample preparation (figure on *left* reprinted with permission from Leap Technologies; figure on *right* reprinted with permission from Mettler Toledo)

The other modules integrated on the system are a vial gripper, a liquid dispenser, a top-loading balance and a vortex. The single PAL arm picks and disposes the Pos-D™ pipette tips for the semi solid dispensing, as well as moves the vials between sample rack, balance, and vortexer. Perez recently presented the application of the Semi Solid Workstation to automated reconstitution of creams (Perez 2010). The viscous cream samples were demonstrated to be routinely aspirated and dispensed with 1 mL positive displacement tip. Sample volumes were checked gravimetrically with the onboard balance and used to scale the reconstitution volume to assure final concentration accuracy and precision. Currently, the Leap system does not include a filter station, but another PAL system with automated filtration has been reported (Leap Technologies 2010b) and can be integrated. The Pos-D™ semi solid pipetting also complements the LUOs offered by TPW-II and APW3. A fully automated semi solid sample preparation station can be foreseen based on the integration of these solutions in the future.

12.5 Conclusions

Sample preparation automation frees pharmaceutical analysts from laborious manual lab operations, which can lead to significant cost savings and efficiency gain. More importantly, lab automation improves consistency in sample preparation, eliminating or greatly reducing out of specification errors due to human error or mix-up. All processing steps during preparation can be well controlled and electronically tracked, which also makes method transfer as straightforward as electronic method cloning. A number of available sample preparation automation solutions were reviewed in this chapter targeting API powder, solid formulation, and semi solid formulations, respectively. Simpler lab operation, tighter analytical precision, more effective process monitoring, and decision making were demonstrated in various applications.

References

Beyer WF (1966) Steroid tablet assay involving automated sample preparation and blue tetrazolium reaction. J Pharm Sci 55(2):200–203

BioDot Inc. (2009) Automated Powder Dosing Technology. http://www.labautopedia.org/mw/index.php/Automated_Powder_Dispensing_Technology. Accessed 8 Nov 2010

Bryant S, Gill I, Edwards D, Smith IJ (2002) Advances in powder-dosing technology. Innov Pharm Technol (Autumn) 95–100. http://www.iptonline.com/pdf_viewarticle.asp?cat=7&article=58

Cherng J-PJ, Gonzalez-Zugasti J, Kane NR, Cima MJ, Lemmo AV (2004) Integration of an opto-mechanical mass sensor with a powder-dispensing device for microgram sensitivity. JALA 8:228–237

Cirillo MG (1986) The Perkin-Elmer masterlab system. J Liq Chromatogr Related Technol 9(14):3185–3190

Cometa G (2009) Performance evaluation of the Metrohm Robotic Soliprep 815. Pfizer, Groton

PA Consulting (2010) PA unveils ground breaking Rapid Tablet Extraction (RTE) sample preparation system for faster solid dose product testing. http://www.paconsulting.com/our-thinking/pa-unveils-ground-breaking-rte-rapid-tablet-extraction-sample-preparation-system-for-faster-solid-dose-product-testing/. Accessed 8 Nov 2010

Fagan PG, Hood MR, Johnston JA, Woolven H (2006) An assessment of the suitability of the XcelodoseTM 600 in the filling of blended products. Paper presented at the 2006 AAPS Annual Meeting and Exposition, San Antonio, 10/28/2006–11/2/2006

Fermier AM, Troisi J, Heritage EC, Drexel MA, Gallea P, Swinney KA (2003) Powder dispensing robot for sample preparation. Analyst 128(6):790–795

Gloekler J, Schuetze T, Konthur Z (2010) Automation in the high-throughput selection of random combinatorial libraries-different approaches for select applications. Molecules 15:2478–2490

Han SM, Munro A (1999) Transfer from manual to automated sample preparation: a case study. J Pharm Biomed Anal 20(5):785–790

Hanna S, Dzus G, Rasero L, Lachman L (1976) Automated analysis of warfarin sodium tablets. J Pharm Sci 65(10):1522–1525

Hariharan M, Ganorkar LD, Amidon GE, Cavallo A, Gatti P, Hageman MJ, Choo I, Miller JL, Shah UJ (2003) Reducing the time to develop and manufacture formulations for first oral dose in humans. Pharma Technol 27(10):68–84

Holler U, Brodhag C, Knobel A, Hofmann P, Spitzer V (2003) Automated determination of selected water-soluble vitamins in tablets using a bench-top robotic system coupled to reversed-phase (RP-18) HPLC with UV detection. J Pharm Biomed Anal 31(1):151–158

Huen JM, Thevenin JP (1979) Sample preparation for HPLC by automatic tablet extraction. Chromatographia 12(6):405–407

Laukart J, Bensel N (2008) Compact weighing system supports routine operations. BIOspektrum 14(7):732–733

Leap Technologies (2010a) Semi-Solid Sample Preparation Workstation. http://www.leapwiki.com/mediawiki/index.php?title=Semi_Solid_Applications. Accessed 8 Nov 2010

Leap Technologies (2010b) Filtering of Samples. http://www.leapwiki.com/mediawiki/index.php?title=Filtering_of_Samples. Accessed 8 Nov 2010

RTS Life Science (2010) PrepStation™ – Automated Sample Preparation System. http://www.rtslifescience.com/solutions/1587_778_prepstation--8482----automated-sample-preparation-system.html. Accessed 8 Nov 2010

Lochmuller CH, Lung KR, Cushman MR (1985) Laboratory robotics: the advent of "soft" instrumentation. J Chromatogr Sci 23:429–436

Lung KR, Ayers J, Yembrick C, Guerrieri P (2005) Laboratory automation in phase I development. Am Pharm Rev 8:62–66

Majors RE (2006) New Sample Preparation Technologies for the Pharmaceutical Industry. LCGC Web Seminar. http://www.chromatographyonline.findanalytichem.com/e-Education. Accessed 19 Nov 2010

Mouro D, Noack R, Musico B, King H, Shah U (2006) Enhancement of xcelodose capsule-filling capabilities using roller compaction. Pharm Technol 30(2):72–81

Opio A, Nickerson B, Xue G, Warzeka J, Norris K (2010) Expanding the application of the tablet processing workstation to support the sample preparation of oral suspensions. JALA (2011) 16(3):229–234

Perez C (2010) A fully automated system for reconstituting semi-solid samples. Paper presented at the Laboratory Robotics Interest Group Mid Atlantic Sixteenth Annual Technology Exhibition and Presentations, New Brunswick

Polesello S (1995) Automated samplers for gas chromatography (GC) and HPLC. Lab (2000) 9(6):34–45. http://www.speciation.net/Database/Journals/Laboratorio-2000-;i729

Reichman M, Gill H (2008) Automated drug screening for ADMET properties, in drug metabolism handbook: concepts and applications (Nassar AF, Hollenberg PF, Scatina J (eds)), John Wiley & Sons, Inc., Hoboken, NJ, USA. doi: 10.1002/9780470439265.ch7

Shamrock WF, Reilly K, Lloyd DK (2000) Automated sample preparation of Roxifiban tablets: transfer of a manual method to an automated workstation. J Pharm Biomed Anal 21(6):1225–1232

VonBehren D (2009) Automated Analysis of Suspensions, Creams, Lotions, and Gels Using SOTAX Assay Workstations. Sotax Corporation http://www.sotax.com/fileadmin/sps/White_Paper___Press_Release/Automated_Analysis_of_Suspensions__Creams__Lotions.pdf. Accessed 8 Nov 2010

Xue G, Brown SM (2009) Powder dosing automation. Paper presented at the Lab Automation 2009, Palm Spring, 1/25–1/28

Zhang W, Hou L, Tian G (2009) A single-cell micro-operation device for automatic microinjection in cell engineering based on flow-field flow and mechanical movement of micro-manipulator. Application: CN, CN Patent 2008–1001909, 101481653

Zinsser NA (2010) REDI 2002 Plus: High Precision Powder Dispensing. http://www.zinsserna.com/redi2002.html. Accessed 8 Nov 2010

Chapter 13
A Systematic Approach for Investigating Aberrant Potency Values

Beverly Nickerson, Ivelisse Colón, Eddie Ebrahimi, Garry Scrivens, and Lin Zhang

Abstract Accurate potency and purity data are critical in the development of drug products. These results are used to make decisions regarding formulation development/selection, formulation stability, and process robustness, and are used to release clinical supplies and set clinical use periods. Sometimes aberrant potency values (e.g., assay values, content uniformity results, stratified core test results) are obtained and significant efforts are spent investigating these issues to identify the root cause, which may be manufacturing or method related. This chapter discusses a systematic approach for investigating aberrant potency values from the analytical method perspective. In addition, several case studies are described.

13.1 Introduction

Accurate potency and purity data are critical in the development of drug products. These results are used to aid formulation development and selection, to assess formulation stability, and to determine process robustness. In addition, potency and purity analysis is an integral component of the release and stability testing of clinical supplies.

Sample preparation and extraction is a critical component of the method used for potency and purity testing of drug products. All of the active ingredient and impurities must be recovered from the dosage form to achieve accurate quantitative results. The sample preparation method should also be robust enough to serve its intended purpose (e.g., use by only the project lab or by multiple labs, low or high sample throughput).

After a sample preparation/extraction method has been developed, validated and is in use, issues, such as obtaining low, high, out of trend, out of specification,

---

B. Nickerson (✉)
Analytical Development, Pfizer Global Research and Development, Pfizer Inc., 558 Eastern Point Road, Groton, CT 06340, USA
e-mail: beverly.nickerson@pfizer.com

B. Nickerson (ed.), *Sample Preparation of Pharmaceutical Dosage Forms: Challenges and Strategies for Sample Preparation and Extraction*, DOI 10.1007/978-1-4419-9631-2_13,

or variable potency values (e.g., assay values, content uniformity results, stratified core test results), may arise. These aberrant values raise questions about the analytical method (e.g., was there insufficient extraction of the API from the dosage form, is the method non-robust) and about the manufacturing process (e.g., was API lost during the process, is the process non-robust). Considerable time and effort can be spent troubleshooting these aberrant potency results to find the root cause. Sometimes these aberrant potency values are method related and may arise when the method has been transferred to a new testing group or when there has been a change in the manufacturing site or process. Sometimes the aberrant potency values are accurate and are the result of a sample, a sampling, or a manufacturing related issue.

A systematic approach for investigating aberrant potency issues, from the analytical method perspective, is discussed in this chapter. In addition, several case studies are described to illustrate the approach. Although the specific examples and case studies involve solid oral dosage forms, the principles can be applied in trouble shooting aberrant assay values for other types of dosage forms as well. Note that these are issues that occur when using a validated method; troubleshooting sample preparation issues during method development is discussed in Chap. 7.

13.2 A Systematic Approach for Investigating Aberrant Potency Values

Investigating aberrant potency values can be a challenging and time consuming effort. Often there are time constraints in performing theses investigations since the results are needed to support such activities as releasing clinical supplies, supporting clinical use periods, or supporting manufacturing scale up. Aberrant potency issues can manifest themselves as obtaining consistently aberrant values or obtaining random or occasional aberrant values during the testing of a sample or product. When random aberrant values are obtained (e.g., some but not all samples for a given lot have low potency values or low values are obtained for some lots and not others), questions are raised regarding the analytical method (e.g., is the method robust/rugged) and the manufacturing process (e.g., nonhomogeneity issues for a given batch or nonrobust manufacturing process leading to batch-to-batch variability). Troubleshooting random aberrant potency values is extremely challenging since it may be difficult to reproduce the issue.

Despite the amount of testing and the number of issues that can occur, there are few resources (Cory et al. 2004; Choi and Dong 2005; Nickerson 2006; Lee 2007) available in the literature that describe investigation of aberrant potency values for solid oral dosage forms. A systematic approach to investigating aberrant potency values is outlined in Table 13.1. Each step of the investigation is described in detail below. Although these steps are presented and described in sequential order, the order in which they are performed should depend on each individual case. In addition, some steps may be omitted if appropriate. It is also useful for the analytical team to periodically meet, review results, and adjust the investigational plan as appropriate based on the results or information obtained during the investigation.

Table 13.1 A systematic approach to investigating aberrant potency values

Step	Considerations
1	Rule out laboratory error • Was the method followed?
2	Review available information; discuss, brainstorm and plan next steps • Involve the appropriate people (e.g., analyst performing the work, analyst who developed/ validated the method, individual(s) with experience investigating aberrant potency issues)
3	Try to confirm aberrant potency results with the current method • Reinject sample solutions • Analyze additional samples along with a control sample • Additional testing performed by another analyst • Analyst who developed method or who has extensive experience with the method performs testing or watches testing being performed
4	See if modifications to the current method will improve potency values (e.g., increase recovery of API) • Let sample solutions sit/stir overnight and retest/reinject • Increase agitation time (e.g., shake time) • Increase extraction solvent volume • Change extraction solvent(s) (e.g., change pH, change organic content)
5	Try to improve results (e.g., increase recovery) using alternative methods (use samples in question along with a control sample) • Dissolution • Near infrared (NIR) spectroscopy – Qualitative analysis comparing aberrant samples vs. "good" samples • Homogenizer (e.g., Polytron®), ball mill or mortar/pestle – Leverage particle size reduction to increase extraction efficiency • Microwave assisted extraction (MAE) or accelerated solvent extraction (ASE) – Leverage elevated temperature to increase extraction efficiency • Water analysis – Is there high or variable water content that results in low or variable potency • CHN analysis – If drug contains nitrogen and excipients do not, compare N content of aberrant sample vs. "good" sample
6	Confirm current method able to extract API from dosage form • Prepare individual tablets with known quantity of API and analyze with the current method to confirm complete extraction of API • Prepare drug-excipient blends and analyze to test for potential drug-excipient interactions

13.2.1 Step 1: Rule Out Laboratory Error

The first step of the investigation should be to conduct a general laboratory investigation to rule out possible laboratory error. The basic question to be answered is, "Was the method followed as written?" Items to check include: use of the correct size volumetric flask(s) and volumetric pipette(s), use of the correct agitation mechanism and time, use of correct filters, assignment of correct API potency factor, etc.

It is important to not dispose of any sample solutions (e.g., stock solutions, final sample solutions) as these can prove valuable if additional testing is needed to try to determine the cause of the aberrant results.

If the aberrant potency values were obtained by a lab that did not develop the method, then it would be useful to have an analyst from the lab that developed the method to be involved in the discussions during the laboratory investigation. The analyst can talk through the method with the analyst who performed the test to see if there were any issues with interpretation of steps written in the method. If there were any unusual or atypical observations, these should be discussed.

Although every step of the sample preparation method is important, a step worth particular attention is ensuring adequate mixing of the sample solution. Samples are often prepared in volumetric flasks. The sample (e.g., tablet) is placed in the volumetric flask along with an approximate volume of dilute and then is shaken or stirred for a specified length of time. The method then typically specifies to dilute to volume and "mix thoroughly." If adequate mixing is not performed, the solution will not be homogeneous and low or variable potency values may result when an aliquot is taken from the flask for analysis or to make a subdilution for analysis. This is especially true for viscous solutions or solutions with insoluble excipients. In addition, when preparing a large number (e.g., >50–100) of samples, if manual mixing is performed, the consistency and/or vigor of mixing may vary from sample to sample and there may be insufficient mixing for some samples. Examples illustrating these issues are discussed in Case Studies 1 and 3 of Sect. 13.3.

13.2.2 Step 2: Review Available Information, Discuss, Brainstorm, and Plan Next Steps

After laboratory error has been ruled out, the analytical team should meet to review the available information and discuss the next steps of the investigation. This discussion should involve all relevant personnel. This includes the analyst who performed the testing and individuals from the laboratory that developed the method or who have extensive experience with the method. It is also useful, if possible, to have individuals involved who have previous experience investigating aberrant results.

The team should review and discuss the issue and any relevant information, observations, and data. This should include information to answer the following types of questions:

- When were the aberrant values observed?
- Were the aberrant values obtained on the first batches (prototypes or lab scale batches) of the formulation? (e.g., there is limited experience with the method.)
- Were the aberrant values obtained on the first batches at a new manufacturing site or post scale-up of the process? (e.g., did something change?)

Table 13.2 Examples of manufacturing changes and potential impact on the sample preparation/extraction method

Manufacturing change	Potential impact on sample preparation/extraction method	Possible actions
Change in tablet hardness	May impact tablet disintegration	Assess ability of method to disintegrate or disperse the tablet
Change in excipient or grade of excipient	May result in drug-excipient interactions	Assess potential for drug-excipient interactions
	May impact disintegration and/or chromatography (e.g., interfering peaks)	Re-evaluate method specificity for excipients
Change in coating thickness for controlled or sustained release dosage forms	May impact tablet dispersion and require longer extraction time	Assess ability of method to disperse the tablet and assess length of time used for extraction of drug
Change in granulation process (wet vs. dry)	May impact tablet dispersion and dissolution/solubilization	Assess extraction time and recovery of drug
Change in API particle morphology or particle size	May impact rate of API dissolution/solubilization	Assess extraction time and recovery of drug

- Were the aberrant values obtained on batches that had a change in the API (e.g., change in particle size) or drug product manufacturing process (e.g., change in tablet hardness)? (e.g., is there a change that may impact the method's ability to extract the sample?)
- Were the aberrant values obtained by a new testing laboratory or new analyst?
- What type of testing or method related issues have been encountered in the past with this method or with this type of dosage form?

Processes such as Kepner-Tregoe problem analysis (Kepner and Tregoe 1976) or DMAIC (Define, Measure, Analyze, Improve, Control – five interconnected phases of a data driven strategy for improving processes in Six Sigma) (Gitlow and Levine 2004) may also be employed.

Analysis of the above information may provide clues as to the cause of the aberrant values. If the incidence of aberrant values correlates with a change in the manufacturing process, this process change and its potential impact on the method should be evaluated. Some manufacturing changes and their potential impact on the sample preparation method are listed in Table 13.2.

Occasional or what appear to be random aberrant values often call into question the robustness of the sample preparation/extraction method (assuming laboratory error is ruled out). In these cases, a sample preparation/ extraction ruggedness or robustness studies should be performed. Possible variables to study to assess the ruggedness and/or robustness of the sample preparation/extraction method are listed

Table 13.3 Possible variables to study to assess sample preparation/extraction method robustness and ruggedness

Method variables for robustness[a]	Variations to study
Shaking	Manual vs. automated
	If manual: shaking by inversion vs. swirling, number of shakes or inversions
	If automated: type of shaker, oscillations per minutes, length of time
Stirring	Stir time
	Stir speed
	Size and shape of magnetic stir bar
	Shape of vessel
Sonication	Probe vs. bath
	Sonication time
	If a bath is used: amount of solution in bath, where flask in placed in bath (on bottom vs. elevated in solution)
	Effect of different sonicators or sonication probes
Grinding	Manual vs. automated
	Potential losses during transfer steps
	Cleaning between samples (i.e., to prevent cross-contamination)
Dissolving/extraction solvent	Amount of solvent
	Slight changes in ratio of solvents if dissolving/extraction solvent is a mixture
	Slight changes in pH
	Changes in reagent grade
Miscellaneous	Temperature
Method variables for ruggedness[b]	**Variations to study**
Variable	Different labs, analysts, instruments, days, etc.

[a]Measure of method's ability to remain unaffected by small, but deliberate variations in method parameters, and provides an indication of its reliability during normal usage
[b]Degree of reproducibility of the results obtained under a variety of conditions

in Table 13.3. An example involving a nonrobust sample preparation/extraction method that resulted in random or occasional low potency values is described in Case Study 3 of Sect. 13.3.

When high variability in potency values are obtained (e.g., higher %RSD than typical for content uniformity), questions are raised regarding the analytical method (e.g., is the method robust/rugged) and the manufacturing process (e.g., nonhomogeneity issues for a given batch or non-robust manufacturing process leading to batch to batch variability). One should ensure that the method is being followed properly (e.g., analyst is adequately mixing, etc). A second area that should be evaluated is method robustness. An example illustrating investigation of low potency and high variability results is discussed in Case Study 1 of Sect. 13.3.

On the basis of the particular details of the case, a plan should be developed for the investigation. These steps can include any or all of Steps 3–6 shown in Table 13.1 and can be performed in parallel or in whatever order is appropriate based on the specific case.

13.2.3 Step 3: Try to Confirm Aberrant Results with Current Method

One should try to confirm the aberrant results (e.g., low potency values) using the current method. This often involves reanalysis/reinjection of the sample solutions (if not previously done in Step 1), analyzing additional samples along with a control sample or having another analyst perform additional testing. At this stage, the analyst who developed the method or who has extensive experience with the method may perform testing or may watch the testing being performed.

If higher potency values are obtained at this stage, then it is likely that the original low potency results are due to either laboratory error, which was not identified during Step 1, or are due to a possible method robustness issue.

13.2.4 Step 4: See if Modifications to the Current Method Will Improve Aberrant Potency Values

One should try modifications to the current method to see if increased potency values can be obtained on the sample(s) in question. Compare methods for similar formulations to get ideas. Method modifications can include, but are not limited to:

- Letting sample solutions sit/stir overnight and then analyzing them
- Increasing agitation time (e.g., shake time)
- Increasing extraction solvent volume
- Changing method of agitation
- Changing extraction solvent(s) (e.g., change pH, change organic content)
- Using increased temperature (e.g., using a stirring hot plate or accelerated solvent extraction [ASE] or microwave assisted extraction [MAE])

If better results are obtained (e.g., original low potency values are now at ~100%) with modifications to the method, then the current method is not providing complete extraction of the API and method modifications are needed.

13.2.5 Step 5: Try to Improve Results Using Alternative Methods

Alternative analysis methods and alternative sample preparation methods should be used to try to improve results (e.g., if original results were low potency then try to increase recovery of samples) or to try to confirm the original aberrant values obtained for the samples. Testing should be performed on the samples in question along with a control sample. The control sample is one with a known potency value, typically close to 100% label claim.

Alternative analysis methods that can be used include dissolution and near infrared (NIR). For solid oral dosage forms, there is typically a dissolution or drug release method that is available. The dissolution or drug release method can be used to see the maximum amount of drug dissolved or released at the end of the test or after an infinity point. NIR can be used as a qualitative method to compare the response of the sample in question vs. a control sample. NIR is a good technique to consider since no sample preparation is required and hence sample extraction is not a factor.

Alternative sample preparation methods that can be used include homogenizer (e.g., Polytron®), ball mill or mortar/pestle; and MAE or ASE. The homogenizer, ball mill, or mortar/pestle can be used to break up the dosage form and reduce the sample particle size to increase extraction efficiency (Majors 1998; Kok and Debets 2001; Nickerson et al. 2008). The homogenizer uses a set of rotating blades combined with wet grinding/shredding/shearing to break up the sample in the extraction solvent. During investigations, a defined volume of diluent and the sample can be placed in an appropriately sized bottle (to avoid subsequent subdilution or the need to quantitatively transfer the solution after homogenization) and then the homogenizer can be inserted into the solution in the bottle to homogenize the sample. If the tablet is larger than the diameter of the homogenizer probe, the homogenizer will not readily grab the tablet and shred it. In these cases, the tablet should be cut into smaller pieces (e.g., cut in half or quarters with a scalpel) or a larger homogenizer probe should be used. For the ball mill, the sample is placed in a stainless steel (or Teflon) chamber with stainless steel (or Teflon) balls, with or without diluent. The chamber is placed in the mill and shaken at high velocity. The balls effectively pulverize and mill the sample. For purposes of the investigation, it is recommended to mill with a defined volume of diluent in the chamber, so that milling and extraction occur simultaneously and the presence of solvent aids in quantitative removal of the sample from the chamber. If no diluent is used, it can be challenging to quantitatively transfer the sample from the chamber after milling. A composite sample can also be milled without diluent and then a portion of the milled material can be removed, weighed, and prepared. Alternatively, manual grinding using a mortar and pestle can be used. Since the homogenizer, ball mill, and mortar/pestle all leverage particle size reduction, typically using only one of these techniques should be sufficient during the investigation.

As described in Chap. 5, MAE and ASE heat the sample to increase extraction efficiency (Eskilsson et al. 1999; Kou and Mitra 2003). In MAE, the sample is enclosed in the extraction vessel with a microwave-absorbing solvent (or a combination of microwave and non microwave-absorbing solvents), and then microwaves are used to heat the solution directly. Direct heating of the solution, as opposed to conductive heating of the extraction vessel, results in reduced temperature gradients, more rapid heating of the sample, and reduced extraction times (on the order of 10 min). Stirring the solution in the extraction vessel helps to decrease the heating up time. Degradation, however, may be an issue with thermally labile analytes of interest. ASE uses elevated temperatures and pressures to extract components from solid and semisolid samples with organic or aqueous solvents. Elevated pressure in itself does not increase API solubility, but is used to prevent vaporization of the solvent under elevated temperatures. There is no sample agitation (e.g., stirring) mechanism in ASE. ASE does have the advantage of allowing one to perform multiple extractions on a single sample, so one can perform several extractions to see if additional drug is extracted after the first extraction.

Additional methods worth considering include water analysis and CHN analysis. Water analysis will reveal if there is more water than expected in the sample, thus leading to lower potency results than expected. Similarly, if solvents are used in the drug product manufacture (e.g., organic solvent used in film-coating) then residual solvent testing may be beneficial. If the API contains nitrogen and the excipients in the formulation do not, then the nitrogen content in the sample, measured by CHN, will be the result of the API. Nitrogen content from the sample in question can be compared with a control sample to see if there is any difference. Alternatively, chemiluminescence nitrogen detection (CLND) instead of CHN analysis can be used to measure nitrogen content.

On the one hand, if better results (e.g., higher potency values) are obtained for the sample using alternative methods, then this suggests that the sample does not have a low potency and that the current method is not adequate to extract all of the API. On the other hand, if comparable low potency values are obtained using alternative methods, then this suggests, although does not prove, that the sample does have a low potency.

13.2.6 Step 6: Confirm Current Method Able to Extract API from Dosage Form

A good way to confirm that the current method is able to completely extract API from the dosage form is to analyze tablets that have been individually made with a known quantity of API. If low recoveries are obtained for these samples, then this suggests that the current method is not extracting all API. Analysis of control samples or previously analyzed samples can also be performed to demonstrate that the method is capable of obtaining reproducible results.

If drug–excipient interactions are suspected, such as adsorption of drug to an insoluble excipient, then one can prepare drug–excipient blends for analysis or wet spike API solution onto dry excipient blend and prepare the sample per the method. If there is drug–excipient interaction, a low recovery will be obtained and the extraction/dissolving solvent or sample preparation method will need to be changed to address this issue.

13.3 Case Studies

Several case studies are detailed in this section to illustrate the investigation strategy outlined in Table 13.1. The case studies highlight the effectiveness of a systematic approach to investigating aberrant potency issues and demonstrate that aberrant values can result from method related issues as well as API and drug product manufacturing issues.

13.3.1 Case Study 1: Low Potency Values Obtained During In-Process Testing of Controlled Release Tablet Cores of Compound A

Background. A modified release osmotic tablet formulation was developed for Compound A. The API is first formulated as a spray dried dispersion (SDD) with a polymer. The film coated tablet formulation consists of 25% SDD and other excipients including polyethylene oxide (PEO) and cellulose acetate. PEO and cellulose acetate can cause challenges during sample preparation (e.g., viscous sample solutions, difficulties with obtaining adequate recovery).

A method for the analysis of the cores and final tablets was developed and validated. In this method, tablets are cut in half for better exposure of the tablet core to the solvent, and transferred to a volumetric flask. Acetonitrile is added to the flask to disperse the film-coating (when present) with the aid of 3.5 h of shaking at ~200 opm. Methanol is then added to the flask to aid in SDD dissolution followed by 3.5 h of additional shaking at ~200 opm. Extracts are then diluted to volume with water, mixed and filtered through a 100k molecular weight cut-off filter using centrifugation. The resulting solutions are analyzed by HPLC. This test method was used throughout formulation development and the average recovery for a development batch was close to 97%.

Low potency and high variability results obtained during IPC testing. A clinical batch of tablet cores was manufactured and 30 cores were sampled for each stratum (i.e., beginning, middle and end) for in-process control (IPC) testing. Ten samples (four from the beginning, two from the middle and four from the end) were tested. Potency values ranging from 77.0 to 97.8% were obtained with an average of 90.9% and with a high variability (6.2% RSD). An investigation involving the testing lab and the lab that developed the method was initiated (Steps 1 and 2 of Table 13.1).

Analytical investigation. After reviewing the data and procedures, it was noted that the cores were not cut in half during the IPC testing. Although significant differences were not expected by omitting this step for analysis of cores, it was deemed as a possible cause for error. The lowest sample (77.0%) was refiltered in triplicate and reanalyzed (Step 3 of Table 13.1) to confirm results and eliminate the possibility of injection error. The %LC increased to 89.6, 89.4, and 88.9%, consistent with the results of other samples, although still lower than the target potency value. The response factors of standards used for quantitation were checked against new standards and found not to be an issue. Also as part of Step 3 of Table 13.1, an analyst from the method development lab with experience with the product and method analyzed $n=5$ for the same tablet core lot (samples taken from a separate sampling of the lot obtained for development purposes), obtaining an average result of 91.1% with a RSD of 1%. Although the reproducibility was better, relatively low potency results were obtained. At this stage, issues with extraction as well as method transferability were suspect. The team then decided to increase the confidence in the current method before doing any additional testing on the IPC samples.

Table 13.4 Potency values obtained for the side-by-side testing by the analyst from the release testing lab (analyst #1) and the analyst who validated the method (analyst #2)

Sample	Assay (% LC)	Average/%RSD
Analyst #1		
200 opm – sample 1	81.5	
200 opm – sample 2	89.3	
200 opm – sample 2	82.2	84.3/5.1
200 opm – sample 1 refiltered	88.9	
200 opm – sample 2 refiltered	89.3	
200 opm – sample 3 refiltered	88.0	88.7/0.7
Analyst #2		
200 opm – sample 1	87.4	
200 opm – sample 2	90.6	
200 opm – sample 3	88.5	88.9/1.8
290 opm – sample 1	86.4	
290 opm – sample 2	90.9	
290 opm – sample 3	88.9	88.7/2.5

Minor modifications to the method were performed to minimize extraction concerns (Step 4 of Table 13.1). The analyst from the method development lab performed several experiments during the troubleshooting exercise including: using larger extraction solvent volumes; leaving samples shaking for a total of 24 h; cutting the tablet cores in half vs. not; and additional shaking (16 h) with 100% organic diluent. All of these experiments resulted in no significant increases in recoveries.

After these results, side-by-side testing was conducted in the testing lab (to account for any differences in equipment). This exercise was performed by the analyst who had tested the IPC samples (analyst #1) and the analyst who had validated the assay method (analyst #2). Each analyst performed triplicate samples following the written procedure. In addition, analyst #2 also tested an additional three samples using a shaker at ~290 opm, as previous development work had occasionally been conducted at this speed. All samples were analyzed in the same HPLC run. The results are summarized in Table 13.4. As shown, analyst #1 still had the lower average and higher % RSD. The shaking speed did not result in significant differences.

After the side-by-side exercise, it was easier for both analysts to review the analysis step by step. Any instrumentation differences were eliminated. After "walking the process," it was realized that mixing of the extracts after diluting to final volume could be the only step where error/differences could have been introduced. After dilution with water, the solutions are highly viscous and "mixing thoroughly" as per the method had a different meaning to the two analysts. Analyst #1 employed slow wrist-action mixing, while analyst #2 mixed by inverting the flask and vigorously shaking for a minimum of 5 times. This root-cause was confirmed when the samples from analyst #1 were thoroughly mixed and refiltered and all recoveries increased to values similar to those obtained by analyst #2 as shown in Table 13.4 (analyst #1, refiltered samples).

After resolving the variability issue, the question related to low potency assay values was still unanswered. Although the confidence in the method highly increased after the troubleshooting exercise, alternative methods were used to further increase confidence in the potency values obtained (Step 5 of Table 13.1).

The first approach taken was through dissolution testing. Cores were tested in duplicate after 24 h using the validated dissolution method. The values obtained were 85 and 86% of label claim. These values are consistent with a potency value in the lower 90s, as it was proven during development that ~5% of the active is retained and not released in the residual sweller layer. A second approach was through the analysis of the active layers, as it is known that the coagulant PEO in the sweller layer makes the extraction much more challenging. For these experiments, the active layers were separated from the sweller layer using a razor blade. The weight of a core vs. the two separate pieces was within 3 mg. Both, the clinical batch and the development batch were used in this experiment. The average % label claim for the development batch active layers was 100.2% and for the clinical batch was 90.7%. Note that this value is close to the average value of the original testing of the cores by the analyst #2 of 91.1%, demonstrating the efficiency of the method and the fact that the two batches are indeed different. A third approach using a Polytron® homogenizer to disperse the tablet and provide agitation during extraction gave an average potency for the clinical batch of 90.5% for $n=4$. Thus, all the analytical data demonstrated that the clinical batch had indeed lower potency than the development batch.

Outcome of analytical investigation. Based on all available data obtained throughout the investigation, it was concluded that thorough mixing of sample extracts prior to filtration is critical to obtaining reproducible results. It is also important that for methods with challenging extraction, the method transfer should not be based only on scientific rationale, but on hands-on familiarization and/or parallel testing involving key stakeholders. After the investigation, the team also had high confidence in the method and that the lower recoveries observed for the clinical batch were not due to the sample preparation.

The cores were coated and the final drug product batch was tested and found to be within specifications with an average potency of 94%. No further issues with the method were encountered after closing the investigation.

13.3.2 Case Study 2: Low Potency Values for Clinical Batches of Tablets of Compound B

Background: Compound B was formulated as an immediate release tablet formulation using standard excipients. The extraction/dissolving solvent used in the tablet assay method is 70% (1%H_3PO_4, v/v)/30% acetonitrile (v/v), in which the solubility of Compound B was determined to be 35 mg/mL. The nominal sample concentration for assay and content uniformity testing is ca. 0.1 mg/mL. Six lots of tablets had been manufactured and tested with average potency values of 97.9–102.1%. Stability for these tablet formulations at room temperature had been demonstrated out to 35 months.

Low potency results obtained during release testing. A new lot of API was manufactured and due to discoloration of the API, recrystallization out of aqueous

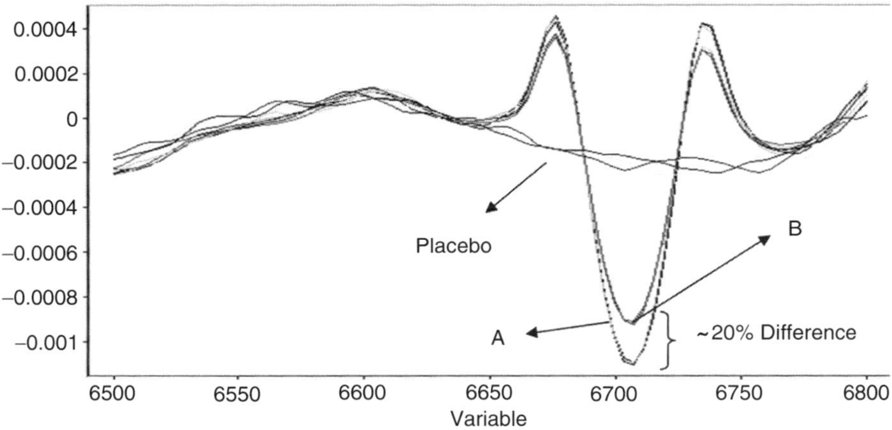

Fig. 13.1 NIR analysis, using reflectance mode and second derivative processing, of placebo tablet (placebo), previously manufactured tablet lot (*A*) and new tablet lot (*B*). Difference in API content between the two active tablets estimated to be 15–20%

media was required. The API was subsequently released for tablet manufacture. One 5 mg tablet lot and one 20 mg tablet lot were manufactured from a common blend and both lots were found to have assay values less than 85%.

Analytical investigation. Laboratory error was ruled out as a cause of the low potency values (Step 1 of Table 13.1). After careful review of the manufacturing documents, it was determined that the correct amount of API was charged. Since the manufacturing process uses vacuum transfer of material with local extract vents (LEVs), it was thought that the finer API particles had been lost in the vents during material transfers. Another batch of blend was made using bin-to-bin transfer to eliminate the LEVs and the blend was tabletted. This tablet lot was also found to have low potency (<85% of label claim). Since no laboratory or manufacturing cause could be assigned for the low potency and the API had passed all release criteria, the analytical method became the focus of the investigation.

The following alternative sample preparation techniques were used to try to increase recovery in the tablets, or confirm the low potency results (Step 5 of Table 13.1): MAE, ball mill, using an ionic liquid extraction/dissolving solvent and Polytron® homogenizer. In addition, modifications to the current method were made (Step 4 of Table 13.1) to try to increase the recovery of the samples by using different solvents, such as DMSO, THF, and DMF, to extract the tablets. No increases in recovery were obtained. In addition, dissolution testing using simulated gastric fluid, simulated intestinal fluid, and surfactants was performed and was found ineffective in extracting additional API.

NIR spectroscopy was then used to determine qualitatively whether the newly manufactured tablet lot had the same amount of API as those previously manufactured and found to have acceptable assay values. NIR does not rely on sample extraction; therefore, it was a valuable tool in this investigation. As shown in Fig. 13.1, when comparing the second derivative preprocessed responses of the new tablet lots to one previously manufactured, it is clear that the new tablet lot contains less API.

A laboratory sample of API (from the API batch used to manufacture the tablets in question) transferred to a scintillation vial from a larger bottle was found to have significant condensation within a few minutes of the transfer. Karl Fischer (KF) testing confirmed the condensation was in fact water, and further KF testing of the API showed approximately 9% water, significantly more than expected. Loss on drying (LOD) experiments were conducted on lab samples and found a loss of approximately 12%. The manufacturing site was asked to conduct LOD experiments on the remaining bulk API utilizing a sampling plan that called for nine samples to be tested. Visual inspection of the drum during sampling showed that there were agglomerates in the bulk drum. LOD results for the samples ranged from 8 to 35%. Samples were also subjected to potency testing, yielding an average of 82% and a range of 74–89%.

Given that the API passed all release criteria, the sampling and drying techniques came into question. This material was tray dried. Conceivably the top layer of material might have been dry, and the release samples could have been taken from the top layer. Another theory is that when dealing with milligrams quantities during analytical testing, the material loses water quickly as the container is continuously opened and closed during testing, which results in acceptable assay values.

Outcome of analytical investigation. Based on the results of the investigation, it was concluded that the low recovery in the tablets was not due to the analytical method, but rather due to high water content in the API. The remaining API was dried fully, tested, and released, and used for subsequent tablet manufacture. These tablets were found to have acceptable potency values. An in-process LOD step was mandated for future API lots to ensure full drying of the material.

This case study demonstrates that low potency values can be accurate and can be the result of a manufacturing related issue. Due diligence, however, is always required to investigate all possible causes of low potency values, including possible method related issues.

13.3.3 Case Study 3: Randomly Low Potency Values for Fixed Combination Tablets of Compounds C and D

Background. An assay method for a fixed combination tablet product had been developed and successfully transferred and used by several laboratories with no issues. The fixed combination tablets consist of two active ingredients, compounds C and D.

Occasional low potency values obtained. During manufacture of scale up and clinical batches where high sample volumes were analyzed (e.g., stratified samples and release samples), occasional low potency values were obtained (~70–80% for each of the two drugs).

Table 13.5 Potential causes of low potency assay values for fixed combination tablets of compounds C and D

Potential causes for low potency results
Potential causes related to the analytical assay
• The tablet disintegration step is insufficient in the tablet assay preparation
– Shaker to shaker variability is the cause (shaker rpm or stroke length discrepancy)
– The shake time is insufficient
– Shaker rpm is insufficient
• A residue in the volumetric flasks (e.g., soap or methanol used for cleaning) is interfering with the tablet disintegration step
• The tablet extraction solvent is not strong enough or optimal to extract the actives
• The pH of the disintegration step is in a non-robust area for disintegration or extraction
• Tablets stick to the bottom of the flasks prior to disintegration by some mechanism not allowing complete disintegration
Tablet related causes and tablet assay causes
• Excipient segregation is generating true low assay tablets
• Hard tablets are not easily disintegrated or extracted by the analytical method
• Tablets with a visible shine are not easily disintegrated or extracted by the analytical method. It was noted that tablets manufactured by Site X have a distinct shine as compared to lab prototypes
• Over lubricated tablets or granules are not easily disintegrated or extracted by the analytical methods

Analytical investigation. Because of the apparent random nature of the low potency values, it was difficult to reproduce the low potency values (Step 3 of Table 13.1). A Kepner-Tregoe analysis was conducted to identify potential causes of the occasional low potency values (Step 2 of Table 13.1). These potential causes are listed in Table 13.5.

The potential causes were evaluated using existing data or through additional experimentation. Data available or generated included the following: analysis of aberrant tablets containing different hardness levels (over and under compressed) and extended lube times; evaluation of the effect of soap residue on the flasks; evaluation of effect of adjusting disintegration solvent pH to 6, 7 (nominal), and 8, and varying the length of time between solvent addition and start of agitation; performing a statistically designed experiment to evaluate effects of shake time, shake speed and disintegration buffer concentration and evaluating the effect of tablet sticking to the bottom of the flask during sample preparation. Based on the initial investigation, it appeared that the tablet disintegration step of the sample preparation method was the cause, specifically that tablets showed a tendency to stick to the bottom of the volumetric flasks during preparation.

Upon additional investigation, however, differences were noted in the way that analysts mixed the solution in the volumetric flasks. It was concluded that insufficient mixing or agitation could lead to a nonhomogeneous solution, which in turn would result in low potency assay values.

Outcome of the analytical investigation. Based on the conclusions, the sample preparation method was revised to detail how the solution in the volumetric flasks should be mixed. The final agitation step of the sample preparation was changed from "Bring each flask to volume with Dissolving Solvent and mix well with several inversions" to "Mix well with 15 inversions allowing the air bubble in the neck of the flask to move to the bottom of the flask, then turn the flask right side up, holding it in this position to allow the solution to settle out of the flask neck. Each inversion should take approximately 3–4 s." Since revising the method to detail how to mix the solutions during the final agitation step, no unexplained low potency values have been obtained with analysis of approximately 3,000 samples.

13.3.4 Case Study 4: Low Potency Value Investigation for Modified Release Tablets of Compound E

Background. Two extemporaneously prepared (EP) 25 mg controlled release (CR) matrix tablet formulations (MR3 and MR4) were developed for compound E. Each individual tablet was compounded by weighing an excipient premix and API into an individual capsule shell and then blending the mixture within the capsule shell. The blended powder mixture from each capsule was then emptied and compressed into a tablet.

The tablet sample preparation procedure uses methanol to disintegrate the tablet with the aid of a 15-min sonication (sample solution concentration is 0.4 mg Active/mL). This is followed by centrifugation and dilution with sample solvent (55% water: 45% methanol).

Low potency values obtained during time zero stability testing. During real time stability assessment of the EP tablets at time zero, low potency values were obtained. Recovery values of 88.8 and 88.7% were found for a composite of two MR3 tablets (MR3 tablet #3, #10) and a composite of two MR4 tablets (MR4 tablet #3, #10), respectively. For an EP formulation, the exact quantity of API spiked into each tablet is known, thus %recovery was calculated against the known spiked API quantity. These low recovery results were not acceptable for formulation nomination.

Analytical investigation. No obvious lab error was identified (Step 1 in Table 13.1). Modifications were made to the potency method to try to increase recovery of the drug (Step 4 of Table 13.1). Only a 15-min sonication is used in the method, thus, an additional shaking step was added by shaking the volumetric flasks containing the tablet and partially filled with methanol for 2 h before diluting to volume. This modified approach was applied to the analysis of one tablet per formulation and 76.8 and 86.4% recoveries were found for the MR3 (tablet MR3 #7) and MR4 (tablet MR4 #7) formulations, respectively.

At this point, a literature search was performed on all historical HPLC assay methods for HPMC containing CR matrix tablet formulations within the company (Step 2 of Table 13.1). These methods all suggest that the modified sample preparation method should suffice in extracting all of the API from the tablet. The fact that more stressed extraction method (e.g., with added shake time) results in lower

Table 13.6 Recovery of Compound E from EP tablet and corresponding capsule shell used to compound the EP tablet

	mg Active			% Recovery		
Sample ID	Shell	Tablet	Theory [a]	Shell	Tablet	Total
MR3 #1	0.94	21.17	25.51	3.69	83.00	86.69 [b]
MR3 #7	4.15	19.67	25.60	16.23	76.82	93.05
MR4 #1	2.72	20.93	25.30	10.74	82.71	93.46
MR4 #7	1.98	22.10	25.57	7.73	86.42	94.16

[a] Theory based on known quantity of Compound E weighed into individual capsule shell during compounding of each individual EP tablet

[b] Lower recovery believed to be the result of not thoroughly rinsing out the capsule shell to recover all the residual powder. (This was the first capsule shell that was rinsed and the technique was not yet perfected)

recovery led the analyst to think that the low recoveries may be due to compounding process variation.

A discussion was held between the formulator and analysts to brainstorm possible causes of potency variation (Step 2 of Table 13.1). The formulator pointed out that there was some small quantity of powder leftover in the capsule shells used for compounding. One additional tablet was analyzed for each formulation (MR3 tablet #1 and MR4 tablet #1) using the original method. The capsule shells used for compounding these tablets (MR3 tablet #1, #7, MR4 tablet #1, #7) were also secured and the powder in the shells was quantitatively rinsed out using acetonitrile, and the sample solutions were analyzed for API content. These results are displayed in Table 13.6. As shown in these results, there was a significant percentage of API left in the capsule shells due to sticking of the powder to the shell. The reason that lower % recoveries were obtained at a more stressed extraction condition is a coincidence that more API was left on capsule for those two tablets.

Outcome of analytical investigation. Based on the results of the investigation, it was concluded that the low recovery in the EP tablets is not due to the analytical method, but rather is due to API loss from sticking to the capsule shell during the compounding process. The EP compounding process was changed to use glass vials instead of capsule shells to blend the excipients and API. In the new process, satisfactory recovery (>96%) was achieved for the tablets.

This is another case study where low potency values were the result of a manufacturing related issue. Again, due diligence is always required to investigate all possible causes of low potency values, including possible method related issues.

13.3.5 Case Study 5: Low Potency Investigation During Release Testing of Tablets of Compound F

Low potency results obtained during release testing. A low potency investigation for an immediate release tablet formulation of compound F was initiated when during the release testing of the first clinical batch of tablets, low potency results were obtained (*ca.* 95%) for the assay and content uniformity tests, but the dissolution results were *ca.* 100%. While the assay and content uniformity results were within specification, they were aberrant results that were unlikely to be product-related and were suspected to be analytical procedure related.

Analytical investigation. Initial investigations (Step 1 of Table 13.1) verified that the standards were prepared correctly and there were no calculation errors, etc. – the extraction procedure appeared to not perform satisfactorily. The sample diluent used in the method is acidified methanol/water, 80/20, v/v. A number of experiments were tried to increase recovery (Step 2 of Table 13.1), such as increasing the acidity of the diluent to 0.5% and doubling the volume of the extraction solution to lower the concentration, but without significant improvement in the potency values obtained.

It was then noticed that the diluent prepared for the release testing appeared not to have the low viscosity and low surface tension that would be expected for a diluent containing 80% methanol – it had the odor of methanol, but appeared in nature to be more like a lower percent organic solution, such as a ~20% methanol solution. The earlier extraction method development work demonstrated that low methanol content in the extraction diluent would lead to low recoveries. The rate of evaporation of a droplet of diluent was compared visually with that of a droplet of a solution of 80% methanol (aq.) and of 20% methanol (aq.); the rate of evaporation of the diluent was the same as the 20% methanol solution and was distinctly different from that of the 80% methanol solution. The shape of the solvent front was also used to provide further evidence that the diluent was prepared incorrectly.

Outcome of analytical investigation. Since an assignable cause was identified (sample diluent prepared incorrectly), the original assay and content uniformity results were invalidated and fresh correctly-prepared diluent was used to provide new assay and content uniformity results; this time all results were *ca.* 100% of nominal.

13.3.6 Case Study 6: Variable Potency Investigation During Site Qualification for Compound G Tablet Manufacture

Variable potency results obtained during site qualification manufacture. Highly variable potency results were obtained for stratified immediate release tablet core samples of compound G collected during site qualification batch manufactures. No anomalies were observed in the granule blend data or in the tablet release data (performed on a composite sample of the film coated tablets); variability was only observed in the stratified tablet core data. Stratified tablet core results for three batches each of two different strengths are shown in Fig. 13.2. The various steps of the analytical investigation are described below.

Initiation of Investigation. A retrospective analytical method investigation was initiated, but the effectiveness of that investigation was severely hampered by the fact that all analytical sample solutions were disposed of before the investigation was

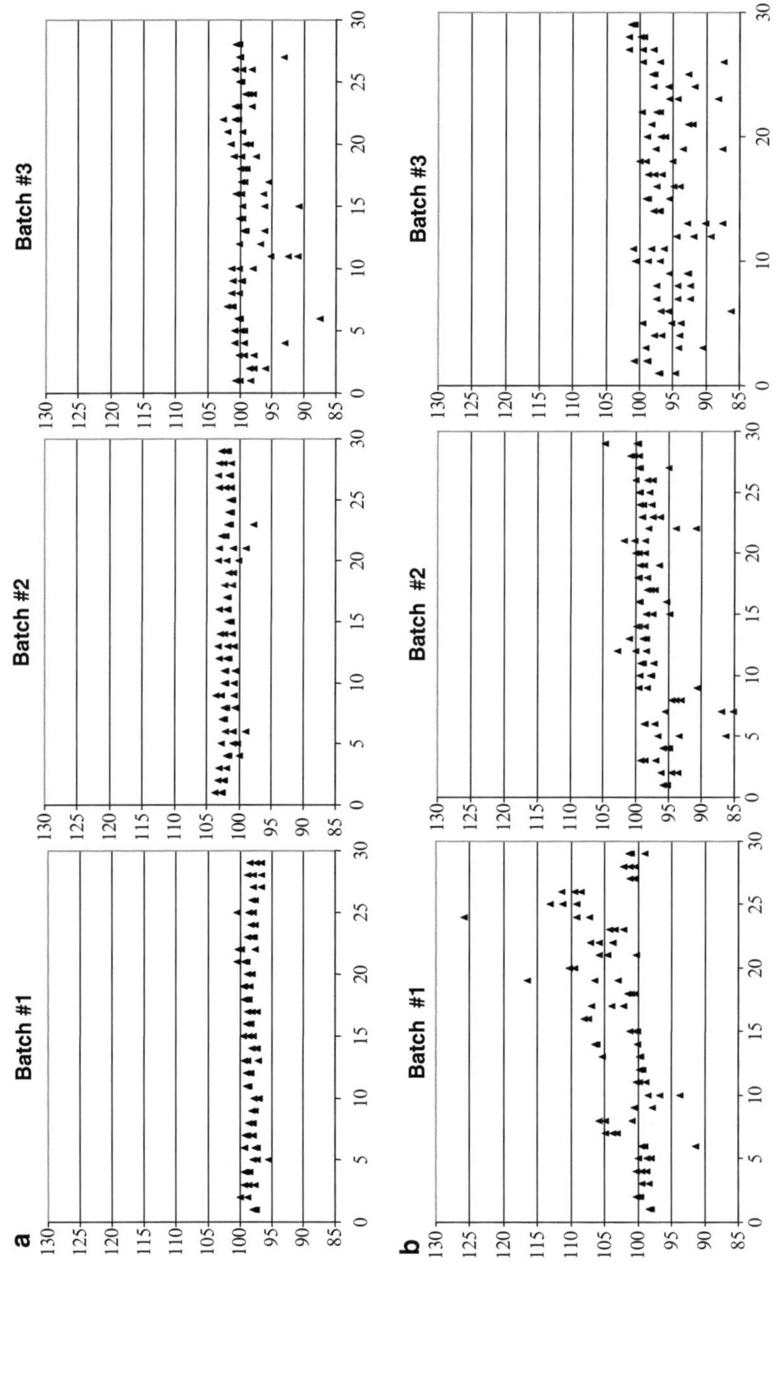

Fig. 13.2 Variable tablet assay results obtained during site qualification batch manufacture. The x-axis corresponds to sampling point during the manufacturing run. The y-axis corresponds to % label claim. **(a)** Results for three batches of the lower strength formulation. **(b)** Results for three batches of the higher strength formulation

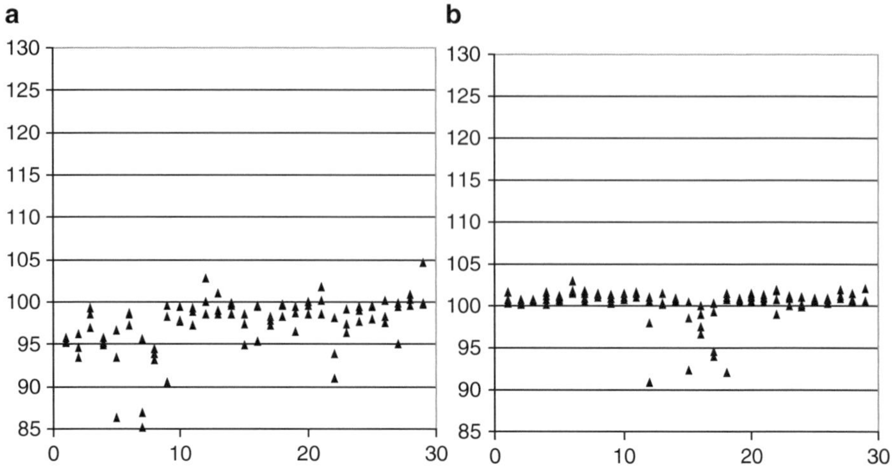

Fig. 13.3 Results for the repeat analysis of batch #2 for the higher strength formulation. The *x*-axis corresponds to sampling point during the manufacturing run. The *y*-axis corresponds to % label claim. (**a**) Original results. (**b**) Results from the repeat analysis

initiated. It was the local site practice to report results and discard all solutions when no errors were found during the check and review of the analytical documentation (as was the case in this case study). The first corrective action by the problem investigation team was to ensure that in the future, analytical solutions are retained until review is complete, and it is confirmed that no analytical investigations are required. Checking and reviewing analytical documentation (e.g., laboratory notebooks) can only reveal chromatographic, calculation, and documentation errors. Many analytical errors (such as those made during sample preparation) are unlikely to be revealed in reviewing analytical documentation. It is therefore critical to retain sample solutions until it is determined that no investigation is required.

Repeat Analysis. The analysis was repeated for one of the batches (Step 3 of Table 13.1) and the results are shown in Fig. 13.3. Statistically different and improved results were obtained, yet there were still several anomalous individual results obtained.

In this case, the analytical solutions were retained from the repeat analysis, which enabled the anomalous samples to be reinjected. In addition, since the sample preparation involved a dilution step, fresh dilute samples were prepared from the stock sample solutions of the samples that gave anomalous results. The reinjected diluted sample solutions confirmed the anomalous result, but the freshly prepared-from-stock solutions produced satisfactory, nominal results as shown in Fig. 13.4.

Cause of Sample Preparation Variability. This method had been used to generate thousands of results, and in the majority of cases satisfactory results were obtained. Therefore, it was concluded that during the original analysis of the site qualification batches, the sample preparation procedure did not perform satisfactorily or it was not carried out correctly for all individual samples. The anomalous result data sets

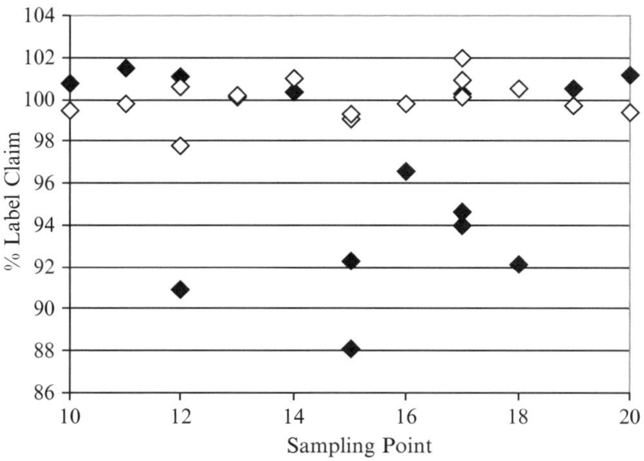

Fig. 13.4 Results obtained for solutions prepared from the stock solutions that produced anomalous results – *black* symbols are original results, *outlined* symbols are results obtained from the redilution of the stock sample solutions

were analyzed statistically to determine whether the variability was normally distributed – a nonnormally distributed variability may be an indication of nonrandom "special cause variability," e.g., caused by a specific erroneous laboratory practice. Figure 13.5 shows the data presented as "normal probability plots" – if the data are normally distributed, all data points would lie approximately on a single straight line. The normal probability plots show that for a significant number of individual samples there is an additional source of error in addition to the expected random normally-distributed variability (i.e., special cause).

Observation of Analytical Practice. The sample preparation for all the lower strength batches and for Batch #3 of the higher strength was carried out by "Analyst A," while Batches #1 and #2 of the higher strength were prepared and analyzed by "Analyst B." Analyst A results are generally less variable than Analyst B, but are better for some batches than others. Anomalous results from Analyst A are only subpotent, while Analyst B produced subpotent and super-potent results (e.g., Batch #1 of the higher strength). This indicates that the analytical technique of the operator is the likely cause of the anomalous results.

Observation of the sample preparation procedure carried out by Analysts A and B (Step 3 of Table 13.1) revealed that the procedure was not carried out precisely as written. The sample preparation procedure is outlined in Table 13.7. Instead of following the steps in the order shown in Table 13.7, both Analysts A and B carried out the procedure in the following order: 1, 2, 3, 5, 6, 4, 7, and then 8–13 as indicated (i.e., Step 4 was carried out after Step 6). This error arose due to familiarity with other methods carried out for other projects within the same laboratory. This error may result in incomplete extraction due to the presence of only about 20 mL of

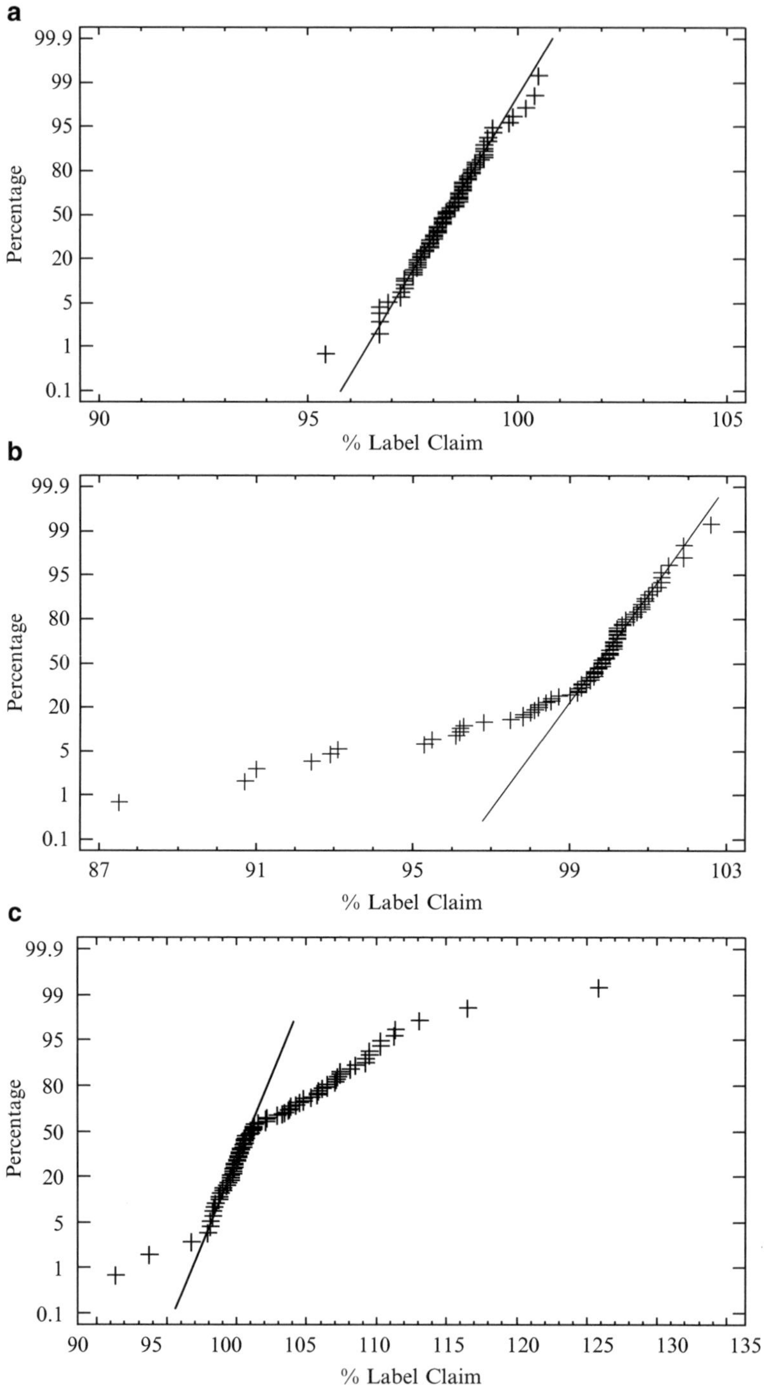

Fig. 13.5 (**a**) Normal probability plot for batch #1 of the lower strength formulation showing reasonably normally-distributed variability. (**b**) Normal probability plot for batch #3 of the lower strength formulation showing a significant nonnormally-distributed "tail" of sub-potent results. (**c**) Normal probability plot for batch #1 of the higher strength formulation showing two significant nonnormally-distributed "tails" of sub-potent and super-potent results

Table 13.7 Sample preparation procedure for tablets of compound G

Step	Procedure
1	Transfer one preweighed tablet to a 100-mL volumetric flask
2	Add approximately 20 mL diluent to the flask
3	Agitate flask to disintegrate the tablet
4	Dilute to volume with diluent
5	Add a magnetic stir bar to the flask
6	Stir for 20 min
7	Manually shake flask to ensure homogeneity
8	Allow contents of flask to settle (to facilitate filtration)
9	Filter a sufficient volume through a 0.45-μm GHP Acrodisc (discarding the first 2 mL)
10	Pipette 2 mL of filtered solution into 100 mL volumetric flask
11	Top up to the mark with diluent
12	Shake to ensure homogeneity
13	Transfer sample solution to HPLC vial and analyze by HPLC

diluent being present during the 20 min stir. However, this error in itself does not fully explain the experimental result anomalies.

An additional potential operator-error considered likely is insufficient agitation at Step 7 of the procedure; the labor-intensive sample preparation procedure coupled with the high number of samples analyzed would suggest that this is quite likely; also some evidence for this was highlighted during the observations of the operator's analytical procedure.

Outcome of analytical investigation. As a result of these analytical investigations, a policy of not discarding solutions until all analytical investigations are complete was instigated. This case study also highlights the limitations of checking and reviewing analytical documentation for investigating anomalous results. The method was rewritten to emphasize and clarify the order of steps in the sample preparation procedure. Analysts A and B were retrained in the procedure. Future methods are now evaluated for their labor-intensiveness since laborious procedures can lead to a lack of diligence by operators and the fewer and simpler the steps are the less likelihood of analytical error.

13.4 Conclusions

A systematic approach to investigating aberrant potency values for solid oral dosage forms is presented and discussed. Application of this approach to several case studies is also presented. The case studies highlight the effectiveness of a systematic approach to investigating aberrant potency issues and demonstrate that aberrant values can result from method related issues as well as API and drug product manufacturing issues.

Acknowledgments The authors thank the following individuals for their efforts in the investigations described in this paper: Wayne Warren and Mike Roy for their efforts on the Case Study 1 investigation; Larry Erhart and Angela Liu for their efforts on the Case Study 2 investigation; Larry Erhart, Hugh Clark and John Larmann for their efforts on the Case Study 3 investigation; and Hannah Jupe for her efforts on the Case Study 5 investigation.

References

Choi CK, Dong MW (2005) Sample preparation for HPLC analysis of drug products. In: Ahuja S, Dong MW (eds) Handbook of pharmaceutical analysis by HPLC. Academic, New York

Cory W, Field K, Wu-Linhares D (2004) Is it the method or the process – separating the causes of low recovery. Drug Dev Ind Pharm 30(8):891–899

Eskilsson CS, Björklund E, Mathiasson L, Karlsson L, Torstensson A (1999) Microwave-assisted extraction of felodipine tablets. J Chromatogr A 840:59–70

Gitlow HS, Levine DM (2004) Six sigma for green belts and champions: foundations, DMAIC, tools, cases, and certification. Prentice Hall, Upper Saddle River

Kepner CH, Tregoe BB (1976) The rationale manager: a systematic approach to problem solving and decision making. Kepner-Tregoe, Princeton

Kok SJ, Debets AJJ (2001) Fast sample preparation for analysis of tablets and capsules: the ball-mill extraction method. J Pharm Biomed Anal 26:599–604

Kou D, Mitra S (2003) Extraction of semivolatile organic compounds from solid matrices. In: Mitra S (ed) Sample preparation techniques in analytical chemistry. Wiley, Hoboken

Lee C (2007) Troubleshooting low potency results in solid oral dosage forms: Is it the method or the formulation? Am Pharm Rev. http://www.americanpharmaceuticalreview.com. Accessed Sept/Oct 2007

Majors RE (1998) Particle-size reduction in solid sample preparation. LC-GC 16(5):436–441

Nickerson B (2006) Troubleshooting sample preparation/extraction methods. Presented at the AAPS Short Course on Sample preparation/extraction methods for solid oral dosage forms, San Antonio, TX on October 29

Nickerson B, Arikpo WB, Berry MR, Bobin VJ, Houck TL, Mansour HL, Warzeka J (2008) Leveraging elevated temperature and particle size reduction to extract API from various tablet formulations. J Pharm Biomed Anal 47:268–278

Chapter 14
Green Chemistry Considerations for Sample Preparation

Paul Ferguson and Mark Harding

Abstract Green chemistry is the design, development and implementation of chemical products and processes to reduce or eliminate the use and generation of substances hazardous to both human health and/or the environment. Although advances in instrumentation have been made to reduce the amount of organic solvents needed for analytical testing (e.g., reduced column i.d. for high performance liquid chromatography analysis, Supercritical Fluid Chromatography, Super Heated Water Chromatography, etc.), generally large quantities of solvent are still used in sample preparation. The green chemistry principles of 'reduce, replace and remove' are discussed with respect to the use of organic solvents in sample preparation. Examples of sample preparation methods that have been developed and have utilized green chemistry principles will also be discussed.

14.1 Introduction

14.1.1 What is Green Chemistry?

All branches of chemistry can adversely affect the natural environment. Increasing and legitimate pressures from environmental agencies obligate chemists to adapt or transform processes that have a detrimental impact on the environment. In 1998, Anastas and Warner (Anastas and Warner 1998) authored the concept of the 'Twelve Principles of Green Chemistry' as a recommendation on how chemists could

P. Ferguson (✉)
Research Analytics, Pfizer Global Research and Development, Pfizer Inc., Ramsgate Road, Sandwich, Kent, UK
e-mail: Paul.Ferguson@Pfizer.com

B. Nickerson (ed.), *Sample Preparation of Pharmaceutical Dosage Forms: Challenges and Strategies for Sample Preparation and Extraction*, DOI 10.1007/978-1-4419-9631-2_14,

Fig. 14.1 The twelve principles of green chemistry. Reproduced from Anastas and Warner (1998) with permission from Oxford University Press

1. Prevent waste.
2. Design safer chemicals and products.
3. Design less hazardous chemical syntheses.
4. Use renewable feedstocks.
5. Use catalysts, not stoichiometric reagents.
6. Avoid chemical derivatives.
7. Maximize atom economy.
8. Use safer solvents and reaction conditions.
9. Increase energy efficiency.
10. Design chemicals and products to degrade after use.
11. Analyze in real time to prevent pollution.
12. Minimize the potential for accidents.

develop and implement synthetic chemistry in a sustainable manner (see Fig. 14.1). Anastas also provided an early definition of Green Chemistry as 'the use of chemistry techniques and methodologies that reduce or eliminate the use or generation of feedstocks, products, by-products, solvents, reagents, etc. that are hazardous to human health or the environment' (Anastas 1999).

Green chemistry has evolved from this core set of principles, as identified by Clark in the journal Green Chemistry (Clark 2006), to focus on a key set of factors that are forecast to have a considerable impact on working practices now and in the future. These factors include the availability of materials that are used in chemical processes; the costs involved in purchasing and disposing of these materials (as well as their products and by-products), and the influence of the regulatory environment. It is important to note that green chemistry is not a specialization within the wider field of chemistry; rather it is concerned with the evolution of chemistry to meet this new set of external realities. The purposes of our chemical processes are unchanged as, by-and-large, is the chemical toolkit available to achieve those aims. Green chemistry is now a well established field of research with some notable industrial applications (Manley et al. 2008).

14.1.2 Green Analytical Chemistry

In recent years, there has been an emphasis on the contribution of the analytical sciences to process understanding and quality-by-design for the manufacture of active pharmaceutical ingredients (API) and drug products (DPs). One desirable consequence of increased process understanding is the potential for more efficient, greener manufacturing processes, which generate less waste. Hence, the analytical sciences can play an important role in furthering the development of green chemistry.

In a situation that parallels with environmental analysis (Anastas 1999), despite the emphasis on the safety, understanding and efficiency of processes and products under analysis, there is a tendency for pharmaceutical analysts to neglect to consider the wider impact of their own methodologies, consumables and waste. Although it is the case that manufacturing processes have much greater potential to consume materials and generate larger volumes of waste than analytical chemistry, with education analysts are increasingly recognizing the need for their discipline to become more environmentally aware. This has resulted in the growth of Green Analytical Chemistry, a subdiscipline of Green Chemistry where the same principles are applied to the analytical sciences, as an emerging area of research (Armenta et al. 2008). In the literature, the use of life-cycle assessment (LCA) has been reported where an attempt to measure the overall environmental impact of an analytical process was made (Keith et al. 2007). Using LCA all materials, energy inputs and outputs to the process are included. However, this approach can be time-consuming and presents some difficulties such as determining the boundaries of the process. In this text, our intention is to focus on solvents, as these often represent the key consumable for analytical chemists; how solvent use can be reduced and which solvents could be replaced or removed entirely by greener alternatives. Furthermore, assessments of desirable and undesirable solvents from a green perspective are available in the published literature (e.g., Alfonsi et al. 2008).

14.1.3 The 3 R's

A common theme within both green chemistry and green analytical chemistry is that of reducing, replacing, or removing solvents from a process. This is often referred to as the '3 R's'. First, the concept is that the easiest adjustment to a method a chemist or analyst can make is to reduce the volume of organic solvent used in an experiment resulting in less wastage (and decreased costs). Second, the next step in optimizing a process would be to change the organic solvent utilized in the experiment to a safer or more environmentally benign one – while still maintaining (or improving) the reaction yield or extraction efficiency. Lastly, where possible, organic solvents should be removed entirely from the process. For a synthetic process, this could be the use of enzymes in aqueous media to afford a synthetic transformation (e.g., like that used in Pfizer's pre-gabalin commercial synthetic route) (Martinez et al. 2008). From an analytical sample preparation perspective, this could be enabled by using super critical fluid extraction (SFE) or super heated water extraction (SHWE), which utilize only carbon dioxide or water, respectively. These techniques will be discussed later in this chapter.

Another reason to embrace the concept of the '3 R's' was highlighted recently. A spotlight was cast onto both Green Chemistry and Green Analytical Chemistry by the acetonitrile shortages experienced in 2008–2009 (Tullo 2008). Acetonitrile is commonly synthesized as an unwanted by-product (approximately 2–3% by weight) in the manufacture of acrylonitrile. Acrylonitrile is a key component in the synthesis of many materials such as plastics and building materials. A number of unfortunate occurrences coincided in 2008 to seriously impact the availability of arylonitrile and thus acetonitrile. These included power-failures and hurricanes damaging major production plants in the US, and the global economic crisis leading to less demand for materials synthesized from acrylonitrile. In the pharmaceutical industry, acetonitrile is commonly used in sample preparation and chromatographic analysis. The shortages impacted many in the analytical community, adversely affecting the cost and availability of this important resource. One key observation made was that while many analysts individually consider the level of resources they use to be low with respect to the processes they are monitoring, the shortage highlighted the significance of analytical chemists collectively as resource consumers (Christopher et al. 2009). The shortage also underlined the importance of understanding where critical resources used in analyses come from and how sustainable and reliable those supplies are. It also prompted many to consider anew the contribution of solvent purchase and, equally as significant, their disposal to their overall operating costs.

A common response to the acetonitrile shortages has been to focus on reducing the amount of solvents used in analyses and preventing waste. For example, there is a steady shift by separation scientists to move from high performance liquid chromatography (HPLC) to ultra-high pressure liquid chromatography (UHPLC). Although often a driver for this shift is increased analytical method performance (in terms of analyte peak efficiency and resolution), there is the serendipitous benefit of a reduction in mobile phase and organic solvent consumption due to the reduction in the analytical column dimensions. Another important way to reduce the amount of solvents used is to move toward online analysis. Processes such as the use of 'on-line' mIR and Raman probe technology completely negates the need for sample preparation and will become much more commonplace in the future (He et al. 2007).

14.1.4 Green Pharmaceutical Sample Preparation and Analysis

There is an increasing awareness that as instrumental analytical techniques move to smaller dimensions, the most environmentally unsound part of an analytical procedure is often the sample preparation step. Sample preparation can still use a large amount of organic solvent to generate appropriate samples compatible with analytical instrumentation. Additionally, large amounts of pollutants can also be generated in sample preparation through vapours, waste reagents, solid sample and consumable wastage. It is therefore increasingly important that we should look holistically at the whole analytical process and in particular sample preparation in a 'green' context (Namiesnik 2001).

Historically one approach to sample analysis has been via 'dilute and shoot', which is solubilizing the drug and its matrix and injecting directly into the analytical system (Choi and Dong 2005). The perception is that sample preparation can often be a time consuming process and one that can be ill-afforded in generating an analytical measurement. However, consideration of the impact of such an approach on consumable usage costs (e.g. HPLC columns) and instrument reliability means that in an increasingly cost conscious industry, appropriate sample preparation methods should be considered as equal a priority as the analytical method development itself.

In the first instance, sample preparation methods that utilize lower volumes of synthetic organic solvents will be more green than many traditional methods.

However, worker safety is also a key component to green practices and should also be considered. It is worth noting that often significantly more solvent and sample manipulation is required to prepare samples from formulated DP than the raw API and for this reason API should be analyzed or used in preference to DP formulations whenever practicable.

In sample preparation, the analyst should also be cognisant of the power of the analytical technique by which the sample will be analyzed. For example, where highly selective and sensitive mass spectrometric techniques are used, it makes little sense to include multiple time consuming and waste generating sample clean-up and preconcentration steps within the sample preparation process as the analytical instrumentation can essentially ‘filter’ many interferences out. Additionally, the sample preparation should ideally be tailored to the scale of the sample volume required for the analytical measurement. It makes little sense to generate large volumes of purified sample when typically only a few microlitres might be used in an analysis. An approach of doing just enough to enable a robust analysis to be performed should always be adopted. Curylo and co-workers have reviewed a number of sample preparation techniques, which speak to these approaches (Curylo et al. 2007).

14.1.5 Types of Analysis

The aim of this chapter is not to provide an in-depth discussion on every sample preparation technique, which could be made more environmentally sound, nor is it possible to do so in a single chapter (see other chapters for more complete guidance on the specific approaches noted below). We will discuss some of the most common experiments that an analyst performs in the laboratory i.e., potency assay, content uniformity and impurity determination. Additionally, we will focus on key techniques – reviewing guidance, offering potential starting points and discussing pertinent examples. It is hoped that this will enable the reader to consider and implement simple adjustments to their sample preparation method, which will minimize the use of solvents (both aqueous and organic), generate less waste, increase analyst safety and be more environmentally sound.

14.2 Potency Assay and Content Uniformity

Potency assay is one of the main analysis types performed to determine the weight fraction of an analyte (typically the API) in a batch of sample or formulated product. The other contents are precursors and impurities from a drug’s synthetic route, solvents (organic and water), residual metals and excipients in formulated products. The assay procedure is typically used to ensure conformance to a label claim (typically 90–110% of label claim) for DPs.

Content uniformity is often performed to assay individual tablets or capsules (i.e., formulated DP) to ensure processing consistency. Typically ten single sample preparations are made with all assay measurements required to fall between a particular range, often 85.0–115.0% of label claim with an appropriate acceptance value (USP General Chapter <905> Uniformity of Dosage Units 2010). There are no sample preparation techniques specifically for content uniformity determination, and any technique discussed in this section could be used.

14.2.1 Liquid–Liquid Extraction (LLE)

As discussed in Chap. 4, liquid–liquid extraction (LLE) or solvent extraction is a commonly used step in sample preparation or isolation. In this technique, compounds are separated based on their relative solubilities in two different immiscible liquids, in which typically one phase is an organic solvent. Generally, the other phase is aqueous and is usually the phase into which DP formulations are initially dissolved prior to extraction. At a simple level, it is the extraction of a substance from one liquid phase into another liquid phase (the analyte(s) usually ending the extraction in the organic phase prior to preconcentration and analysis). This technique often uses large volumes of solvents and many historical methods utilize chlorinated or fluorinated solvents, which are highly detrimental to the environment.

Two of the main requirements of LLE are that the liquids used must be immiscible and be chemically inert to each other (and to the analytes that will partition between the liquids). In the work-up of synthetic organic reactions, many types of LLE are utilized by chemists, which are typically based on the solvent a particular reaction step is performed in.

In reference to the 12 rules of Green Chemistry noted in Fig. 14.1, to make LLE processes greener we should consider the following:

1. Renewable feedstocks – from a green perspective it is preferable to use solvents derived from renewable sources. However, at this time organic solvents are limited to bioethanol and potentially biomethanol.
2. Worker safety – replacement of a compatible solvent with a less toxic alternative is highly desirable.
3. Environmental considerations – even though the majority of solvents utilized in the pharmaceutical industry are either recovered or disposed of via incineration, it is desirable to minimize any potential contamination of the environment.

Figure 14.2 provides a guide developed at Pfizer, which categorizes the solvents that are desirable to use through those that should be avoided for synthetic chemistry transformations. It should also be possible to use this approach to identify potentially greener alternatives for LLE. For example, toxic hexane could be replaced with the more benign solvent heptane. As both of these solvents are straight-chain alkanes, there is unlikely to be significant difference in the extraction selectivity and could be replaced directly into any existing established LLE protocol. Additionally, chloroform, dichloroethane, or carbon tetrachloride (all potentially mutagenic or carcinogenic solvents), which were commonplace in many historical LLE methods, can

Preferred	Usable	Undesirable
Water	Cyclohexane	Pentane
Acetone	Heptane	Hexane(s)
Ethanol	Toluene	Di-isopropyl ether
2-Propanol	Methylcyclohexane	Diethyl ether
1-Propanol	Methyl *t*-butyl ether	Dichloromethane
Ethyl actetate	Isooctane	Dichloroethane
Isopropyl acetate	Acetonitrile	Chloroform
Methanol	2-MethylTHF	Dimethyl formamide
Methyl ethyl ketone	Tetrahydrofuran	*N*-Methylpyrrolidinone
1-Butanol	Xylenes	Pyridine
t-Butanol	Dimethyl sulfoxide	Dimethyl acetate
	Acetic acid	Dioxane
	Ethylene glycol	Dimethoxyethane
		Benzene
		Carbon tetrachloride

Fig. 14.2 Pfizer solvent selection guide for green chemistry (Alfonsi et al. 2008). Reproduced by permission of The Royal Society of Chemistry

Table 14.1 Suggested acids (upper part of table) and bases (lower part of table) for adjusting the pH of the aqueous solvent to aid drug extraction when used in LLE

Preferred	Usable	Undesirable
Acetic acid	Formic acid	Trifluoroaceticacid
Phosphoric acid	Hydrochloric acid	Methane sulfonic acid
	Sulfuric acid	
Sodium hydroxide	Triethylamine	Ammonium hydroxide
Potassium hydroxide	Diethylamine	
Triethanolamine		

be replaced with dichloromethane, which will provide similar extraction efficiency. Although dichloromethane is still undesirable, it is significantly safer than the other halogenated solvents mentioned.

In choosing more environmentally benign solvents for LLE, an understanding of the physicochemical properties in switching to greener solvents is required. Green solvent application in synthetic chemistry will correlate strongly with their suitability for LLE and these principles should be used in choosing suitable solvents.

As mentioned earlier, in addition to consideration of the organic solvent in LLE, the use of a water-based partitioning solvent is often required to disperse and/or solubilize the DP matrix. The aqueous extraction solvent is often pH adjusted to optimize extraction of the drug from the matrix, and it is therefore also worth considering which acid or base to use, and again its potential impact on the environment. At Pfizer Global Research and Development, an internal green chemistry guide outlining which acids and bases should be used in synthetic reactions is available. These are ranked according to ecotoxicity, biodegradation, bioaccumulation and volatility potential. Some of this information pertinent to sample preparation has been tabulated in Table 14.1. Obviously the pH of the liquid phase will be dependent on which acid or base is used and at what concentration. For example, if an aqueous partitioning solvent requiring acidification is used, it is preferable to use, e.g., phosphoric acid in place of say trifluoroacetic acid (but is also dependent on acid solubility in a particular liquid). The analyst should also be cognizant when using organic acids or bases that these may be soluble in the organic extraction solvent and an appropriate one used.

LLE is currently engrained within analytical practices arising through historical usage and vast analyst experience. For all its limitations (long extraction times, use of large solvent volumes, worker safety, etc.), LLE is still a widespread and commonplace sample extraction technique. However, from a green perspective, LLE should be avoided whenever possible and alternative techniques allowing smaller solvent volumes and lower toxicity solvents employed.

14.2.2 Solid Phase Extraction (SPE)

One of the most commonly used extraction techniques is solid phase extraction (SPE) and was detailed extensively in Chap. 4. Like LLE, this technique involves the transfer of analytes from a weak to a strong solvent. With SPE, a drug formulation would typically be solubilized and dissolved into a primarily aqueous solution (tablet formulations possibly having being ground initially) where the pH of the solution may have been modified to aid analyte extraction (again the acids or bases in Table 14.1 could be utilized for the SPE process too). An appropriate sorbent chemistry is selected to retain the desired analytes and pretreated as necessary. The analyte solution is then forced through the SPE cartridge by gravity, manual force, or through vacuum suction. The SPE cartridge is then flushed with an appropriate solvent to elute the analyte(s) into a collection vessel. This solution may be analyzed directly or may undergo further concentration steps as necessary to enable the analytical measurement. Note that the elution solvent may not necessarily be organic in nature. If the sorbent is for example a cation-exchange phase, then an ionic solution of say sodium chloride could be used to elute the cationic analytes from the cartridge. However, depending on the analytical technique to be employed for the measurement, high ionic strength solutions may be an issue, but from a green perspective it is worth considering.

One advantage of SPE is that there are a large range of chemistries available as sorbent, which can be utilized to selectively 'tune' extractions to retain particular analytes or be selective for analyte classes, e.g., positively charged analytes over anionic species. This can greatly reduce sample preparation time, sample work-up complexity and can offer a greater range of versatility over the numerous solvents available for LLE. Other benefits of SPE over LLE include the following:

1. Improved sample preconcentration – samples are usually eluted into smaller volumes of solvent than LLE and samples are therefore more concentrated
2. Uses less organic solvent – mL instead of typically hundreds of mL
3. Increased worker safety – from exposure to smaller volumes of solvent
4. Not being limited to solvents immiscible with water
5. No emulsion formation (which are common in LLE and difficult to separate)
6. Can be readily automated for high throughput

From a green perspective, SPE should be used in place of LLE whenever possible for the above reasons. Further information on SPE sorbents and approaches to sample preparation are available from many of the manufacturer's websites, e.g., Waters, Varian, Phenomenex, etc.

14.2.3 Microwave-Assisted Extraction (MAE)

Microwave-assisted extraction (MAE) was discussed in Chap. 5. MAE benefits from the ability to rapidly heat a sample solvent to high temperatures, which accelerates mass transfer of analytes from a sample matrix into the extraction solvent when performed in a closed vessel. The technique typically uses much smaller volumes of solvent than traditional LLE experiments (typically 10–30 mL of solvent is used in MAE compared with often 10 times that or more in LLE). Several extractions may be performed in one heating cycle and the capacity is only limited by the volume of the microwave extractor. One example of the use of MAE is the extraction of an API in a chewable pharmaceutical tablet formulation (Hoang et al. 2002). The samples were extracted with water and small volumes of methanol (approximately 13 mL per sample), and this approach demonstrated higher extraction efficiency compared with a traditional mechanical method.

Another example citing the use of MAE was for the content uniformity determination of a proprietary Merck immediate release tablet formulation (Hoang et al. 2007). Although a relatively high volume of pure acetonitrile was used (50 mL per sample) during the procedure, the extraction time was 4 times faster than the traditional LLE procedure with much lower solvent consumption.

14.2.4 Ultrasound-Assisted Extraction (UAE)

As the name suggests, this approach to sample preparation utilizes ultrasonic energy to aid sample dissolution. High frequency acoustic waves are generated to create microscopic bubbles in liquids. When these bubbles collapse, small shockwaves and cavitations are produced, which promote dissolution of solids (Tadeo et al. 2010) and is applicable to pharmaceutical sample preparation. A more detailed discussion of this approach may be found in Chap. 3. UAE is often used as part of a process, which may also include the likes of a filtration step to remove excipients or SPE to preconcentrate analytes prior to analysis. There are few discussions of this technique's application in the literature from a green perspective for pharmaceutical preparations. However, one example described the extraction of nicotine from chewing gum and transdermal patches where the analyte was extracted into heptane prior to analysis by gas chromatography (GC). This approach reduced the solvent consumption by approximately 80% relative to the standard method (Zuo et al. 2004).

14.2.5 Supercritical Fluid Extraction (SFE)

Although this technique was discussed extensively in Chap. 5, it is worth noting it in a green context. As discussed, the technique typically utilizes carbon dioxide, which can be 'tuned' to modify its solubilizing and extraction properties (the critical temperature and pressure of CO_2 being 31°C and 73.8 bar, respectively). By adjusting the temperature or pressure of the extraction system, the diffusion rate and thus the solvating properties of CO_2 may be easily manipulated. Typically the system temperature is kept low (30–40°C) and the pressure is varied. By keeping the temperature low in this manner, analyte degradation is minimized. Although carbon dioxide has a reasonable solvating strength (comparable to hexane) and can extract a diverse range of polarity analytes by itself, it is often necessary to add a small amount of organic modifier, particularly if the analytes are polar in nature. In this case, methanol tends to be used, which is also considered a green solvent (Jiménez-González et al. 2004; Alfonsi et al. 2008).

Carbon dioxide can be extracted from air, although it is often formed as a by-product of catalytic cracking of methane gas to form hydrogen. When it is sourced from the atmosphere, SFE can be classified as a very 'green' technique as there is no net increase in carbon dioxide released into the environment. Some examples of SFE being employed in the extraction of analytes from drug-product formulations include extraction of felodipine from sustained release tablets (Howard et al. 1994) and isolations of vitamins from tablets (Scalia et al. 1995).

Although SFE has its place in pharmaceutical sample preparation, the technique is being steadily replaced by accelerated solvent extraction (ASE) (Slack and Snow 2007), which utilizes similar instrumentation, but does not have the added complexity of working with a compressible gas as an extraction medium.

14.2.6 Accelerated Solvent Extraction (ASE)

ASE or pressurized fluid extraction (PFE) was discussed at length in Chap. 5. Again, the principles of the technique are the same as for SHWE. The use of higher temperatures in ASE increases the capacity of solvents to solubilize analytes and promotes desorption of analytes from a matrix. Increased temperature decreases the viscosity of solvents allowing increased penetration into matrices. Additionally, the increased pressure used in this technique forces solvent into matrix pores, which may be inaccessible under standard extraction conditions. Its historical development was primarily for use in the environmental analysis field, but is now becoming more prevalent in pharmaceutical analysis.

The technique utilizes much smaller volumes of solvent, has faster sample processing time (compared to for example LLE) and can be automated for high throughput. The basic instrument configuration is essentially the same as for SHWE (see Sect. 14.2.7). Typically, static conditions are used where the extraction cell is filled

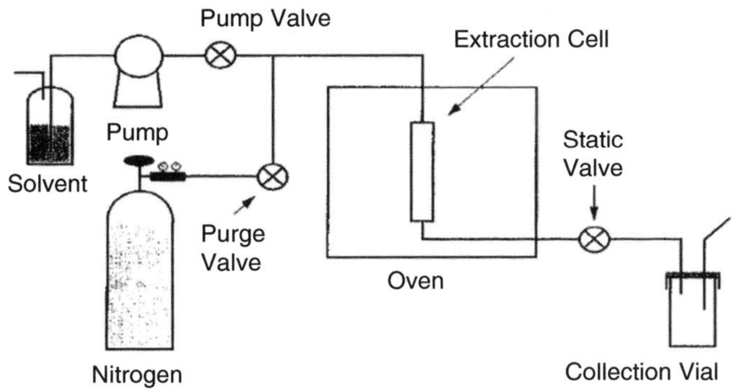

Fig. 14.3 Schematic of an ASE instrument. Reprinted with permission from (Richter et al. 1996). Copright 1996 American Chemical Society

with an appropriate solvent and heated under pressure (to maintain the solvent in a liquid state) for a short period of time (e.g. 5–10 min). The cell is then purged into a sample vessel using a compressed gas. Multiple extractions can be performed on a given sample. Absorbent materials may also be added to the extraction cell to simultaneously perform sample clean-up alongside extraction, e.g., C18 polymeric resins or activated carbon can be included to retain highly lipophilic material. It is also possible to use a flow-through technique, if a pumping system is connected to the extraction cell. A schematic of a typical ASE instrument is illustrated in Fig. 14.3.

An example of the application of ASE for DP formulations was demonstrated by Abend and co-workers (Abend et al. 2003). Their study describes a procedure to extract and quantitate a canine/feline antiparasitic in a meat-based formulation, although some sample pretreatment was required (grinding and blending of the samples). When compared with the traditional extraction technique of platform shaking and sonication, up to 5% higher recoveries were observed while much lower solvent volumes were used. Another application outlined the extraction of feldopine from tablet formulations using ASE (Bjorklund et al. 1998).

Commercial systems are available from Dionex, which allow full automation of this technique, e.g., the ASE 100–350 systems, and multiple extraction solvents can be used to selectively extract analytes of differing polarities from the same sample. An additional benefit resulting from these automated platforms is a reduction in worker exposure to the solvents applied in this approach.

14.2.7 Super Heated Water Extraction (SHWE)

SHWE or pressurized hot water extraction (PHWE) utilizes pressurized water at elevated temperatures (between its boiling point at 100°C and its critical point at 374°C) and pressures (to keep water in its condensed phase) to extract analytes from complex matrices. It can be considered as a polar alternative to SFE

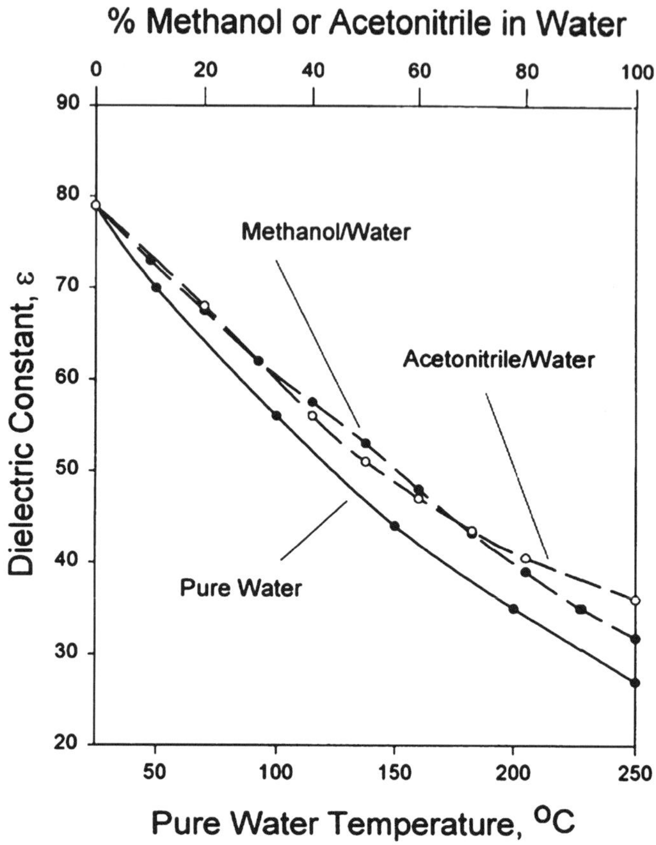

Fig. 14.4 Control of solvent polarity (dielectric constant) by changing temperature (at 50 bar) with pure water and aqueous methanol and acetonitrile blends. Reprinted from Yang et al. (1998) with permission from Elsevier

(Keith et al. 2007). It is most commonly used in the analysis of foodstuffs (Hassas-Roudsari et al. 2009) or in the environmental monitoring area (Zbiral and Nemec 2009) and was first reported in 1994 for the extraction of polar and nonpolar analytes from soil samples (Hawthorne et al. 1994). The technique can readily dissolve a range of low and medium polarity analytes from difficult matrices (Dong et al. 2007; Teo et al. 2008). As water is nontoxic and can be disposed of with minimal environmental issues, the technique can exhibit broad extraction applicability in an environmentally sound manner. A diagram exemplifying the potential of super-heated water for extraction is illustrated in Fig. 14.4, where the dielectric constant is a reflection of the polarity of the solvent. As can be seen, increasing the temperature of water from approximately 35 to 200°C decreases the polarity of water by half due to hydrogen bonds being broken and increasing water molecule disorder. At 200°C, water has a similar extraction power as 80% (v/v) aqueous methanol or acetonitrile.

The hardware required to utilize SHWE is similar to that used for ASE (see Sect. 14.2.6) such as those made by Dionex. The basic system consists of a water reservoir, a pump, a preheating coil, a heater (e.g., a GC oven), an extraction cell, a back-pressure regulator and a sample collection vessel. A very thorough review of the technique across all its application areas and the parameters affecting extraction has recently been reported (Teo et al. 2010).

One example of the application of this approach was demonstrated by Richter and co-workers (Richter et al. 2006) for the determination of nifedipine in tablets (both assay and content uniformity). While the extraction was performed in pure water, some of the 'green' benefits were negated by sample receipt into methanol to aid quantitation, but is still a pertinent example of this approach.

Obviously, one of the limiting factors of SHWE is the potential for thermal or hydrolytic degradation of analytes at these high temperatures. However, studies have been cited where no evidence was observed of thermal degradation of temperature sensitive compounds (Richter et al. 1996). While noting that the technique uses an environmentally benign solvent in pure water, the extracted sample is a cold and dilute aqueous sample and often requires further steps prior to analysis, e.g., analyte preconcentration using SPE or solid phase micro extraction (SPME) (Smith 2006).

14.2.8 Liquid-Phase Microextraction (LPME)

Liquid-phase microextraction (LPME) is a sample preparation technique that is rapid, inexpensive and utilizes small volumes of solvent. It is often used where the sample matrix is very dirty, e.g., river water, sewage, biological fluids where a particular analyte in the matrix is desired for extraction, but is equally applicable to DP formulations. It can be classed as a counterpart to SPME (see Sect. 14.3.1), but utilizes smaller solvent volumes (in the order of microlitres). The technique is essentially a derivative of LLE but on a much smaller scale. As sample volumes are large relative to the extraction solvent, very high analyte enrichment is possible (Rasmussen et al. 2000). The technique is also compatible with a range of analytical techniques, e.g., HPLC, GC and capillary electrophoresis (CE) (Pedersen-Bjergaard and Rasmussen 2008). Additional information on this technique may be found in Chap. 4.

The technique works by utilizing a porous hollow fiber that is immersed in a sample solution. Inside the hollow fiber is a desired extraction solvent (e.g., octanol). The fiber is left in the solution for the desired extraction time and the extraction solvent then removed from the fiber and collected in a suitable receptacle, for example a microvial. As the fibers are essentially disposable, cross-contamination or sample carry-over can be eliminated. A good example of this approach was demonstrated for the analysis of β2-agonists salbutamol and terbutaline in tablet formulations (Yamini et al. 2006). Payán and co-workers (Payán et al. 2009) illustrated the utility of this technique for the measurement of ibuprofen in pharmaceutical preparations and urine. With chemiluminescent detection they showed linearity and sub ppm limits of detection with good reproducibility. Another good example of this technique was for the trace-level determination of triphenylphosphine oxide albeit in API samples (Richoll and Colón 2006).

One group has taken this approach a step further in terms of its environmental friendliness and investigated the use of naturally occurring oils such as almond and olive oil as the acceptor solution in the hollow fiber (Pedersen-Bjergaard and Rasmussen 2004) thus totally removing the use of synthetic organic solvents. They demonstrated that these oils were the equal of synthetic solvents for the extraction of a range of pharmaceutical compounds from water, plasma and urine. While these are obviously not DP matrices, this approach offers the possibility of extension to DP formulations as well.

14.3 Impurity Determination

The monitoring of process related impurities and degradants in pharmaceutical compounds is important to assure drug purity and stability. Regulatory and guidance bodies such as the Food and Drug Administration (FDA) and the International Conference on Harmonisation (ICH) provide guidelines and validation criteria for such methods. Typically process-related impurities are not measured in DP formulations as long as they have been controlled in the API. However, this is not always the case. Degradents are always monitored in the formulated product. From a DP sample preparation perspective, it is important to ensure complete extraction of the API and degradants while minimizing sampling of contaminants such as excipients and preservatives (which may require reporting, identification and possibly qualification).

Many of the techniques noted in Sect. 14.2 could equally be applied for impurity or degradant determination, but may require additional steps to achieve the desired method selectivity or sensitivity.

14.3.1 Solid Phase Micro Extraction (SPME)

This technique was developed by Pawliszyn and co-workers in 1990 at the University of Waterloo in Canada (Arthur and Pawliszyn 1990) and was discussed in Chap. 4. They demonstrated by using a fused silica fiber that was coated with a polymeric stationary phase, and they could directly extract a variety of common water contaminants by immersing the fiber in groundwater samples. The technique can significantly reduce total analysis time as sampling, extraction, concentration and compatibility with the analytical technique can all be combined into a single process for gaseous, liquid and solid samples. This technique has numerous features which make it a 'green' approach to sample preparation including the following:

- Simplicity of application
- Speed of extraction (typically <15 min)

- Generic applicability
- Elimination of solvents
- Small sample volumes
- High sensitivity
- Low cost
- Simple automation

The technique can be coupled with GC (via immersion or headspace analysis), HPLC, or CE. Analyte preconcentration depends on a number of factors such as the type of stationary phase (e.g., polydimethylsilane [PDMS] or divinylbenzene-polydimethylsilane [DVB-PDMS]), thickness of the fiber and extraction time. Head space extraction prolongs fiber lifetime compared to immersion in solution, although sample carry-over can be an issue. One example of the use of this approach was demonstrated by Loknauth and Snow (Lokhnauth and Snow 2005), who utilized SPME for the determination of paraben preservatives in a range of pharmaceutical formulations, e.g., creams, lotion and ointment with ion-mobility spectrometry (IMS) detection. They compared this approach with a generic HPLC determination and showed excellent agreement in the results. The SPME approach used no organic solvent in the sample preparation or analysis. The comparative HPLC approach used significant volumes of organic solvent in both the sample preparation and HPLC analysis. A similar sample preparation approach was utilized for the analysis of phthalate plasticizers in liquid medicines and intravenous injection solutions (Mitani et al. 2003, 2004).

An additional example of this technique was for the determination of residual solvents in pharmaceutical preparations using a range of SPME coated fibers, e.g., PDMS and PDMS/DVB (Coran et al. 2001; Raghani 2002). A comprehensive review and discussion of SPME and related extraction techniques was recently published by Kataoka (Kataoka 2005).

14.4 Some Comparison of Extraction Techniques from a Green Perspective

As discussed in this and other chapters, sample preparation is dominated by manual and often repetitive operations, be it weighing, grinding, pipetting, filtering, or any number of other processes. As a result of the often 'large' scale of these operations, many of these steps unduly utilize larger solvents volumes than are really required and could be optimized by using smaller sample weights or volumes. To expedite sample preparation processes, it is highly likely that automation will play a more prominent role in making sample preparation greener in the future. An article by Lee et al. compared an automated tablet processing workstation (TPW), with MAE and ASE for a spray-dried dispersion (SDD) of a pharmaceutical compound (Lee et al. 2007). They demonstrated that the automation option required shorter extraction times per sample than MAE and ASE and gave much better recoveries

Table 14.2 Comparison of a number of sample preparation techniques for the analysis of an API in an SDD formulation

Technique	Recovery (%)	Temperature	Total prep time (per tablet)/min	Organic solvent (acetonitrile) used (per tablet)/mL
Manual	100.0	Ambient	330	200
TPW	101.4	Ambient	30	120
MAE	100.7	40°C	70	40
ASE	91.1	70°C	70	9

Adapted from Lee et al. (2007) with permission from Elsevier

Sample techniques	Extraction constituents	Organic solvent	Extraction time (min)	Sample amount (g)	Special equipment	Quantitative analysis
HS-SPME	VC	NR	15–60	1–5	NR	SAQ
HS-LPME	VC	NR	15–60	1–5	NR	SQA
MD	VC	NR	30–60	20–40	R	QA
MAE	VC and NVC	R	30–120	5–40	R	QA
PLE	VC and NVC	R	5–30	0.5–5	R	QA
SFE	VC and NVC	NR	120–240	20–40	R	QA

VC, volatile constituents; NVC, non-volatile constituents; NR, no requirement; R, requirement; SQA, semi-quantitative analysis; QA, quantitative analysis.

Fig. 14.5 Comparison of sample preparation techniques used to analyse traditional Chinese medicines. Reprinted from Deng et al. (2007) with permission from Elsevier

than ASE. MAE and ASE utilize much lower volumes of organic solvent (acetonitrile) however by a factor of 3 (MAE) to approximately 13 (ASE) – see Table 14.2. All three techniques were significantly superior in terms of speed and solvent usage to manual sample preparation, which was used as a reference.

Another paper presented a review of sample preparation techniques for traditional Chinese medicines, e.g., herbs (Deng et al. 2007). A number of techniques were reviewed and their relative merits discussed. Depending on the nature of the analysis and result required, any one of these techniques could be utilized illustrating the diversity of the sample preparation options available. A summary of this is shown in Fig. 14.5.

14.5 Other Potential Options for Green Sample Preparation

There are numerous options available for sample preparation of DP formulations, only a few of which have been discussed in this chapter. There are many other techniques often employed in other fields or other processes that could become more prevalent in the future for the preparation of analytes from DP formulations. Some suggestions for techniques which are currently not used routinely in DP extractions are noted below.

14.5.1 Membrane Extraction Techniques

Supported-liquid membrane (SLM) is an integrated two-step LLE. The technique utilizes a porous synthetic hydrophobic membrane in which the pores are filled with an appropriate organic solvent (such as long chain hydrocarbons or dihexyl ether for extraction of more polar analytes), which acts as a 'barrier' to two liquid phases either side of the membrane. The liquid matrix containing the sample is either flowed past one side of the membrane in a continuous manner (dynamic extraction) or is static. The analytes are essentially transported from one side of the membrane to the other aqueous solution through the organic solvent medium with mass transport dependent on analyte solubilities and diffusion coefficients. The technique is often used for the extraction of analytes from biological fluids (Kataoka and Lord 2002) or environmental samples (Prieto et al. 2010). This technique has many beneficial aspects over other techniques such as LLE including:

- Utilizing small volumes of typically non-toxic solvents
- High enrichment factors due to the high sample/solvent ratio
- Excellent sample clean up
- No solvent evaporation (safety aspect)
- Can be automated in-line with some analytical techniques such as GC

While no current examples of the application of this technique to DP formulations were identified, that is not to say it is not a feasible technique and may become more common place in the future.

14.5.2 Stir Bar Sorptive Extraction (SBSE)

Stir bar sorptive extraction (SBSE) was first reported by Baltussen and co-workers in 1999 (Baltussen et al. 1999). In this technique, analytes are extracted from a matrix into a nonmiscible liquid phase. The nonmiscible liquid-phase is almost always polydimethylsiloxane (PDMS) coated onto the type of standard stirrer bars, which can be found throughout chemistry laboratories. Inorganic absorbents have been investigated, but often interact too highly with trapped compounds and can be difficult to release. The stirrer bar may be either immersed in a solution containing the analyte, or into a head space autosampler for volatile analytes. Because of the low phase ratio (volume of aqueous solution relative to the volume of the stir bar stationary phase) high recoveries are obtainable. SBSE has higher sample capacity and recoveries than SPME as a result of higher coating levels on the stirrer bar (Kataoka 2005). Following immersion in a liquid sample, the bar is rinsed to remove unwanted contaminants, dried and then submitted for liquid or thermal desorption (assuming the analytes are thermally stable). In a headspace mode (head space sorptive extraction – HSSE as developed by Bicchi et al. (Bicchi et al. 2002)), the bar is suspended in a headspace vial, although this approach is less utilized than the aforementioned immersion approach. The number of researchers utilizing these approaches to sample preparation has steadily increased over the last 10 years to around 400 publications in 2009 alone (Prieto et al. 2010). A thorough review of the published literature and further guidance on optimizing an SBSE sample preparation approach may be found in the literature (Prieto et al. 2010).

The absorption of analytes onto PDMS film correlates well with the octanol-water partition coefficient K_{ow} (David et al. 2003). The fact that absorption can be predicted rather than having to be measured means that less experimentation is required and therefore less waste generated. A software package is available to calculate theoretical recoveries from the sorbent (KowWIN software, Syracuse Research Corporation, Syracuse, New York, USA) that utilizes this relationship. Alternatively, partition factors for neutral or charged species (log P or Log D, respectively) may be calculated by other commercially available software packages such as ACDLabs physical chemistry suite (Toronto, Canada). Additionally, the volumetric scale of the extraction is relatively small – typically a few mL of methanol, acetonitrile, or aqueous buffers (or organic solvents mixed with aqueous buffer), which is also beneficial. When used in the head-space mode and then thermally desorbed in a GC, no organic solvent is used at all.

One example where an SBSE approach was demonstrated for a formulated 'drug' product (Ochiai et al. 2002) was for the determination of preservatives in various beverages including 'quasi-drug drinks' (as defined by the Japanese Ministry of Health, Labor and Welfare). The drink was a mix of vitamins, minerals and caffeine (amongst other ingredients). The investigators found that this approach when coupled with thermal desorption GC-MS was very sensitive (low ng/mL limits of detection) with good linearity. No organic solvent was used at any step in the sample preparation or analysis.

An interesting application reported in the environmental field by Rodil and Popp (Rodil and Popp 2006) was the trace level analysis of pesticides in soil samples. In this work they utilized SHWE to extract the pesticides and then preconcentrate the analytes via SBME prior to analysis by GC/MS. This allowed the quantitation of analytes at sub nanogram/gram levels. This approach was compared with more conventional approaches utilizing acetone-hexane solvent blends for extraction. While results were comparable, the combination approach utilizing SHWE and SBME used a more environmentally friendly and safer solvent (acetonitrile). While the soil matrix is obviously more complex than many drug-product formulations, this process could easily be modified for trace analysis in pharmaceutical formulations.

14.5.3 QuEChERs

QuEChERS (the acronym being formed from 'Quick, Easy, Cheap, Effective, Rugged and Safe' and pronounced 'Catchers') is a streamlined approach to sample preparation (Schenck and Hobbs 2004; Payá et al. 2007). The approach was originally developed for the analysis of pesticide residues in food stuffs (initially in fruits and vegetables, but extended to meat products), but is now seeing some application in pharmaceutical analysis (primarily in bioanalysis). In this technique, the sample is homogenized, placed in a (typically) Teflon™ tube with reagents and agitated for a short time. The single-phase extraction solvent used is compound dependent, but is often acetonitrile, which is a reasonably green solvent (Alfonsi et al. 2008). Care must be taken to select the best extraction solvent and one that minimizes degradation of potentially labile analytes. This is followed by liquid–liquid partitioning via the addition of anhydrous $MgSO_4$ and NaCl. The final clean-up step is solid-phase extraction prior to analysis, which is typically a hyphenated mass spectrometry technique.

This approach has numerous 'green' characteristics compared with conventional sample preparation methods. Small amounts of reagents are used, the technique is broadly applicable ('generic'), utilizes a limited amount of glassware and is automatable. Stepan and co-workers (Stepan et al. 2008) described a QuEChERS approach for the analysis of steroids in nutritional supplements. While in theory such an approach might be adopted for other pharmaceutical formulations such as tablet or creams, it is likely that the main application for this approach in a pharmaceutical/nutraceutical setting would be for the analysis of nutritional supplements, or food-like formulations such as those sometimes used in canine and feline veterinary medicines.

14.6 Summary and Conclusions

In this chapter, we hope that we have provided the reader with a flavor of potential options for making their approaches to sample preparation more environmentally sound and safer for the analyst. There are numerous alternative approaches to commonly employed LLE, which in addition to being greener are more cost effective, require less manual manipulation and offer faster throughput. However, from a green perspective, the ideal situation is one where sample preparation prior to analysis is not required, and this is discussed in Chap. 16. The use of on-line analytical techniques such as near and mid-infrared spectroscopy is becoming more prevalent and worthy of note, but detailed discussion is beyond the scope of this text.

Many of the instrumental techniques noted above enabling greener sample preparation are currently not widely used. This is likely the result from a number of factors including use of historical or pharmacopeial methods, cost, additional training burden and a reluctance to utilize new approaches to name a few. However, as these techniques become more prevalent in the literature and in the laboratory, it is envisaged that adoption will increase and sample preparation will become inherently greener.

Acknowledgments We acknowledge the useful discussions with Dr Pete Dunn (Head of Green Chemistry, Pfizer Global Research & Development, Sandwich, UK) in the preparation of this manuscript.

References

Abend AM, Chung L, McCollum DG, Wuelfing WP (2003) Development and validation of an automated extraction method (accelerated solvent extraction®) and a reverse-phase HPLC analysis method for assay of ivermectin in a meat-based chewable formulation. J Pharm Biomed Anal 31(6):1177–1183

Alfonsi K, Colberg J, Dunn PJ, Fevig T, Sandra J, Johnson TA, Kleine HP, Knight C, Nagy MA, Perry DA, Stefaniak M (2008) Green chemistry tools to influence a medicinal chemistry and research chemistry based organisation. Green Chem 10(1):31–36

Anastas PT (1999) Green chemistry and the role of analytical methodology development. Crit Rev Anal Chem 29(3):167–175

Anastas PT, Warner JC (1998) Green chemistry: theory and practice. Oxford University Press, Oxford

Armenta S, Garrigues S, de la Guardia M (2008) Green analytical chemistry. TrAC Trends Anal Chem 27(6):497–511

Arthur CL, Pawliszyn J (1990) Solid phase microextraction with thermal desorption using fused silica optical fibers. Anal Chem 62(19):2145–2148

Baltussen E, Sandra P, David F, Cramers C (1999) Stir bar sorptive extraction (SBSE), a novel extraction technique for aqueous samples: theory and principles. J Microcolumn Sep 11(10):737–747

Bicchi C, Iori C, Rubiolo P, Sandra P (2002) Headspace sorptive extraction (HSSE), stir bar sorptive extraction (SBSE), and solid phase microextraction (SPME) applied to the analysis of roasted arabica coffee and coffee brew. J Agric Food Chem 50(3):449–459

Bjorklund E, Jaremo M, Mathiasson L, Karlsson L, Strode JT, Eriksson J, Torstensson A (1998) Determination of felodipine in tablets using accelerated solvent extraction. J Liq Chromatogr Relat Technol 21(4):535–549

Choi CK, Dong MW (2005) Sample preparation for HPLC analysis of drug products. In: Ahuja S, Dong MW (eds) Handbook of pharmaceutical analysis by HPLC, 1st edn. Elsevier, Amsterdam, pp 123–143

Christopher JW, Tanja B, Wes S, Xiaoyi G (2009) Performance to burn? Re-evaluating the choice of acetonitrile as the platform solvent for analytical HPLC. Green Chem 11(8):1232–1238

Clark JH (2006) Green chemistry: today (and tomorrow). Green Chem 8(1):17–21

Coran SA, Giannellini V, Furlanetto S, Bambagiotti-Alberti M, Pinzauti S (2001) Improving gas chromatographic determination of residual solvents in pharmaceuticals by the combined use of headspace solid-phase microextraction and isotopic dilution. J Chromatogr A 915(1–2):209–216

Curylo J, Wardencki W, Namiesnik J (2007) Green aspects of sample preparation – a need for solvent reduction. Polish J Environ Stud 16(1):5–16

David F, Tienpont B, Sandra P (2003) Stir-bar sorptive extraction of trace organic compounds from aqueous matrices. LC GC Eur 16:108–113

Deng C, Liu N, Gao M, Zhang X (2007) Recent developments in sample preparation techniques for chromatography analysis of traditional chinese medicines. J Chromatogr A 1153(1–2):90–96

Dong L, Wang J, Deng C, Shen X (2007) Gas chromatography-mass spectrometry following pressurized hot water extraction and solid-phase microextraction for quantification of eucalyptol, camphor, and borneol in chrysanthemum flowers. J Sep Sci 30(1):86–89

Hassas-Roudsari M, Chang PR, Pegg RB, Tyler RT (2009) Antioxidant capacity of bioactives extracted from canola meal by subcritical water, ethanolic and hot water extraction. Food Chem 114(2):717–726

Hawthorne SB, Yang Y, Miller DJ (1994) Extraction of organic pollutants from environmental solids with sub- and supercritical water. Anal Chem 66(18):2912–2920

He Y, Tang L, Wu X, Hou X, Lee Y-I (2007) Spectroscopy: the best way toward green analytical chemistry? Appl Spectrosc Rev 42(2):119–138

Hoang TH, Farkas R, Wells C, McClintock S, Di Maso M (2002) Application of pressurized liquid extraction technology to pharmaceutical solid dosage form analysis. J Chromatogr A 968(1–2):257–261

Hoang TH, Sharma R, Susanto D, Di Maso M, Kwong E (2007) Microwave-assisted extraction of active pharmaceutical ingredient from solid dosage forms. J Chromatogr A 1156(1–2): 149–153
Howard AL, Shah MC, Ip DP, Brooks MA, Strode JT III, Taylor LT (1994) Use of supercritical fluid extraction for sample preparation of sustained-release felodipine tablets. J Pharm Sci 83(11):1537–1542
Jiménez-González C, Curzons AD, Constable DJC, Cunningham VL (2004) Expanding GSK's solvent selection guide – application of life cycle assessment to enhance solvent selections. Clean Technol Environ Policy 7(1):42–50
Kataoka H (2005) Recent advances in solid-phase microextraction and related techniques for pharmaceutical and biomedical analysis. Curr Pharm Anal 1:65–84
Kataoka H, Lord HL (2002) Sampling and sample preparation for clinical and pharmaceutical analysis. In: Pawliszyn J (ed) Sampling and sample preparation in field and laboratory, Vol 37. Comprehensive analytical chemistry. XXXVII Elsevier Science, Amsterdam
Keith LH, Gron LU, Young JL (2007) Green analytical methodologies. Chem Rev 107(6):2695–2708
Lee C, Gallo J, Arikpo W, Bobin V (2007) Comparison of extraction techniques for spray dried dispersion tablet formulations. J Pharm Biomed Anal 45(4):565–571
Lokhnauth JK, Snow NH (2005) Determination of parabens in pharmaceutical formulations by solid-phase microextraction-ion mobility spectrometry. Anal Chem 77(18):5938–5946
Manley JB, Anastas PT, Cue BW Jr (2008) Frontiers in green chemistry: meeting the grand challenges for sustainability in r&d and manufacturing. J Cleaner Prod 16(6):743–750
Martinez CA, Hu S, Dumond Y, Tao J, Kelleher P, Tully L (2008) Development of a chemoenzymatic manufacturing process for pregabalin. Org Process Res Dev 12(3):392–398
Mitani K, Narimatsu S, Izushi F, Kataoka H (2003) Simple and rapid analysis of endocrine disruptors in liquid medicines and intravenous injection solutions by automated in-tube solid-phase microextraction/high performance liquid chromatography. J Pharm Biomed Anal 32(3):469–478
Mitani K, Izushi F, Kataoka H (2004) Analysis of phthalate contamination in infusion solutions by automated on-line in-tube solid-phase microextraction coupled with high-performance liquid chromatography. J Anal Toxicol 28:575–580
Namiesnik J (2001) Green analytical chemistry – some remarks. J Sep Sci 24(2):151–153
Ochiai N, Sasamoto K, Takino M, Yamashita S, Daishima S, Heiden A, Hoffmann A (2002) Simultaneous determination of preservatives in beverages, vinegar, aqueous sauces, and quasi-drug drinks by stir-bar sorptive extraction (SBSE) and thermal desorption GC-MS. Anal Bioanal Chem 373(1):56–63
Payá P, Anastassiades M, Mack D, Sigalova I, Tasdelen B, Oliva J, Barba A (2007) Analysis of pesticide residues using the quick easy cheap effective rugged and safe (QuEChERS) pesticide multiresidue method in combination with gas and liquid chromatography and tandem mass spectrometric detection. Anal Bioanal Chem 389(6):1697–1714
Payán MR, López MÁB, Fernández-Torres R, Navarro MV, Mochón MC (2009) Hollow fiber-based liquid-phase microextraction (HF-LPME) of ibuprofen followed by fia-chemiluminescence determination using the acidic permanganate-sulfite system. Talanta 79(3):911–915
Pedersen-Bjergaard S, Rasmussen KE (2004) Liquid-phase microextraction utilising plant oils as intermediate extraction medium – towards elimination of synthetic organic solvents in sample preparation. J Sep Sci 27(17–18):1511–1516
Pedersen-Bjergaard S, Rasmussen KE (2008) Liquid-phase microextraction with porous hollow fibers, a miniaturized and highly flexible format for liquid-liquid extraction. J Chromatogr A 1184(1–2):132–142
Prieto A, Basauri O, Rodil R, Usobiaga A, Fernández LA, Etxebarria N, Zuloaga O (2010) Stir-bar sorptive extraction: a view on method optimisation, novel applications, limitations and potential solutions. J Chromatogr A 1217(16):2642–2666
Raghani AR (2002) High-speed gas chromatographic analysis of solvents in pharmaceuticals using solid phase microextraction. J Pharm Biomed Anal 29(3):507–518

Rasmussen KE, Pedersen-Bjergaard S, Krogh M, Grefslie Ugland H, Grønhaug T (2000) Development of a simple in-vial liquid-phase microextraction device for drug analysis compatible with capillary gas chromatography, capillary electrophoresis and high-performance liquid chromatography. J Chromatogr A 873(1):3–11

Richoll SM, Colón I (2006) Determination of triphenylphosphine oxide in active pharmaceutical ingredients by hollow-fiber liquid-phase microextraction followed by reversed-phase liquid chromatography. J Chromatogr A 1127(1–2):147–153

Richter BE, Jones BA, Ezzell JL, Porter NL, Avdalovic N, Pohl C (1996) Accelerated solvent extraction: a technique for sample preparation. Anal Chem 68(6):1033–1039

Richter P, Ines Toral M, Toledo C (2006) Subcritical water extraction and determination of nifedipine in pharmaceutical formulations. J AOAC Int 89(2):365–368

Rodil R, Popp P (2006) Development of pressurized subcritical water extraction combined with stir bar sorptive extraction for the analysis of organochlorine pesticides and chlorobenzenes in soils. J Chromatogr A 1124(1–2):82–90

Scalia S, Ruberto G, Bonina F (1995) Determination of vitamin A, vitamin E, and their esters in tablet preparations using supercritical fluid extraction and HPLC. J Pharm Sci 84(4):433–436

Schenck FJ, Hobbs JE (2004) Evaluation of the quick, easy, cheap, effective, rugged, and safe (QuEChERS) approach to pesticide residue analysis. Bull Environ Contam Toxicol 73(1):24–30

Slack GC, Snow NH (2007) HPLC sample preparation. In: Ahuja S, Rasmussen H (eds) HPLC method development for pharmaceuticals, vol separation science and technology. Elsevier, Amsterdam, pp 237–268

Smith R (2006) Superheated water: the ultimate green solvent for separation science. Anal Bioanal Chem 385(3):419–421

Stepan R, Cuhra P, Barsova S (2008) Comprehensive two-dimensional gas chromatography with time-of-flight mass spectrometric detection for the determination of anabolic steroids and related compounds in nutritional supplements. Food Addit Contam A Chem Anal Control Expo Risk Assess 25(5):557–565

Tadeo JL, Sánchez-Brunete C, Albero B, García-Valcárcel AI (2010) Application of ultrasound-assisted extraction to the determination of contaminants in food and soil samples. J Chromatogr A 1217(16):2415–2440

Teo CC, Tan SN, Yong JWH, Hew CS, Ong ES (2008) Evaluation of the extraction efficiency of thermally labile bioactive compounds in gastrodia elata blume by pressurized hot water extraction and microwave-assisted extraction. J Chromatogr A 1182(1):34–40

Teo CC, Tan SN, Yong JWH, Hew CS, Ong ES (2010) Pressurized hot water extraction (PHWE). J Chromatogr A 1217(16):2484–2494

Tullo A (2008) A solvent dries up: acetonitrile is in short supply, and chemists are concerned. Chem Eng News 86(47):27

USP General Chapter <905> Uniformity of Dosage Units (2010) In: United States Pharmacopeia Vol 33 Reissue. USP-NF Online. http://www.uspnf.com. Accessed July 2010

Yamini Y, Reimann CT, Vatanara A, Jönsson JÅ (2006) Extraction and preconcentration of salbutamol and terbutaline from aqueous samples using hollow fiber supported liquid membrane containing anionic carrier. J Chromatogr A 1124(1–2):57–67

Yang Y, Belghazi M, Lagadec A, Miller DJ, Hawthorne SB (1998) Elution of organic solutes from different polarity sorbents using subcritical water. J Chromatogr A 810(1–2):149–159

Zbiral J, Nemec P (2009) Comparison of some soil extractants for determination of boron. Commun Soil Sci Plant Anal 40:96–105

Zuo Y, Zhang L, Wu J, Fritz JW, Medeiros S, Rego C (2004) Ultrasonic extraction and capillary gas chromatography determination of nicotine in pharmaceutical formulations. Anal Chim Acta 526(1):35–39

Chapter 15
No Sample Preparation

Yang (Angela) Liu, George L. Reid, and Zhongli Zhang

Abstract Sample preparation for pharmaceutical dosage forms can be a time-consuming and labor-intensive task. One option to reduce or eliminate this work is to use an analysis method that requires no or minimal sample preparation. This chapter discusses uses of vibrational spectroscopy (e.g., infrared, Raman) and mass spectrometry techniques to analyze dosage forms with no or minimal sample preparation to obtain identification, polymorph, water content, potency, and purity information. A high-level description of these techniques will be presented along with example applications.

15.1 Introduction

The analysis of pharmaceutical dosage forms typically involves three separate functions: (1) sample preparation, (2) sample analysis, and (3) data analysis and result generation. Multiple approaches toward minimizing the overall sample to results time exist, with advantages and disadvantages of each. In a traditional analytical work process, the sample is prepared (e.g., dissolved/sonicated/shaken/filtered) and the filtrate is analyzed by an end analysis technique such as chromatography. The time to obtain the filtrate can be lengthy and require significant amounts of organic solvents, and approaches to minimize this have been discussed in previous chapters. Although significant progress has been made in the reduction of sample analyses times via the introduction of ultra high-pressure liquid chromatography, the use of sample preparation and organic solvent is still required. In this chapter, the use of techniques that require no to minimal sample preparation

G.L. Reid (✉)
Analytical Development, Pfizer Global Research and Development, Pfizer Inc.,
558 Eastern Point Road, Groton, CT 06340, USA
e-mail: george.l.reid@pfizer.com

B. Nickerson (ed.), *Sample Preparation of Pharmaceutical Dosage Forms: Challenges and Strategies for Sample Preparation and Extraction*, DOI 10.1007/978-1-4419-9631-2_15,

will be discussed. Additional benefits of these reduced sample preparation techniques typically include rapid sample analysis time, dosage form often left intact for subsequent testing, and green techniques. Two main approaches to reduced sample preparation will be discussed in this chapter, vibrational spectroscopy and mass spectrometry.

Vibrational spectroscopic techniques probe the sample at the molecular level. Different molecular structures and chemical environments can give unique spectral features across the spectrum, thus selectivity is achieved without separation or preparation of the sample matrix. Multivariate data analysis is often performed to extract the qualitative or quantitative information of interest. Mid-infrared (MIR), near-infrared (NIR), and Raman spectroscopy are well-established spectroscopic techniques in the pharmaceutical industry and are described in pharmacopoeias (Ph.Eur. 2005a; Ph.Eur. 2005b; Ph.Eur. 2005c). Terahertz pulsed spectroscopy (TPS) is a more recently introduced technique for solid-state analysis, and some interesting applications have been reported in recent years (Taday and Newnham 2004). These spectroscopic techniques cover different ranges of the electromagnetic spectrum, have complementary activation kinetics, and allow flexible sampling approaches that are appropriate for a wide range of pharmaceutical applications. Spectroscopic methods are also information rich and can often provide a holistic view of the sample (chemical and physical information such as particle size). As a result of reduced sample preparation and analysis speed, vibrational spectroscopic techniques have gained an important place as process analytical technology (PAT) tools and have become important enablers for real-time release testing.

Mass spectrometry (MS) is an analytical technique used for determining the elemental composition of a sample and for elucidating the chemical structures of molecules. MS is one of the core technologies in support of structure characterization, identification, and quantification of drug entities in pharmaceutical analyses. The MS principle consists of ionizing the sample to generate charged molecules or molecular fragments and measurement of their mass-to-charge ratios (m/z). The mass-to-charge ratio of the parent ion and fragments is used to identify and quantitate components in the sample.

The use of MS in the pharmaceutical industry is mainly as a hyphenated technique in combination with chromatography, e.g., liquid chromatography and mass spectrometry (LC-MS). LC-MS offers good selectivity for complex mixture analyses and provides multidimensional structural information about the analytes. However, this technique cannot be used directly on most pharmaceutical dosage forms without appropriate sample preparation. Often the sample preparation of the dosage form is the key determiner in the outcome of the LC-MS analysis. Researchers have worked to develop new technologies to meet this requirement; however, most efforts have focused on acceleration of the separation method instead of sample preparation procedures (Chen and Horvath 1995; Swartz 2005; Wang and Zhang 2007; Wu and Clausen 2007). The recent development of open air ionization techniques has allowed for the routine analysis of materials with minimal or no sample preparation.

15.1.1 *Vibrational Spectroscopy Overview*

15.1.1.1 Mid Infrared (MIR)

MIR spectroscopy frequency is defined from 4,000 to 400 cm^{-1} in the electromagnetic spectrum. MIR radiation probes the intramolecular vibrations of functional groups in the test material. A change in the dipole moment needs to occur for a molecular vibration to be infrared active. Polar groups, such as C-F, Si-O, C=O, and C-O, absorb the infrared energy strongly, and antisymmetric stretches correspond with high intensity in the spectrum. Because of the high absorption intensity, the sample often needs to be diluted (e.g., in KBr for Diffuse Reflectance Infrared Fourier Transform Spectroscopy (DRIFTS)). With an attenuated total reflection (ATR) accessory, powder and liquid phase samples can be analyzed directly without sample preparation. However, the sampling volume is very limited and for finished drug product, some simple sample preparation procedures such as peeling off the film coating, grinding, or opening a capsule are still needed. As a result, MIR is often used for pure material analysis when high specificity is required, and has also gained use and acceptance in the drug substance development and manufacture area.

15.1.1.2 Near Infrared (NIR)

NIR spectroscopy covers the frequency range of 12,000 to 4,000 cm^{-1}. It is one of the most established spectroscopic tools for PAT in the pharmaceutical industry. The peaks observed in NIR spectra are mostly due to the absorption bands of overtones and combinations of the fundamental modes of –CH, –NH, –OH (and –SH) functional groups (Siesler et al. 2002). The intensity of NIR spectrum is on the order of 10^3 weaker than that of MIR spectrum. The weak absorption enables NIR to analyze the sample directly without any dilution or sample preparation. The penetration depth is also significantly deeper in comparison with MIR, which allows for a larger interrogation of the dosage form's volume. Although the feasibility and the performance of NIR methods are formulation and case-dependant, there are many successful applications for low-dose products that have as low as a 1 wt% drug loading. A typical solid oral dosage form 3–4 mm in thickness can often be analyzed in transmission mode. Diffuse reflectance mode is also an option for thicker tablets and powders. Transdermal patches may be analyzed in reflectance mode with the thin release liner on. Lyophiles in glass vials or even thin plastic bottles can be analyzed through the container without impacting the integrity of the package. Liquid samples can be analyzed in transmission mode.

Signal transmission through fiber optic cables is very efficient in NIR region; therefore, processes can be monitored in situ through fiber optic probes. NIR spectroscopy reveals information about the physical properties (particle size, density, morphology, temperature) besides the chemical properties. On one hand, the rich information can provide more understanding of the sample. On the other hand, the physical information can become interference for chemical property analysis. Chemometrics, i.e., multivariate data analysis, is often required to reduce the physical interference and resolve the overlapping bands for quantitative and qualitative analysis.

15.1.1.3 Raman

Raman spectroscopy typically involves the frequency region between 4,000 and 200 cm^{-1}. Raman spectra correspond to the polarizability changes in the sample induced by the incident monochromatic light. It can be considered as a complementary technique to infrared. The radiation is more effectively scattered back from nonpolar groups. Symmetric stretches of the molecules respond to higher intensity values in Raman spectrum. Raman spectroscopy offers sharp spectral features and has good specificity. The physical properties of the sample have minimal impact on the spectrum. Raman spectroscopy is good for solid-state analysis, such as polymorphic analysis, because of its nondestructiveness and high specificity (Lewis and Edwards 2001). Water is Raman inactive; therefore, Raman is advantageous for process monitoring in aqueous system. In recent years, transmission Raman has been developed. Compared to the conventional back scattering Raman spectroscopy, transmission Raman has an increased sampling volume and is very promising for tablets or capsules analysis with a more representative sampling (Johansson et al. 2007).

Raman spectra can be obtained directly on the final drug product such as tablets. Raman bands used for identification and quantification of drug products can often be found in the spectral region 1,750 to 1,500 cm^{-1}. These bands are due to vibrations of aromatic group in the APIs, which are present in the majority of drugs. There are generally no interfering Raman bands in this spectral range from excipients, as most excipients do not contain aromatic functionality (e.g., lactose, cellulose, phosphates, silicon dioxide, magnesium stearate). This selectivity of Raman spectroscopy allows identification and quantification of even low-dosage drug products (Saal 2006).

15.1.1.4 Terahertz

Terahertz radiation lies between the microwave and infrared regions, frequencies between 130 and 2 cm^{-1}. It is a part of the far-infrared region. TPS detects low-frequency molecular flexing and intermolecular vibrations in the solid state. In crystals, phonon modes are also detected and therefore, the information is largely gathered from the intermolecular level (Taday and Newnham 2004) and crystal lattice changes are directly probed (Day et al. 2006). The peaks in TPS spectra cannot easily be assigned, therefore the interpretation of spectra is more difficult compared to MIR spectroscopy. Historically, the terahertz region has been used much less than the NIR region to study the pharmaceutical materials. In recent years, applications regarding TPS have been predominantly aimed toward obtaining the physical information of solid-state properties of APIs, especially of polymorph recognition, characterization, and quantification (Strachan et al. 2006). Terahertz pulse imaging has been applied to characterize different coating structures and coating thicknesses (Fitzgerald et al. 2005). Pharmaceutical packaging materials such as plastic are transparent at terahertz frequencies, therefore, in-package analysis can be conducted.

15.1.1.5 New Sampling Techniques and New Instrumentation

In recent years, advancements have been made on sampling techniques and automation to fully exploit spectroscopy's rapid analysis times and ability to make measurements with no sample preparation. For lab-based instrumentations, automated sample wheels or stages have been developed to reduce resources required for analyzing multiple samples and to aid fast screening tests. Representative sampling for less homogeneous samples is now feasible using transmission Raman (Johansson et al. 2007), and spatial offset spectroscopy (Eliasson and Matousek 2007) can be used to evaluate samples through thicker packages. Online NIR tablet analyzers have been developed for real-time content uniformity analysis right at the production line, and fiber optic probes have enabled remote monitoring of processes. The maturing of Micro-Electro-Mechanical Systems (MEMS) technology has enabled the miniaturization of spectrometers. Handheld NIR (e.g., Polychromix Phazir) and handheld Raman (e.g., Ahura TruDefense) spectrometers are employed in agile material identification, including counterfeit detection in the field. Wireless process NIR spectrometers are widely used for blending and drying monitoring.

15.1.2 Multivariate Data Analysis

Data analysis is a key component of spectroscopic analysis. Vibrational spectroscopic data are multivariate by nature. The data for complex sample matrix are collinear, thus spectroscopy is usually combined with chemometric models to provide either qualitative (e.g., classification, identification) or quantitative information (Martens and Naes 1989). Spectroscopic methods are often considered secondary for quantitative analysis. The method itself is capable of qualitative trending. However, in order to do a quantitative analysis, a calibration model needs to be constructed between the spectra of the calibration standards and their properties of interest tested by primary method(s). For identification, a spectral library followed by classification model is typically needed. There are various algorithms for model construction, among these algorithms, principal component analysis (PCA) and partial least squares (PLS) regression being the most widely used linear methods.

One of the critical elements for a successful and robust calibration model is the choice of a representative calibration sample set. Ideally, the calibration sample set should cover the same variance space that future samples will fall in. Another aspect is spectral preprocessing. Before calibration modeling, spectral data, especially NIR data, often need to be preprocessed to remove or reduce baseline variation and filter interferences to enhance the performance and the robustness of the model. Standard normal variate (SNV) or multiplicative scattering correction (MSC) are often used to correct the scattering effect caused by the sample physical properties when chemical information is of interest. Derivatives are often applied to the spectral data when band overlapping and baseline variations exist.

The development of a robust model usually requires a large number of samples and analysis from the primary method. The quantitative model is generally specific to the formulation and spectrometer used in the development of the calibration model. Therefore, significant efforts in the development, maintenance, and method transfer are typically required for quantitative methods.

15.1.3 Ambient Desorption Ionization Mass Spectrometry Overview

A new mass spectrometry field of ambient desorption ionization mass spectrometry (also known as open air ionization mass spectrometry) has become very active. This is because this technology allows for the direct in situ mass spectral analysis of samples in all the physical phases with minimum or no sample preparation (Arnaud 2007). Numerous open air ionization techniques have been widely applied in pharmaceutics, food science, and forensic science (Ratcliffe et al. 2007; Harper et al. 2008; Venter et al. 2008; Ma et al. 2009; Yu et al. 2009).

Table 15.1 categorizes some recent ambient ionization mass spectrometry techniques based on their desorption and ionization mechanisms (Chen et al. 2009). Scientists can choose appropriate combinations according to the chemical property of the analyte and the available mass spectrometer model. For example, for compounds that lack proton acceptor or donor sites in their structures, atmospheric pressure chemical ionization (APCI) related techniques (DART, DAPCI, DAPPI, PADI, DBDI, etc.) are appropriate. Desorption atmospheric pressure chemical ionization (DAPCI) offers better ionization efficiency than direct analysis in real time (DART) or desorption electrospray ionization (DESI) on compounds of moderate to low polarity (Williams et al. 2006). For analytes that are thermally labile and sensitive to degradation during mass spectral analysis, low-temperature plasma probe (LTP) may be considered (Harper et al. 2008). If dealing with volatile or semivolatile solid or liquid samples, atmospheric solids analysis probe (ASAP) may be the preferred choice (McEwen et al. 2005). For high-throughput analysis and user friendly automation set-up, laser diode thermal desorption (LDTD) is an option (Fayad et al. 2010). Among all these ambient desorption ionization techniques, DESI (Prosolia Inc.), ASAP (Waters Crop.), LDTD (Phytronix Inc.), and DART (IonSense Inc.) are well commercialized and have been widely used in drug discovery, counterfeit medicine screening, and other pharmaceutical research and development areas (Arnaud 2007). Table 15.2 provides references and abbreviations for recently developed ambient desorption ionization mass spectrometry techniques.

Table 15.1 Classification of ambient ionization methods according to the dominant desorption/volatilization and ionization mechanisms

	Dominant ionization/postionization method				
Dominant desorption/ volatilization method	Direct	ESI Spray	Chemical post-ionization	Plasma, Penning, electrons	Laser or lamp post-ionization
(Gas-Phase introduction)		SESI		FAPA	
		EESI			
(Aerosol introduction)		EESI			
		SESI			
Liquid spray/nebulization		EESI			
Momentum transfer (liquid or gas jet)		ND-EESI, JEDI			DAPPI
ESI (including sonic spray)	DESI, EASI				
	DeSSI				
Laser desorption		LAESI, ELDI, (IR-)LADESI	LD/APCI	LA-FAPA	
Energetic particles	DART			PADI, FAPA (?)	
Plasma	LTP		DART, DAPCI, DBDI, LTP, PADI, DAPI, APGDDI	PADI, DART (?)	
Thermal desorption		APTDI	DART, ASAP, TD-APCI, LDTD, DAPCI	DART, FAPA, APGD	DAPPI
Acoustic desorption		UA-EESI, RADIO	LIAD		

Boxes shaded in *gray* indicate clusters of acronyms that could be replaced by a single acronym

Reprinted from Chen et al. (2009), with permission from Elsevier

Table 15.2 Recently developed ambient desorption ionization mass spectrometry techniques

Technique full name	Acronym
Desorption electrospray ionization	DESI (Takats et al. 2004)
Electrospray laser desorption ionization	ELDI (Shiea et al. 2005)
Surface sampling probe	SSP (Ford and Van Berkel 2004)
Fused droplet electrospray ionization	FD-ESI (Shieh et al. 2005)
Desorption sonic spray ionization	DeSSI (Haddad et al. 2006)
Easy ambient sonic spray ionization	EASI (Haddad et al. 2008)
Extractive electrospray ionization	EESI (Chen et al. 2006b)
Secondary electrospray ionization	SESI (Wu et al. 2000)
MALDI-assisted electrospray ionization	MALDESI (Sampson et al. 2006)
Neutral desorption extractive electrospray ionization	ND-EESI (Chen et al. 2007)
Laser ablation electrospray ionization	LAESI (Nemes and Vertes 2007)
Infrared laser-assisted desorption electrospray ionization	(IR)-LADESI (Rezenom et al. 2008)
Direct analysis in real time	DART (Cody et al. 2005)
Ambient solid analysis probe	ASAP (McEwen et al. 2005)
Desorption atmospheric pressure chemical ionization	DAPCI (Takats et al. 2005a)
Plasma-assisted desorption ionization	PADI (Ratcliffe et al. 2007)
Low-temperature plasma probe	LTP (Harper et al. 2008)
Dielectric discharge barrier ionization	DBDI (Na et al. 2007)
Helium atmospheric pressure glow discharge ionization	HAPGDI (Andrade et al. 2006)
Atmospheric pressure thermal desorption ionization	APTDI (Chen et al. 2006a)
Desorption atmospheric pressure photo ionization	DAPPI (Haapala et al. 2007)
Laser-induced acoustic desorption	LIAD (Cheng et al. 2009)
Flowing atmospheric pressure afterglow	FAFA (Schilling et al. 2010)
Laser ablation flowing atmospheric pressure afterglow	LA-FAPA (Shelley et al. 2008)
Laser diode thermal desorption	LDTD (Fayad et al. 2010)

15.1.3.1 Desorption Electrospray Ionization (DESI) Mass Spectrometry

DESI mass spectrometry was introduced by Graham Cook's group (Takats et al. 2004).The DESI process can be split into two steps, liquid–solid extraction and gas phase transition/ionization. The sample is placed on a planar surface, which has good interaction with the analyte, followed by the bombarding with a stream of nebulized charged aqueous droplets. The resulting desorbed ions are introduced into the mass spectrometer analyzer for mass spectral analysis (Takats et al. 2004).

DESI has the features of electrospray ionization (ESI) and generates a spectrum resembling ESI. In addition to no sample preparation, another advantage of DESI includes generation of multiply charged ions. This feature enables scientists to analyze biological macromolecules, such as proteins and polypeptides (Shin et al. 2007). Another advantage of DESI is the wide variety of solvent applicability (including those commonly used reversed phase HPLC mobile phase) allowing great flexibility during methods development to optimize the ionization signal, such as the proton affinity of the solvent, acidity of the solvent, addition of complexation, or chemical reagents (Kauppila et al. 2006). As such, DESI can be used to analyze broad types of compounds.

Even though there is nearly no sample preparation requirement, matrix effects are the major concern for DESI as this technique is not entirely immune to matrix interference, such as alkali metal adducts formation that may result in analyte signal suppression. To achieve the best sensitivity and reproducibility, it is necessary to optimize the planar surface–analyte interaction, the geometry alignment among spray emitter tip, sample surface and mass spectrometer inlet orifice, DESI jet plume impact region size, sample spot size, physical distance between adjacent spots, sample surface concentration, composition of spray solvent, spray solvent flow rate, nebulization gas flow rate, surface scan rate, and spray probe moving speed (Takats et al. 2005b). The optimization process is necessary for the analysis of low-level analytes or the analyte in complex matrices. This procedure may be time consuming and cannot be easily automated.

15.1.3.2 Direct Analysis in Real-Time (DART) Mass Spectrometry

DART technique was invented by Robert Cody from Jeol USA, Inc. and James Laramee from EAI Corp (Cody et al. 2005). This technique is APCI related, and the analyte is placed in the space between the DART gas flow outlet and the sample cone orifice of a mass spectrometer. A stream of heated helium or nitrogen gas flux containing metastable species that are generated by electrical discharge passes through the DART gas chamber and desorbs the analyte from the solid surface. After ionization (e.g., penning ionization, energy transfer, proton transfer for positive ions, or electron capture for negative ions), the ions drift to the mass spectrometer via vacuum and electrical potential differences (Cody et al. 2005).

Compared with DESI, in which the analytes are usually placed on a planar surface, samples can be introduced into DART in more flexible ways. For example, in addition to applying the sample on the planar surface, such as glass, TLC plate, etc., a solid powder sample can be sampled with a glass rod or a liquid sample can be sampled by dipping the glass rod into the sample solution, and then manually positioning the glass rod in front of the DART gas flow exit. Sample position is critical and it determines the quality of the resulting mass spectrum. Unlike DESI, the optimization procedure in DART is much simpler, which makes this technique suitable for routine operation. Another advantage of DART is that this technique has high tolerance for interference resulting from high concentration of alkali metal cations, enabling DART to analyze the compounds in matrices containing mass spectrometer-incompatible materials without sample pretreatment.

DART has been used to analyze a wide range of drug compounds in different dosage forms. However, it cannot generate multiply charged ions because of its APCI or APPI like ionization mechanism and is not used for biological macromolecule analysis.

15.2 Identification Testing

Vibrational spectroscopic techniques are widely used for identification of incoming raw materials such as excipients and APIs along with finished drug products. The identification typically involves library construction of the materials, optimization of the data preprocessing, and multivariate classification algorithms.

The testing of raw materials by spectroscopy has moved from the laboratory with the introduction of handheld spectrometers. These handheld Raman (e.g., Polychromix Phazir) and NIR (e.g., Ahura TruDefense) analyzers are targeted for fast material identification, including counterfeit detection. The equipment comes with common pharmaceutical raw material libraries and built-in classification algorithms. The ID test of the raw materials can be done within seconds in the drum at the loading dock or just prior to charging on the manufacturing floor.

MS is also used for identification purposes, via ambient desorption ionization, solubilization, and infusion and via hyphenation with chromatography. Ambient desorption ionization has been reported for applications such as identification of the API in authentic dosage forms and as a rapid counterfeit detection method.

15.2.1 Raw Material Identification

MIR is a well-established and accepted tool for raw materials identification. The finger print region offers great specificity. The identity test is often conducted through spectra matching against the spectral library. Extensive spectral libraries are commercially available for common raw materials in many industries (Wang 2008). Because of the strong absorption in the MIR region, the interrogated sample volume is very small. It is useful for precious samples or for forensic analyses where only limited sample is available. However, for testing a large sample, obtaining a representative sample or subsampling needs to be considered.

Absorptions in the NIR region are much less intense than in the MIR region, which allows for deeper penetration of the source light into the material. This means a larger amount of material is investigated so the sampling error is reduced. Also, the glass vials and plastic films used as containers or packaging materials produce less spectroscopic interference. Therefore, the sampling can be done without removing the material from the original container or exposing the operator to the material. Besides chemical information, NIR spectra also contain some physical information of the sample. For powder materials, the same compound with different particle size distributions and pack densities gives different scattering effects, which appear as a baseline shift along the spectrum. This physical information can be useful for grade identification. For example, hydroxypropylcellulose (HPC) EXF and EF are two common grades used for tablet formulation, and the main difference is particle size. HPC EXF has much finer particle size than HPC EF, and this difference can be reliably distinguished by NIR spectra. Particle size distribution of the excipients can impact the process ability and the final tablet attributes.

The combination and overtone bands in NIR region are broad and overlapping. Therefore, the specificity is weaker than MIR or Raman. It is often used as identification confirmation or a conformity test, but is not recommended for unknown sample structure characterization by itself. Accordingly, chemometrics is needed for reliable identification by NIR. Data preprocessing is often used to deconvolute the overlapping band and to remove or reduce the sample physical information when needed. Distance-based algorithms are often used to compare the incoming sample against the known excipient groups in the preestablished spectral library (Blanco et al. 1998; Forbess and Shukla 1998).

Raman is fundamentally complementary to MIR and NIR. Polar molecular bonds with dipole moment changes when vibrating are MIR and NIR active. By contrast, the molecular bonds associated with Raman scattering are nonpolar. Therefore, it is common for Raman to provide information about the carbon–carbon bonds along the backbone of organic raw materials (Al-Zoubi et al. 2002). The spectrum generated by a Raman instrument generally has sharper and better resolved peaks than NIR, which can provide more chemical information of unknown samples. The Raman excitation laser can also penetrate optically clear materials such as glass or plastic, which also limits the spectroscopic interference from the sample containers.

15.2.2 Finished Dosage Form Identification

15.2.2.1 Spectroscopy

Spectroscopic techniques are quite capable of nondestructive drug product identification and counterfeit drug detection. Many dosage form types can be facilely tested, and include solid dosage forms (tablets, capsules) and parenteral dosage forms (oral solutions, lyophiles and transdermal patches). NIR and Raman are the most popular techniques because of the versatile sampling capability and the spectral selectivity of the active ingredients in the full formulation matrix. Validation guidelines exist from different regulatory authorities, and spectroscopic methods for identification of commercial dosage forms have been approved.

NIR can penetrate through common film coatings and obtain tablet core composition information. A model has been reported that was able to correctly identify clinical study tablets with an API content of 1, 2, and 3% (w/w) using NIR transmittance spectra (De Maesschalck and Van den Kerkhof 2005). They used a modified supervised classification approach based on PLS beta classification (PLSBC) as the identification algorithm. This new classification approach is also able to quantify the probability of misclassification, giving a measure of robustness. A validated reflectance NIR method for a 100% capsule identity check on the packaging line was proposed (Herkert et al. 2001).

A systematic and rational development and validation strategy is also very important for a robust method. Typically, the dosage form identity test focuses on an API spectral feature, rather than the excipients to prove the identity of the finished product. A validation strategy for film-coated thiamazole tablets following ICH guidelines has been reported (Alvarenga et al. 2008). The method presented was successfully validated and was approved by European regulatory authorities. An approach for constructing more robust classification models with few batches available by using the same method together with the principle of data augmentation using both reflectance and transmittance modes was suggested (Van den Kerkhof et al. 2006).

The same classification development principle applies to capsules and other dosage forms. As an example (Liu et al. 2008), NIR was used for active identification for an oral solution by collecting the transmission NIR spectrum of the solution through a glass vial. The active band intensity was sufficiently strong for the given concentration range. A spectral distance-based algorithm was used for the classification. The method was able to correctly identify the active product and distinguish closely related products at the manufacturing site for release testing.

Identification in the package by conventional NIR or Raman spectroscopy is possible for thin plastic blister package or some less dense plastic bottles and glass vials. Clinical trial tablets assessed directly through the blister packaging by NIR reflectance mode have been presented (Aldridge et al. 1994). However, in many cases, especially with dark-colored coatings or capsules, or thicker and opaque packaging, Raman or NIR can be completely blocked or the signal of API can be heavily polluted with the signals from the coating, capsule, blister pack, or plastic bottle itself. For these types of samples, spatial offset spectroscopy (SOS) can be a potential solution without sample preparation. Rather than detecting the signal directly at the point of laser illumination, SOS detects the signal at a distance from the incident position thus reducing the contribution of the surface layers in the spectrum and much clearer spectral features of sublayers are observed. Spatially offset Raman spectroscopy (SORS) was applied to the noninvasive identification of APIs in various blister-packed and bottled pills (Eliasson and Matousek 2007). They compared the conventional back scattering Raman and SORS spectra of commercial, packaged formulations that included the pain relievers ibuprofen or paracetamol with reference Raman spectra of the interior surfaces of broken tablets. The spectra were contaminated with Raman signals from the packaging materials; however, the researchers found the contaminating spectra to be less pronounced with SORS analysis.

15.2.2.2 Mass Spectrometry

The ability of ambient desorption ionization mass spectrometer to perform in situ analysis of sample in all physical phases without any pretreatment enables this technique to rapidly identify APIs, excipients, impurities, or degradants in various drug products. DESI has been utilized to analyze the single active ingredient in folic acid, acetaminophen, and aspirin tablets (Chen et al. 2005). Plasma-assisted desorption ionization (PADI) has been used for the direct identification of single API in generic Mefenamic, Ibuprofen, and Aspirin tablets (Ratcliffe et al. 2007). Similar studies were performed in the author's lab using Cephradine, Cimetidine, and Ofloxacin capsules and DART ionization source. In the mass spectra of Cimetidine and Ofloxacin capsules, the protonated molecular ions are the predominant species (Fig. 15.1, top and middle panels). The ion at m/z 318 is confirmed to be the degradant of Ofloxacin as shown in middle panel of Fig. 15.1. The molecular ion of Cephradine was not detected in the spectrum, even with additional experimental studies. The predominant peak at m/z 306 is the degradation product of Cephradine resulting from the loss of carboxylic acid (Fig. 15.1, bottom panel).

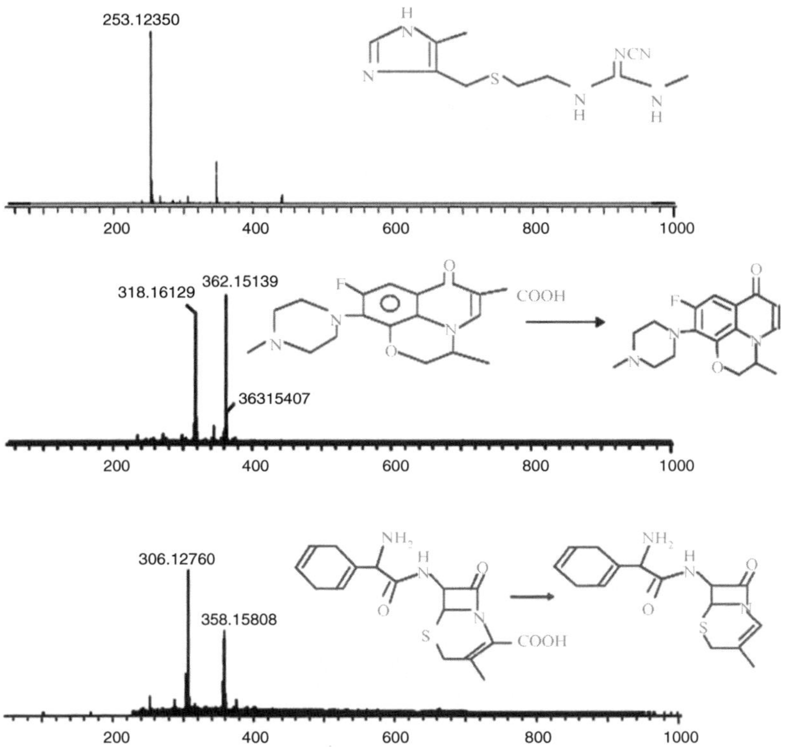

Fig. 15.1 Positive DART mass spectrum of single active ingredient in Cimetidine (*top panel*), Ofloxacin (*middle panel*), and Cephradine capsule (*bottom panel*)

Ambient desorption ionization mass spectrometry has been applied to identify multiple active ingredient components in tablets. Using DESI, all three APIs were identified (acetaminophen, acetylsalicylic acid, and caffeine) in Excedrin tablets (Chen et al. 2005). Additionally, CID MS/MS fragmentation was combined with DESI to detect almost all the vitamin components in Centrum tablets, including lycopene and lutein, present at a level of 0.01% (w/w) in the tablet. DAPCI, DESI, and DART were used to identify the multiple APIs in Anadin Extra tablets containing caffeine, paracetamol, aspirin and Solpadeine Max tablets containing paracetamol and codeine phosphate (Williams et al. 2006). The author has used DART to identify four APIs in an imported cold tablet. The tablet composition was listed as containing paracetamol, amantadine, caffeine, and chlorpheniramine. Positive DART mass spectrum showed the $[M+H]^+$ ions of caffeine (m/z 195), chlorpheniramine (m/z 275), and other peaks (Fig. 15.2). An enlargement of the spectrum around at m/z 152 indicates the presence of two isobaric ions, paracetamol (m/z 152.07), and the free base of amantadine (m/z 152.14) (confirmed by a high-resolution time of flight (TOF) mass spectrometer).

Ambient desorption ionization mass spectrometry has been applied to identify components in liquid formulation. Multiple APIs (acetaminophen, doxylamine

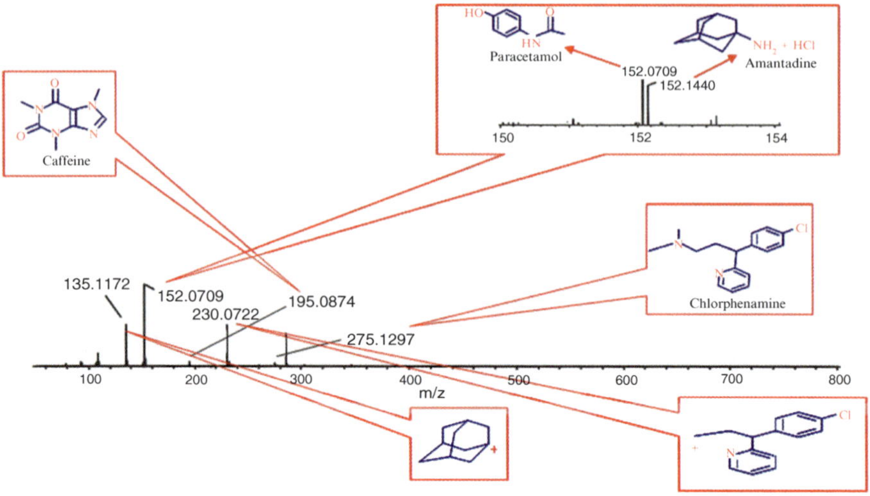

Fig. 15.2 The main APIs in anticold tablet include paracetamol (66%), amantadine (27%), caffeine (4%), and chlorpheniramine (0.5%)

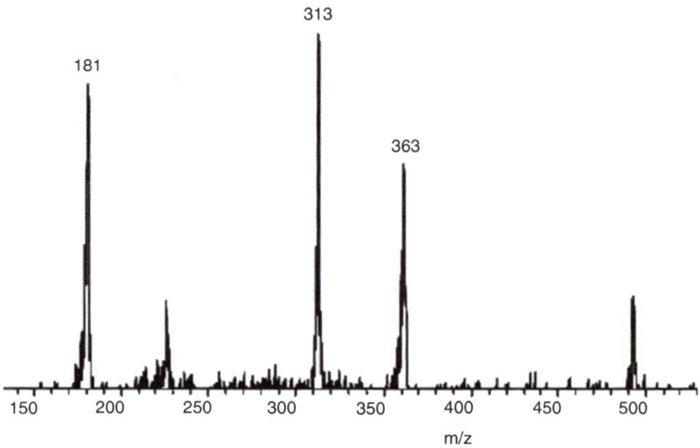

Fig. 15.3 Negative DART mass spectrum of Valdecoxib formulation

succinate, and dextromethorphan) were identified in liquid cold medication with IR LADESI (Rezenom et al. 2008). The author used DART to perform the in situ detection of Valdecoxib in a povidone/mannitol formulation. Figure 15.3 shows the DART mass spectrum that gives a negative singly charged ion $[M-H]^-$ at m/z 313, which is consistent with the deprotonated ion of Valdecoxib. Additionally, mannitol (m/z 181) and its dimer (m/z 363) were detected (confirmed by CID MS/MS fragmentation). Ibuprofen in a gel was detected with DESI within seconds (Williams et al. 2006). Clotrimazole and ketoconazole creams were analyzed by DESI, and the APIs in both ointment samples were identified by molecular mass and CID MS/MS fragmentation data (Chen et al. 2005). Hydrocortisone cream was evaluated by IR LADESI and the result demonstrated that hydrocortisone, which has been observed to have poor mass spectral response in DESI, gave a predominant peak at m/z 363 that is consistent with the protonated hydrocortisone (Rezenom et al. 2008). DESI analysis combined with ion mobility (IMS) separation was used to detect nicotine ($[M+H]^+$, m/z 163) in a slow-release nicotine skin patch (Weston et al. 2005).

The use of mass spectrometry to identify and characterize dosage forms containing low levels of API can be challenging due to the level of the API and matrix effects. The mass spectrum can be very complicated and the signal of the analyte of interest may be suppressed due to the interference from other excipients in the drug formulation, such as alkali salts, some surfactant, polyethylene glycol (PEG), etc. In Rezenom's IR LADESI work on liquid-formulated cold medicine, the signals of acetaminophen, doxylamine succinate, and dextromethorphan were to some extent suppressed by cluster peaks of PEG (Rezenom et al. 2008). A similar observation was obtained in Weston's investigation on nicotine skin patch and chlorhexidine gluconate cream. In the combined full scan mass spectrum of nicotine patch, the relative abundance of positive nicotine peak (m/z 163) was only about 5% due to the interference from other patch excipients (Weston et al. 2005).

The limit of detection and the selectivity of the analyte are critical concerns when analyzing dosage forms containing low levels of APIs in complex matrices using ambient desertion ionization mass spectrometry. One way to improve the sensitivity of a low-level analyte is to use selected ion monitoring (SIM), a mass spectral scanning mode, in which only the ions of the analyte of interest are recorded and all the other ions are filtered out. An immediate release (IR) tablet containing approximately 0.1% API ($[M-H]^-$, m/z 483) was analyzed by DART in the author's lab. The film-coated tablet was cut into half and the inside portion was exposed to the DART. The API was almost absent in the combined full scan spectrum, which shows a lot of peaks representing the excipients in that tablet (Fig. 15.4, bottom panel and insert). When using SIM, the deprotonated molecular ion of the API was detected with a good signal noise ratio (Fig. 15.4, top panel and insert). The well-resolved chlorine isotope peak (m/z 485) confirms the identity of this compound.

Multiple reaction monitoring (MRM) or selected reaction monitoring (SRM) can be used to increase selectivity and enhance detection. In MRM, instead of the selected ion, it is the fragmentation that is monitored. DART in combination with MRM has been used in quantitation of trace level samples in biological matrices. Without any sample clean-up manipulation, verapamil at 0.1 ng/ml in rat plasma was detected with a good signal noise ratio (Yu et al. 2009). The best limit of detection (LOD) of the low-level analyte depends on the optimized conditions of ambient desorption ionization source, the model of the chosen mass spectrometer instrument, the composition of the drug formulation, and the chemical structure of the analyte of interest. The sensitivity achieved using various ambient desorption ionization techniques including DESI, DART, ELDI, SSP, DBDI, LAESI, and DAPPI has been studied (Venter et al. 2008). The sensitivity of DESI MRM directly applied on solid sample is comparable to that achieved using LC-MS full scan on the same sample prepared in dissolving solvent (Garcia-Reyes et al. 2009).

Another approach to improve the selectivity of the analyte without sacrificing the time and small sample consumption advantages of ambient desorption ionization technique is to add an additional dimension of rapid separation other than LC or

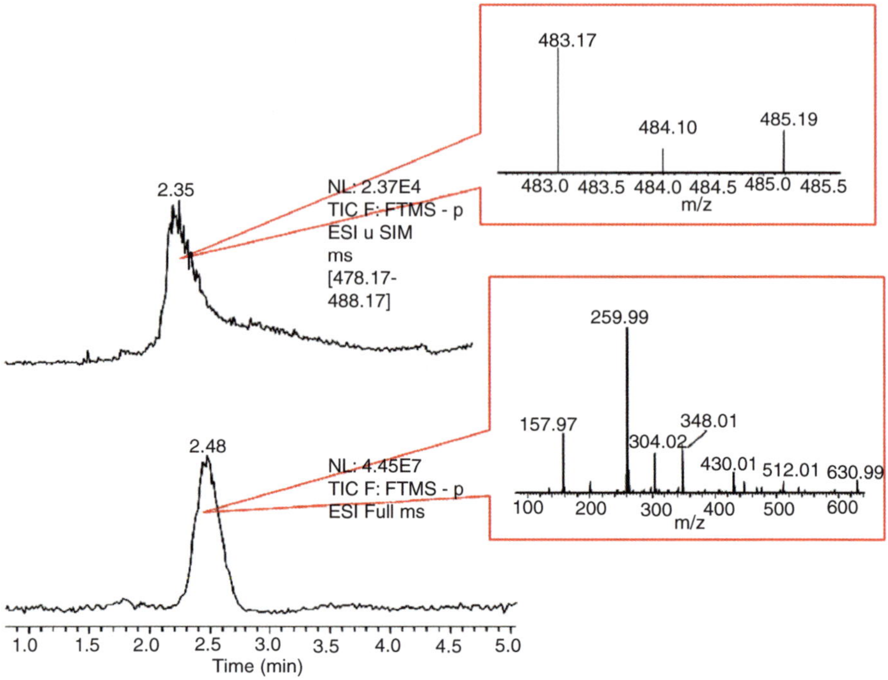

Fig. 15.4 Negative DART selected ion monitoring (SIM) chromatogram (*top panel*) and the total ion current (TIC) chromatogram (*bottom panel*) of noncommercial IR tablet. The inserts are the corresponding mass spectra

GC. In Weston's works on nicotine skin patch and chlorhexidine cream, ion mobility spectrometry (IMS), which separates compounds based on their different crossing sections in the IMS drifting tube determined by their charge state, molecular size, and shape, was hyphenated into DESI mass spectrometer. As a result, the API ions (nicotine and chlorhexidine, respectively) generated in the DESI source were separated from the other excipients, showing overwhelming mass spectral response and thus gave much better signal noise ratio in the spectrum (Weston et al. 2005).

15.2.3 Counterfeit Drug Identification

Counterfeit drugs are defined by the World Health Organization as "a medicine, which is deliberately and fraudulently mislabeled with respect to identity and/or source. Counterfeiting can apply to both branded and generic products and counterfeit products may include products with the correct ingredients or with the wrong ingredients, without active ingredients, with insufficient active ingredients or with fake packaging." (Forzley 2005)

During the last few years, counterfeiters have become increasingly sophisticated in making dosage forms and packaging look authentic. Counterfeit drugs include those that do not contain the labeled active ingredients, those that contain an active ingredient that is not declared, and others that are subpotent or adulterated with structurally similar compounds to the drug. Spectroscopy and MS are widely applied to counterfeit detection.

The use of different excipients, API concentrations, and/or polymorphic forms in counterfeit drugs can present different spectral features in comparison with the genuine drug. Both NIR and Raman have been used as effective counterfeit detection tools. Raman has better specificity, especially for polymorph detection and has more potential on the complex counterfeit detection due to sharper and more resolved bands. Raman spectroscopy is proposed as a fast and reliable method for the detection of counterfeit Viagra® tablets (De Veij et al. 2008a) and other erectile dysfunction tablets (De Veij et al. 2008b). With multivariate analysis, a spectroscopy method can be automated for counterfeit screening by customs or used in the field to identify counterfeit tablets without the involvement of specialists.

In addition to the API identity and concentration, homogeneity of the drug product is also a key indicator used to identify counterfeits. The use of chemical imaging can provide chemical information and also the spatial distribution of the components; therefore, it can be used to detect counterfeits. NIR chemical imaging has been successfully applied to the identification and characterization of counterfeit drug products (Wolff et al. 2008). In order to better characterize the NIR imaging data, a four-stage concept and a new image linear signature for counterfeit drug identification were developed (Pucherta et al. 2010).

Ambient ionization mass spectrometry (DESI and DART) has been applied for screening of counterfeit malaria tablets without any additional sample manipulation (Fernandez et al. 2006). The absence of the drug Artesunate ($[M-H]^-$, m/z 383) in counterfeit tablets was determined within seconds. ASAP was used to study the adulteration of some herbal dietary supplements for the treatment of erectile dysfunction (Twohig et al. 2010). The results indicated that samples were adulterated with commercially available phosphodiesterase type-5 (PDE-5) drugs including Tadalafil, Sildenafil, and a combination of the two. Additionally, a sample gave a predominant singly charged ion peak at m/z 505. Based on the results from the following elemental composition determination, collision induced dissociation (CID) MS/MS fragmentation experiments, and similar PDE-5 analogues studies performed by other research groups (Venhuis et al. 2008; Zou et al. 2008; Reepmeyer and d'Avignon 2009; Singh et al. 2009), this unknown compound was proposed as PDE-5 analogue, Thiohomosildenafil. DART was used to determine adulterated materials containing structural analogues of the active along with subpotent levels of the active. DART mass spectrometry on malarial drugs successfully detected the counterfeit drugs spiked with Artemisinin, a precursor of API- Artesunate, Chloramphenicol, and Metronidazole (Fernandez et al. 2006). The counterfeit drug contained around 20% of expected API, and acetaminophen was also detected in that study.

15.3 Polymorph Analysis

Polymorphism can affect the quality, safety, and efficacy of a drug product. Properties such as the solubility, release profile, and bioavailability of the API and the stability of the formulated product under storage are directly related to the physiochemical properties of the polymorph present. Polymorphs may have different physical properties that cause flowability or compatibility differences, which can influence the drug product. As a result, crystalline forms and all solid-state manifestations of the API, including amorphous forms, solvates, and hydrates fall under regulatory scrutiny. It is important to monitor the polymorphic form of the drug during the formulation development and manufacturing to account for any impact of particle processing including crystallization, granulation, milling, and compaction, and while on stability [Hilfiker 2006].

Several techniques are widely used for polymorph analysis. The most definitive one is single crystal X-ray diffraction; however, this is not applicable to the polymorph determination of formulated drug products. X-ray powder diffraction and other methods that include solid-state NMR, microscopy, differential scanning calorimetry, thermal gravimetric analysis, and vibrational spectroscopy are also widely used. Compared to other commonly used solid-state techniques, vibrational spectroscopy techniques such as Raman, MIR and NIR are rapid and sensitive. The flexible sampling formats of vibrational spectroscopy techniques can accommodate sample sizes that range from single crystals to bulk material. Multiwell plates or autosamplers are particularly useful for high-throughput screening. Microscope slides and glass vials can also be used. Finished dosage drug products such as tablets can be directly analyzed. With imaging techniques, the analysis can be conducted on a single crystal or a microscopic area in the sample. Remote sampling using fiberoptic probes for Raman and NIR allows in situ monitoring of the processing-induced polymorphic transformations. In recent years, TPS has also been developed for polymorph detection.

15.3.1 Raman Applications

Raman spectroscopy has excellent selectivity for polymorphism. Many analytes of pharmaceutical interest contain aromatic functionalities, which typically possess a very large Raman-scattering cross section in comparison to aliphatic molecules. The long range order crystalline materials typically demonstrate sharp spectral features (often with the spectra subtly offset for different polymorphs) with high intensity compared to the amorphous materials. Therefore, Raman is suitable for the differentiation of different polymorphic and amorphous forms (Anderton 2004).

Raman imaging has been utilized for detection of polymorphs microscopically allowing enhanced insight of polymorphic distribution in dosage forms. Raman imaging to detect undesired polymorphs in a very low-dosage tablet has been reported. The results demonstrated that Raman imaging was able to detect 5%

undesired polymorph in the 1% drug loading control tablets. Therefore, the method can be used for stability sample screening during dosage form development or on formal stability studies (Sasic et al. 2009).

15.3.2 MIR Applications

Similar to Raman spectroscopy, the high specificity of MIR allows for small and sometimes subtle spectral differences to be detected and has been used for polymorph identification or quantitation (Bertacche et al. 2006; Hakkinen et al. 2005). MIR does require some sample preparation such as KBr disks, DRIFTS (Pollanen et al. 2005), or the use of ATR accessory (Helmy et al. 2003). The KBr disks or DRIFTS methods require the dilution of the sample. The ATR accessory can analyze the pure materials in their native state (Salari and Young 1998), but for finished dosage forms such as tablets, grinding the tablets is still required and the subsampling can be a more pronounced concern for low-dosage drug products. The direct analysis of polymorphic form in a drug product such as in tablets cannot be achieved. It is also important to note that during the preparation of the KBr disks or using the ATR, the sample can be subjected to significant pressure with the danger of polymorphic conversion. Koradia et al. have used a KBr disc method for quantification of clopidogrel bisulphate polymorphs, and it is recognized that KBr disc compression could lead to conversion (Koradia et al. 2004).

15.3.3 NIR Applications

Differences in molecular networks (crystal lattices) result in shifts in the fundamental bands in the MIR region. These shifts, especially the ones due to the hydrogen-bonded networks, are also visible in the overtone and combination bands in the NIR region. Although the NIR features are broad compared to Raman and MIR, the classification and even quantitation of different forms are often feasible by using data preprocessing such as derivative and multivariate classification/regression analysis.

NIR for polymorph detection of an API has been reported (Aldridge et al. 1996). The NIR spectra showed differences among the desired form and four additional forms including a solvate, a hydrate, a free acid form, and an undesired form. Because of the flexible sampling ability, NIR is often used for in-process polymorph conversation monitoring or off-line process understanding for unit operations such as milling, wet granulation (Li et al. 2005), and drying (Davis et al. 2004). Off-line NIR was used to study sulfathiazole polymorph conversation for the milling process (Aaltonen et al. 2003). The difference of the amorphous and crystalline form can be clearly differentiated by the NIR spectra. They found that the milling process increased the amount of amorphous sulfathiazole. The crystallinity decrease can be followed in the scores plot by conducting a PCA analysis of the second derivatized

NIR spectra. The scores of the samples with the highest milling speeds and the longest milling times moved close to each other as the relative amount of the amorphous component grew.

NIR spectroscopy is very sensitive to the hydration state of the compound (Higgins et al. 2003). It can easily distinguish hydrates from anhydrous forms. NIR has been used to study the transformation of anhydrous theophylline to theophylline monohydrate during wet granulation (Rantanen et al. 2001). Their results showed that NIR spectroscopy was able to detect different states of water molecules during the wet granulation process. In the higher water content stages of granulation, the increasing absorption maxima occurred at 1,410 and 1,905 nm due to OH vibrations of free water molecules. Jorgensen et al. (2002) compared NIR and Raman in terms of hydrate form study for theophylline and caffeine during the wet granulation. Although Raman is able to provide molecule structural change and therefore provide hydrate form information, NIR is more suitable for water–solid interaction study. With a multivariate calibration model, a quantitative method can be established for the quantification of hydrate and anhydrate forms in the powder mixture (Rantanen et al. 2005).

15.3.4 TPS

TPS has been used to distinguish amorphous, crystalline, hydrate, solvate, and liquid crystalline solid states in a number of drug molecules. It is now commonly accepted that terahertz spectra provide sufficient information to distinguish subtle differences in condensed matter properties. As terahertz spectra are reflecting intermolecular vibrations, TPS is an excellent tool for differentiating amorphous systems from their respective crystalline counterparts. Because of diminished long-range order in the amorphous state, the amorphous materials show smooth, featureless, and increasing terahertz absorption spectra. The capability of TPS was demonstrated for polymorph recognition in drug product using tablets of ranitidine hydrochloride (Zantac and apo-ranitidine) (Taday et al. 2003). Pronounced differences were found in the terahertz transmission spectra of the different tablet samples, and it was clearly shown that Zantac contained polymorph II while apo-ranitidine consisted of polymorph I.

TPS is also effective for solid-state quantification. The clear differences generated between different crystalline forms in the terahertz spectra can be exploited for quantitative analysis. A study was conducted on binary mixtures of a number of different forms of polymorphic drug materials (carbamazepine and enalapril maleate), crystalline in amorphous form (indomethacin), and crystalline in liquid crystalline form (fenoprofen calcium), and was quantified in combination with multivariate analysis using PLS algorithm (Strachan et al. 2005). A limit of detection of one polymorphic form in the other as low as 1.80% and crystallinity limit of detection as low as 1.05% were observed.

15.4 Water Content Analysis

NIR spectroscopy is extremely sensitive to the state of water and the water content. The combination of O-H stretching and bending occurs at 1,900–1,950 nm, and the first overtone and second overtone of O-H stretching are at 1,400–1,450 nm and 900 nm respectively. NIR is widely used for rapid water content quantitation, as well as identification and quantitation of the hydrate form of drug compounds forming both single and multiple hydration states (see Sect. 13.3.3). The nondestructiveness of NIR allows the kinetics of water uptake/loss or hydrate form conversions to be measured for the pure materials or directly in formulated products.

NIR for water content measurements has been applied to both active pharmaceutical ingredients (Liu et al. 2009), excipients (Beyer and Steffens 2003) and finished dosage drug products including solid oral dosage forms (Burns and Ciurczak 2008), lyophiles (Kamat et al. 1989; Cao et al. 2006; Zheng et al. 2008), etc. Compared to conventional moisture methods such as loss-of-drying or Karl Fischer titrimetry, NIR has the advantage of speed, nondestructiveness, and no/little sample preparation, and the disadvantage of being a nonprimary method, requiring a reference method to develop a calibration curve for quantitative measurements. A typical measurement takes under a minute and samples can be scanned in the original bottles, packages, or with a simple sample transfer. Consequently NIR results typically have superior repeatability and less variability. In terms of method sensitivity, the water content can go down to 0.1% or lower with a NIR calibration model.

The quantitation of water content by NIR is achieved by establishing a multivariate calibration model of the correlation between the NIR spectra and the reference method (typically Karl Fischer titrimetry). In order to construct the calibration model, a calibration sample set can be prepared by equilibrating the samples at different humidity conditions or by adding varying amounts of water to span the target moisture range. Figure 15.5 shows the NIR water content calibration for a very low water content API (0.02–0.25%). The materials were conditioned in different RH chambers ranging from 11 to 98% RH. The NIR spectra were collected before conducting the Karl Fischer test on the same sample. A one latent variable PLS model was constructed, and the method was tested by additional stability samples. The predictions have shown good correlation with KF results with a RMSEP of 0.009%.

The advantages of NIR are more apparent and significant when analyzing materials that have rapid water sorption/desorption kinetics. Figure 15.6 shows a water uptake kinetic profile for an API obtained by NIR analysis. The sample was exposed to ambient (RH = 31%) conditions. The NIR spectra were collected at different time intervals for 6 h, and a preestablished NIR model was used to predict the %water across the exposure time. Based on the profile, the material started to uptake water rapidly and had a fast uptake rate in the first 20 min. This study showed that sample exposure to ambient conditions before the KF test may impact the water concentration result.

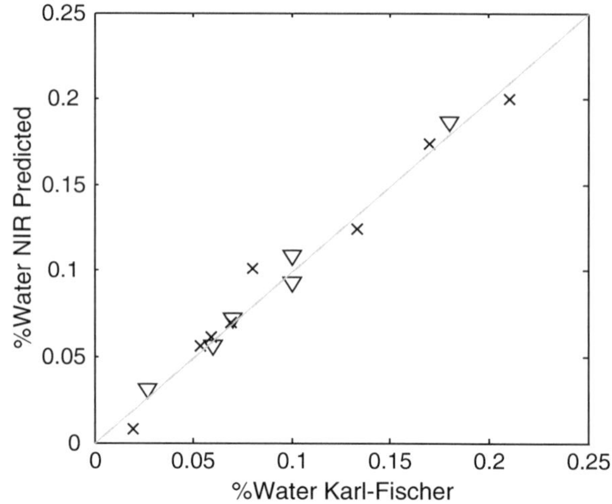

Fig. 15.5 NIR prediction vs. Karl Fischer for an API sample. It is a 1 latent variable PLS model, with a RMSEC=0.010% and a RMSEP=0.009%. *Cross*: Calibration samples. *Triangles*: external validation samples

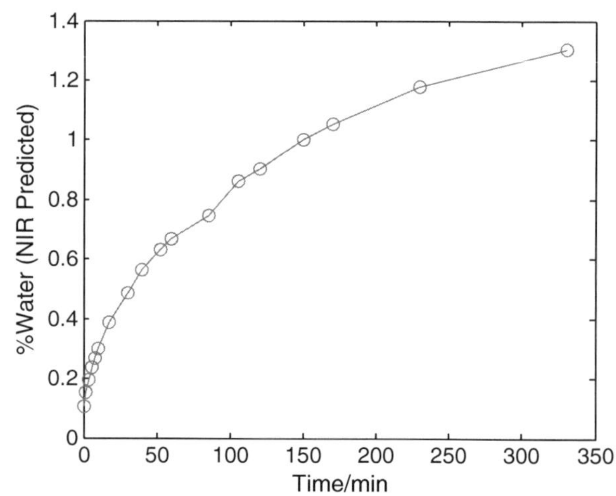

Fig. 15.6 Water uptake profile for an API sample obtained through NIR nondestructive analysis and a preestablished %water PLS model. The compound can be well dried under anhydrous conditions; however, once exposed to the ambient humidity (RH=31%), the material uptakes water rather quickly

Lyophiles are moisture sensitive. Clear borosilicate glass commonly used as the package in ampoules and serum vials is transparent in the near-infrared region. Due to this, the measurement may be performed through the bottom of the vial via a diffuse reflectance mode (Leasure and Gangwer 2002). For biologic lyophile samples, plastics bottles are commonly used. Although the plastic bottles have characteristic absorption, the NIR light may still penetrate through thin clear or less opaque PE or PET bottles.

NIR is capable of water content determination for in-process testing for blends (Gupta et al. 2005) and tablet cores. With a calibrated and validated model, NIR can be used for water content release testing for finished dosage forms such as capsules (Berntsson et al. 1997) and film-coated tablets provided the film coat is not totally reflective.

15.5 Content Uniformity Analysis for Finished Dosage Forms

Potency or content uniformity of the finished dosage form is a key attribute and quality test for process development and product release. In recent years, spectroscopic analysis has become a popular approach for fast content uniformity tests aided by the influence of FDA PAT initiative and real-time release paradigm. NIR is the most mature and commonly used of these spectroscopic techniques.

15.5.1 NIR

A quantitative NIR content uniformity analysis requires a preestablished chemometric model, such as a PLS model, constructed by a calibration sample set with the potency values provided by a reference method (e.g., HPLC). In order to have a robust calibration model, the calibration set needs to have a sufficient variance space to cover the variance during regular product manufacturing. In addition to the concentration in the specification range, the hardness of the tablets, particle size, multiple lots of raw materials, and manufacturing experience at scale are typical common factors included in the model calibration. The design of the calibration set is critical for a robust model, and it is important to fully evaluate the interferences from the formulation matrix.

The limit of quantitation for NIR methods is formulation-dependent. For common immediate release tablet dosage forms (with proper calibration set design and data analysis), NIR can be successfully applied to formulations containing as low as 0.5% w/w API (Alcala et al. 2008). NIR can be used for potency analyses of blends and parenteral dosage form products such as lyophiles (Hirsch 2006).

NIR spectra contain both physical and chemical information about the sample. There are many papers about the impact of tablet hardness on content uniformity models (Saeed et al. 2009). In order to minimize the physical interferences, spectral preprocessing is often applied before the model construction. SNV and second derivative are common algorithms to reduce, but not eliminate, the spectral variation caused by sample physical differences. Based on the understanding of the potential risk and inferences, the calibration set should be designed with consideration of avoiding potential chance correlations to improve the robustness of the model (Xiang et al. 2009).

Depending upon the dosage form, dimension, and composition, transmission or reflectance mode can be used for spectral acquisition. Transmission NIR is typically recommended for tablets and capsules due to more representative sampling. In cases when the NIR light is not able to penetrate through the sample due to the thickness or too strong absorption of the formulation, reflectance NIR becomes a viable option for sample analysis. A comparison study of tablet content uniformity (Schneider and Kovar 2003) has shown that when the sample homogeneity is an issue, transmittance mode offers a smaller prediction error than reflectance mode. However, if the sample is considered homogeneous, both reflectance and transmission can provide sufficiently good content uniformity results.

In addition to the lab-based or at-line bench top instrument, most NIR vendors have developed (or are developing) fully automated online analysis systems for solid dosage forms such as tablets and capsules. An example is Tandem online tablet characterization tool by Bruker Optics. The system consists of a Bruker MATRIX™ near-infrared spectrometer, a Dr. Schleuniger 10X-T tablet testing system, and a tablet handling unit. The equipment can be connected with any tablet press to analyze the samples in near real-time during production. With a preestablished model, content uniformity for stratified cores can be obtained in addition to parameters including weight, size, thickness, hardness, and diameter. Systems like Tandem enable large numbers of samples to be evaluated automatically. Additionally, the analysis information can be used immediately to adjust production parameters to improve product uniformity (feedback control).

15.5.2 Raman

Raman spectroscopy has been evaluated for solid dosage form potency analysis. A major advantage is that Raman has better selectivity and less interference from the physical properties of the sample in comparison with NIR spectroscopy. Therefore, Raman has the potential for simpler calibrations, improved robustness, and shorter method development time. A major disadvantage is potential fluorescence, initially or after storage. Although the Raman effect is very weak and was associated with long integration times and noisy spectra, recent technology developments such as charge-coupled device (CCD) detectors and holographic notch filters have made high quality Raman analysis of tablets in less than a second possible. Besides the conventional back scattering Raman spectroscopy, transmission Raman is evolving for more representative sampling of the tablets or capsules compared with the backscatter mode. A comparative study between backscatter Raman and transmission Raman for quantitative analysis of tablets and capsules was conducted (Johansson et al. 2007). The study results demonstrated that the prediction errors for independent test sets of tablets were found to be 25–30% lower for the transmission mode compared with the backscatter mode. The calibration models tended to be simpler and easier to interpret for the transmission mode, with a lower rank than those for the backscatter mode.

15.5.3 Ambient Desorption Ionization Mass Spectrometry

There are minimal literature references on the use of ambient desorption ionization mass spectrometry for the direct quantitative analysis of pharmaceutical dosage forms. In DESI, signal reproducibility, and hence quantitative reproducibility, is highly dependent on several experimental factors including spray incident angle, collection angle, and spray tip to dosage form distance. Fixing the geometry of the probe can minimize this irreproducibility (Venter and Cooks 2007).

The use of reactive DESI for the direct quantitative analysis of Artesunate in tablets has been reported (Nyadong et al. 2008). A deuterated reference standard was homogeneously dispersed on the tablet surface prior to analysis. DESI variability was reduced by taking the ratio of the analyte to reference standard signals; however, physical attributes of the tablet (e.g., hardness, dosage form geometry) impacted signal and quantitative capabilities of the approach. The accuracy and precision of this application was acceptable for rapid screening, but would not be sufficient for routine dosage form development or manufacturing applications.

15.6 Purity Analysis by Ambient Desorption Ionization Mass Spectrometry

Ambient desorption ionization mass spectrometry has some demonstrated advantages for the direct analysis of complex mixtures, including specificity, speed, and sensitivity. Multiple components in a mixture can be detected in the spectrum as long as they can be efficiently desorbed from the sample holding surface or matrix and be ionized well under the ambient condition. Accordingly, impurities can be potentially detected with ambient desorption ionization mass spectrometry even though no separation is applied. The confidence in impurity identification and characterization can be enhanced if using both positive and negative controls, as well as high-resolution mass spectrometer, which can provide the elemental composition information.

Ambient desorption ionization mass spectrometry may have some limits in trace level impurity identification due to the interference from either matrix effects or the open air environment. SIM or MRM scanning mode can minimize those interferences and improve the LOD, however, in both scanning modes, the identity of the target analyte is known. For unknown impurities, it would be difficult to determine the molecular ion in a complicated spectrum without any structure information. Unlike the traditional mass spectrometries, which operate either in the vacuum (EI, CI, FAB) or in the sealed environment (API), ambient desorption ionization mass spectrometry operates in the open air region. As a result, artifact impurities resulting from the ambient environment can be an issue. Because of the potential variability in the open air region surrounding the ambient desorption ionization source and the ion entrance of mass spectrometer, sometimes the spectrum subtraction of negative/positive control may not always work appropriately. Representative sampling due to homogeneity in the dosage form or other distributions of the analyte of interest in the dosage form can be another concern.

15.7 Conclusions

Vibrational spectroscopy and mass spectrometry can be routinely employed to reduce or eliminate sample preparation during dosage form analysis. Ambient desorption ionization mass spectrometry can be easily applied independent of development stage.

Early in the development process, vibrational spectroscopy can be used for trend analyses or qualitative assessments, whereas later in development with the building of a model, quantitative assessments of API in formulations can be accomplished. These techniques have precedence for obtaining a variety of data including identification, polymorphic form, water content, potency, and purity information. Additional benefits of this approach typically include rapid sample analysis time, intact dosage forms available for subsequent testing, and use of "green" techniques. These techniques will become further utilized and embedded in pharmaceutical workflows as faster analyses become the norm and as real-time release is enabled by applying the appropriate technology to obtain the sought information in the required timeframe.

References

Aaltonen J, Rantanen J, Siiria S, Karjalainen M, Jorgensen A, Laitinen N, Savolainen M, Seitavuopio P, Louhi-Kultanen M, Yliruusi J (2003) Polymorph screening using near-infrared spectroscopy. Anal Chem 75:5267–5273

Alcala M, Leo J, Ropero J, Blanco M, Romanach R (2008) Analysis of low content drug tablets by transmission near infrared spectroscopy: selection of calibration ranges according to multivariate detection and quantitation limits of PLS models. J Pharm Sci 97:5318–5327

Aldridge PK, Mushinsky RF, Andino MM, Evans CL (1994) Active ID through package (blister). Appl Spec 48:1272–1276

Aldridge PK, Evans CL, Ward HW II, Colgan ST, Boyer N, Gemperline PJ (1996) Near-IR detection of polymorphism and process-related substances. Anal Chem 68:997–1002

Alvarenga L, Ferreira D, Altekruse D, Menezes JC, Lochmann D (2008) Tablet identification using near-infrared spectroscopy (NIRS) for pharmaceutical quality control. J Pharm Biomed Anal 48:62–69

Al-Zoubi N, Koundourellis JE, Malamataris S (2002) FT-IR and Raman spectroscopic methods for identification and quantitation of orthorhombic and monoclinic paracetamol in powder mixes. J Pharm Biomed Anal 29(3):459–467

Anderton C (2004) A valuable technique for polymorph screening. Eur Pharm Rev 9:68–74

Andrade FJ, Wetzel WC, Chan GC-Y, Webb MR, Gamez G, Ray SJ, Hieftje GM (2006) A new, versatile, direct-current helium atmospheric-pressure glow discharge. J Anal At Spectrom (21): 1175–1184

Arnaud CH (2007) Taking mass spec into the open–open-air ionization methods minimize sample prep and widen range of mass spectrometry applications. C&EN 85(41):13–18

Berntsson O, Zackrisson G, Ostling G (1997) Determination of moisture in hard gelatin capsules using near-infrared spectroscopy: applications to at-line process control of pharmaceutics. J Pharm Biomed Anal 15:895–900

Bertacche V, Pinin E, Stradi R, Stratta F (2006) Qualitative determination of amorphous cyclosporine in crystalline clorosporine sample by Fourier transform infrared spectroscopy. J Pharm Sci 95:159–166

Beyer J, Steffens JK (2003) Calibration models for determination of water content in pharmaceutical excipients using near-infrared spectroscopy (NIRS). Pharm Ind 65:186–192

Blanco M, Coello J, Iturriag H, Maspoch S, De la Pezuela C (1998) Near-infrared spectroscopy in the pharmaceutical industry. Analyst 123:135R–150R

Burns DA, Ciurczak EW (2008) Handbook of near-infrared analysis. CRC Press, Boca Raton

Cao WJ, Mao C, Chen W, Lin H, Krishnan S, Cauchon N (2006) Differentiation and quantitative determination of surface and hydrate water in lyophilized mannitol Using NIR Spectroscopy. J Pharm Sci 95:2077–2086

Chen H, Horvath CS (1995) High speed high performance liquid chromatography of peptides and proteins. J Chromatogr A 705(1):3–20
Chen H, Talaty NN, Takats Z, Cooks RG (2005) Desorption electrospray ionization mass spectrometry for high-throughput analysis of pharmaceutical samples in the ambient environment. Anal Chem 77(21):6915–6927
Chen H, Ouyang Z, Cooks RG (2006a) Thermal production and reactions of organic ions at atmospheric pressure. Angew Chem Int Ed Engl 45:3656–3660
Chen H, Venter A, Cooks RG (2006b) Extractive electrospray ionization for direct analysis of undiluted urine, milk and other complex mixtures without sample preparation. Chem Commun 19:2042–2044
Chen H, Yang S, Wortmann A, Zenobi R (2007) Neutral desorption sampling of living objects for rapid analysis by extractive electrospray ionization mass spectrometry. Angew Chem Int Ed Engl 46(40):7591–7594
Chen H, Gamez G, Zenobi R (2009) What can we learn from ambient inoization techniques. J Am Soc Mass Spectrom 20(11):1947–1963
Cheng SC, Cheng TL, Chang HC, Shiea J (2009) Using laser-induced acoustic desorption/electrospray ionization mass spectrometry to characterize small organic and large biological compounds in the solid state and in solution under ambient conditions. Anal Chem 81:868–874
Cody RB, Laramee JA, Durst HD (2005) Versatile new ion source for the analysis of materials in open air under ambient conditions. Anal Chem 77(8):2297–2302
Davis TD, Peck GE, Stowell JG, Morris KR, Byrn SR (2004) Modeling and monitoring of polymorphic transformations during the drying phase of wet granulation. Pharm Res 21:860–866
Day GM, Zeitler JA, Jones W, Rades T, Taday PF (2006) Understanding the influence of polymorphism on phonon spectra:lattice dynamics calculations and terahertz spectroscopy of carbamazepine. J Phys Chem B 110(1):447–456
De Maesschalck R, Van den Kerkhof T (2005) Implementation of a simple semi-quantitative near-infrared method for the classification of clinical trial tablets. J Pharm Biomed Anal 37:109–114
De Veij M, Deneckere A, Vandenabeele ML (2008a) Detection of counterfeit Viagra® with Raman spectroscopy. J Pharm Biomed Anal 46:303–309
De Veij M, Vandenabeele P, Moens L (2008b) Detection of counterfeit erectile dysfunction drugs with Raman spectroscopy. Spectrosc Eur 20(3):7–10
Eliasson C, Matousek P (2007) Noninvasive authentication of pharmaceutical products through packaging using spatially offset raman spectroscopy. Anal Chem 79:1696–1701
Fayad PB, Prevost M, Sauve S (2010) Laser diode thermal desorption/atmospheric pressure chemical ionization tandem mass spectrometry analysis of selected steroid hormones in wastewater: method optimization and application. Anal Chem 82(2):639–645
Fernandez FM, Cody RB, Green MD, Hampton CY, McGready R, Sengaloundeth S, White NJ, Newton PN (2006) Characterization of solid counterfeit drug samples by desorption electrosray ionization and direct-analysis-in-real-time coupled to time-of-flight mass spectrometry. ChemMedChem 2006(1):702–705
Fitzgerald AJ, Cole BE, Taday PF (2005) Nondestructive analysis of tablet coating thicknesses using terahertz pulsed imaging. J Pharm Sci 94:177–183
Forbess RA, Shukla AJ (1998) Use of near-infrared spectroscopy (NIRS) for qualitative analysis of pharmaceutical raw materials and dosage forms. Amer Pharma Rev 1(1):67
Ford MJ, Van Berkel GJ (2004) An improved thin-layer chromatography/mass spectrometry coupling using a surface sampling probe electrospray ion trap system. Rapid Commun Mass Spectrom 18(12):1303–1309
Forzley M (2005) Combating counterfeit drugs: a concept paper for effective international collaboration. World Health Organization (WHO) Health Technology and Pharmaceuticals
Garcia-Reyes JF, Jackson AU, Molina-Diaz A, Cooks RG (2009) Desorption electrospray ionization mass spectrometry for trace analysis of agrochemicals in food. Anal Chem 81:820–829
Gupta A, Peck GE, Miller RW, Morris KR (2005) Real-time near-infrared monitoring of content uniformity, moisture content, compact density, tensile strength, and Young's modulus of roller compacted powder blends. J Pharm Sci 94:1589–1597

Haapala M, Pol J, Saarela V, Arvola V, Kotiaho T, Ketola RA, Franssila S, Kauppila TJ, Kostiainen R (2007) Desorption atmospheric pressure photoionization. Anal Chem 79(20):7867–7872
Haddad R, Sparrapan R, Eberlin MN (2006) Desorption sonic spray ionization for (high) voltage-free ambient mass spectrometry. Rapid Commun Mass Spectrom 20(19):2901–2905
Haddad R, Sparrapan R, Kotiaho T, Eberlin MN (2008) Easy ambient sonic spray ionization-membrane interface mass spectrometry for direct analysis of solution constituents. Anal Chem 80:898–903
Hakkinen A, Huhtanen M, Reinikainen SP, Karjalainen M, Rantanen J, Louhi-Kultanen M, Nyström L (2005) DRIFT-IR for quantitative characterisation of polymorphic composition of sulfathiazole. Anal Chim Acta 544:108–117
Harper JD, Charipar MA, Mulligan CC, Zhang X, Cooks RG, Ouyang Z (2008) Low-temperature plasma probe for ambient desorption ionization. Anal Chem 80:9097–9104
Helmy R, Zhou GX, Chen YW, Crocker L, Wang T, Wenslow RM Jr, Vailaya A (2003) Characterization and quantitation of aprepitant drug substance polymorphs by attenuated total reflectance Fourier transform infrared spectroscopy. Anal Chem 75:605–611
Herkert T, Prinz H, Kovar KA (2001) One hundred percent online identity check of pharmaceutical products by near-infrared spectroscopy on the packaging line. Eur J Pharm Biopharm 51:9–16
Higgins JP, Arrivo AM, Reed RA (2003) Approach to the determination of hydrate form conversions of drug compounds and solid dosage forms by near-infrared spectroscopy. J Pharm Sci 92:2303–2316
Hilfiker R (2006) Polymorphism in the pharmaceutical industry. Wiley-VCH, Weinheim, Germany
Hirsch J (2006) Near-infrared analysis of critical parameters in lyophilized materials. BioPharm International 19:2
Johansson J, Sparen A, Svensson O, Folestad S, Claybourn M (2007) Quantitative transmission Raman spectroscopy of pharmaceutical tablets and capsules. Appl Spec 61:1211–1218
Jørgensen A, Rantanen A, Karjalainen M, Karjalainen M, Khriachtchev L, Räsänen E (2002) Wet granulation studied by spectroscopic methods and multivariate analysis. Pharm Res 19:1285–1291
Kamat MS, Lodder RA, Deluca PP (1989) Near infrared spectroscopic determination of residual moisture in lyophilized sucrose through intact glass vials. Pharm Res 6:961–965
Kauppila TJ, Wiseman JM, Kelola RA, Kotiaho T, Cooks RG, Kostianen R (2006) Desorption electrospray ionization mass spectrometrty for the analysis of pharmaceutical and metabolites. Rapid Commun Mass Spectrom 20(3):387–392
Koradia V, Chawla G, Bansal AK (2004) Qualitative and quantitative analysis of clopidogrel bisulphate polymorphs. Acta Pharm 54:193–204
Leasure RM, Gangwer MK (2002) Near-Infrared spectroscopy for in-process moisture determination of a potent active pharmaceutical ingredient. Amer Pharm Rev 5(1):103–109
Lewis IR, Edwards HG (2001) Handbook of Raman spectroscopy: from the research laboratory to the process line. Marcel Dekker, New York
Li W, Worosila GD, Wang W, Mascaro T (2005) Determination of polymorph conversion of an active pharmaceutical ingredient in wet granulation using NIR calibration models generated from the premix blends. J Pharm Sci 94:2800–2806
Liu Y, Leasure R, Carroll F, Berry M, Ferraina R, Lukaszewicz J (2008) Online PAT monitoring for parenteral process understanding and design space mapping, 35th FACSS Reno NV
Liu Y, Pelletier M, White LJ, Reid G, Timpano R (2009) Near infrared spectroscopy for water analysis: beyond a rapid quantitation, 2009 AAPS annual conference and exhibition, Los Angeles, CA
Ma X, Zhang S, Lin Z, Liu Y, Xing Z, Yang C, Zhang X (2009) Real-time monitoring of chemical reactions by mass spectrometry utilizing a low-temperature plasma probe. Analyst 134:1863–1867
Martens H, Naes T (1989) Multivariate calibration. Wiley, Chichester, UK
McEwen C, Mckay R, Larsen B (2005) Analysis of solids, liquids, and biological tissues using solid probe introduction at atmospheric pressure on commercial LC/MS instrumnets. Anal Chem 77(23):7826–7831
Na N, Zhao M, Zhang S, Yang C (2007) Development of a dielectric barrier discharge ion source for ambient mass spectrometry. J Am Soc Mass Spectrom 18(10):1859–1862

Nemes P, Vertes A (2007) Laser ablation electrospray ionization for atmospheric pressure, in vivo, and imaging mass spectrometry. Anal Chem 79(21):8098–8106

Nyadong L, Late S, Green MD, Banga A, Fernandez FM (2008) Direct quantitation of active ingredients in solid Artesunate antimalarials by noncovalent complex forming reactive desorption electrospray ionization mass spectrometry. Am Soc Mass Spectrom 19:380–385

Ph.Eur. (2005a). 5th ed, 2.2.24. Absorption spectrophotometry, infrared. EuropeanPharmacopoeia 5.08. Strasbourg France, Council of Europe: 01/2005:20224

Ph.Eur. (2005b). 5th ed, 2.2.40. Near-infrared spectrophotometry. European Pharmacopoeia 5.08. Strasbourg France, Council of Europe: 01/2005:20240

Ph.Eur. (2005c). 5th ed, 2.2.48. Raman spectrometry. European Pharmacopoeia 5.08. Strasbourg France, Council of Europe: 01/2005:20248

Pollanen K, Hakkinen A, Reinikainen SP, Rantanen J, Karjalainen M, Louhi-Kultanen M, Nystrom L (2005) IR spectroscopy together with multivariate data analysis as a process analytical tool for in-line monitoring of crystallization process and solid-state analysis of crystalline product. J Pharm Biomed Anal 38(2):275–284

Pucherta T, Lochmannb D, Menezesc JC, Reich G (2010) Near-infrared chemical imaging (NIR-CI) for counterfeit drug identification – a four-stage concept with a novel approach of data processing (Linear Image Signature). J Pharm Biomed Anal 51:138–145

Rantanen E, Rantanen J, Jørgensen J, Karjalainen M, Paakkari T, Yliruusi J (2001) Novel identification of pseudopolymorphic changes of theophylline during wet granulation using near infrared spectroscopy. J Pharm Sci 90:389–396

Rantanen J, Wikstrom H, Rhea FE, Taylor LS (2005) Improved understanding of factors contributing to quantification of anhydrate/hydrate powder mixtures. Appl Spec 59:942–951

Ratcliffe LV, Rutten FJM, Barrett DA, Whitmore T, Seymour D, Greenwood C, Aranda-Gonzalvo Y, Robinson S, McCoustra M (2007) Surface analysis under ambient conditions using plasma-assisted desorption/ionization mass spectrometry. Anal Chem 79(16):6094–6101

Reepmeyer JC, d'Avignon DA (2009) Structure elucidation of thioketone analogues of sildenafil detected as adulterants in herbal aphrodisiacs. J Pharm Biomed Anal 49(1):145–150

Rezenom YH, Dong J, Murray KK (2008) Infrared laser-assisted desorption electrospray ionization mass spectrometry. Analyst 133:226–232

Saal C (2006) Quantification of polymorphic forms by Raman spectroscopy. Amer Pharm Rev 9:76–81

Saeed M, Saner S, Oelichmann J, Keller H, Betz G (2009) is. J Pharm Sci 98(12):4877–4886

Salari A, Young RE (1998) Application of attenuated total reflectance FTIR spectroscopy to the analysis of mixtures of pharmaceutical polymorphs. International J of Pharm 163:157–166

Sampson JS, Hawkridge AM, Muddiman DC (2006) Generation and detection of multiply-charged peptides and proteins by matrix-assisted laser desorption electrospray ionization (MALDESI) Fourier transform ion cyclotron resonance mass spectrometry. J Am Soc Mass Spectrom 17(12):1712–1716

Sasic S, Whitlock M, Mehrens S (2009) statistically optimized Raman microscopy for detecting low levels of undesired polymorphic forms in low dose active tablets. 36th FACSS Louisville, KY

Schilling GD, Shelley JT, Barnes JH IV, Sperline RP, Benton MB, Barinaga CJ, Koppenaa DW, Hieftje GM (2010) Detection of positive and negative ions from a flowing atmospheric pressure afterglow using a mattauch-herzog mass spectrograph equipped with a faraday-strip array detector. J Am Soc Mass Spectrom 121(1):97–103

Schneider RC, Kovar KA (2003) Analysis of ecstasy tablets: comparison of reflectance and transmittance near infrared spectroscopy. Forensic Sci Int 134:187–195

Shelley JT, Ray SJ, Hieftje GM (2008) Laser ablation coupled to a flowing atmospheric pressure afterglow for ambient mass spectral imaging. Anal Chem 80(21):8308–8313

Shiea J, Huang MZ, Hsu HJ, Lee CY, Yuan CH, Beech I, Sunner J (2005) Electrospray-assisted laser desorption/ionization mass spectrometry for direct ambient analysis of solids. Rapid Commun Mass Spectrom 19(24):3701–3704

Shieh IF, Lee CY, Shiea J (2005) Eliminating the interferences from TRIS buffer and SDS in protein analysis by fused-droplet electrospray ionization mass spectrometry. J Proteome Res 4(2):606–612

Shin YS, Drolet B, Mayer R, Dolence K, Basile F (2007) Desorption ionization-mass spectrometry of proteins. Anal Chem 79:3514–3518

Siesler HW, Ozaki Y, Kawata S, Heise HM (2002) Near-infrared spectroscopy. Principles, instruments, applications. Wiley-VCH, Weinheim

Singh S, Prasad B, Savaliya AA, Shah RP, Gohil VM, Kaur A (2009) Strategies for characterizing sildenafil, vardenafil, tadalafil and their analogues in herbal dietary supplements, and detecting counterfeit products containing these drugs. Trends Anal Chem 28(1):13–28

Strachan CJ, Taday TF, Newnham DA, Gordon KG, Zeitler JA, Pepper M, Rades T (2005) Using terahertz pulsed spectroscopy to quantify pharmaceutical polymorphism and crystallinity. J Pharm Sci 94:837–846

Strachan CJ, Ho L, Zeitler JA, Rantanen J, Rades T, Gordon KC (2006) Terahertz applications for the analysis of solid dosage forms. Pharm Tech Eur 18(11):27–33

Swartz ME (2005) UPLC: an introduction and review. J Liq Chromatogr Relat Technol 28(7–8):1253–1263

Taday PF, Newnham DA (2004) Technological advances in terahertz pulsed systems bring far-infrared spectroscopy into the spotlight. Spec Eur 16:22–25

Taday PF, Bradley IV, Arnone DD, Pepper M (2003) Using terahertz pulse spectroscopy to study the crystalline structure of a drug: a case study of the polymorphs of ranitidine hydrochloride. J Pharm Sci 92:831–838

Takats Z, Wiseman JM, Gologan B, Cooks RG (2004) Mass spectrometry sampling under ambient conditions with desorption electrospray ionization. Science 306:471–473

Takats Z, Cotte-Rodriguez I, Talaty N, Chen HW, Cooks RG (2005a) Direct, trace level detection of explosives on ambient surfaces by desorption electrospray ionization mass. Chem Commun 15:1950–1952

Takats Z, Wiseman JM, Cooks RG (2005b) Ambient mass spectrometry using desorption electrospray ionization (DESI): instrumentation, mechanisms and applications in forensics, chemistry, and biology. J Mass Spectrom 40:1261–1275

Twohig M, Skilton SJ, Fujimoto G, Ellor N, Plunb RS (2010) Rapid detection and identification of counterfeit of adulterated products of synthetic phosphodiesterase type-5 inhibitors with an atmospheric solids analysis probe. Drug Test Analysis 2(2):45–50

Van den Kerkhof T, Maesschalck R, Vanhoutte K, Coene MC (2006) Augmentation of near infrared diffuse reflectance and transmittance spectral data for the development of robust PLSBC models for classifying double blind clinical trial \ts. J Pharm Biomed Anal 42:517–522

Venhuis BJ, Blok-Tip L, de Kaste D (2008) Designer drugs in herbal aphrodisiacs. Forensic Sci Int 177(2–3):e25–e27

Venter A, Cooks RG (2007) Electrospray ionization in a small pressure-tight enclosure. Anal Chem 79:6398–6403

Venter A, Nefliu M, Cooks RG (2008) Ambient desorption ionization mass spectrometry. Trends Anal Chem 27(4):284–290

Wang PG (2008) High-throughput analysis in the pharmaceutical industry. CRC Press, Boca Raton

Wang D, Zhang M (2007) Rapid quantitation of testosterone hydroxl metabolites by ultra-performance liquid chromatography and mass spectrometry. J Chromatogr B 855(2):290–294

Weston DJ, Bateman R, Wilson ID, Wood TR, Creaser CS (2005) Direct analysis of pharmaceutical drug formulations using ion mobility spectrometry/quadrupole-time-of-flight mass spectrometry combinjed with desorption electrospray ionization. Anal Chem 77:7572–7580

Williams JP, Patel VJ, Holland R, Scrivens JH (2006) The Use of Recently Described Ionisation Techniques for the Rapid Analysis of Some Common Drugs and Samples of Biological Origin. Rapid Commun Mass Spectrom 20:1447–1456

Wolff JC, Warrack JK, Kidder L, Lewis EN (2008) NIR-based chemical imaging as an anticounterfeiting tool. Pharm Manuf 7:27–31

Wu N, Clausen AM (2007) Fundamental and practical aspects of ultrahigh pressure liquid chromatography for fast separations. J Sep Sci 30(8):1167–1182

Wu C, Sims WF, Hill HH (2000) Secondary electrospray ionization ion mobility spectrometry/mass spectrometry of illicit drugs. Anal Chem 72:396–403

Xiang D, Berry J, Buntz S, Gargiulo P, Cheney J, Joshi Y, Wabuyele B, Wu H, Hamed M, Hussain AS, Khan MA (2009) Robust calibration design in the pharmaceutical quantitative measurements with near-infrared (NIR) spectroscopy: avoiding the chemometric pitfalls. J Pharm Sci 98(3):1155–1166

Yu S, Crawford E, Tice J, Musselman BWuJ (2009) Bioanalysis without sample cleanup or chromatography: the evaluation and initial implementation of direct analysis in real time ionization mass spectrometry for the quantification of drugs in biological matrixes. Anal Chem 81:193–202

Zheng Y, Lai X, Bruun SW, Ipsen H, Larsen JN, Lowenstein H, Sondergaard L, Jacobsen S (2008) Determination of moisture content of lyophilized allergen vaccines by NIR spectroscopy. J Pharm Biomed Anal 46(3):592–596

Zou P, Hou P, Oh SS, Chong YM, Bloodworth BC, Low MY, Kou HL (2008) Isolation and identification of thiohomosildenafil and thiosildenafil in health supplements. J Pharm Biomed Anal 47(2):279–284

Index

B. Nickerson (ed.), *Sample Preparation of Pharmaceutical Dosage Forms: Challenges and Strategies for Sample Preparation and Extraction*, DOI 10.1007/978-1-4419-9631-2,